Pólya Urn Models

CHAPMAN & HALL/CRC
Texts in Statistical Science Series

Analysis of Failure and Survival Data
P. J. Smith

The Analysis of Time Series — An Introduction, Sixth Edition
C. Chatfield

Applied Bayesian Forecasting and Time Series Analysis
A. Pole, M. West and J. Harrison

Applied Nonparametric Statistical Methods, Fourth Edition
P. Sprent and N.C. Smeeton

Applied Statistics — Handbook of GENSTAT Analysis
E.J. Snell and H. Simpson

Applied Statistics — Principles and Examples
D.R. Cox and E.J. Snell

Bayesian Data Analysis, Second Edition
A. Gelman, J.B. Carlin, H.S. Stern and D.B. Rubin

Bayesian Methods for Data Analysis, Third Edition
B.P. Carlin and T.A. Louis

Beyond ANOVA — Basics of Applied Statistics
R.G. Miller, Jr.

Computer-Aided Multivariate Analysis, Fourth Edition
A.A. Afifi and V.A. Clark

A Course in Categorical Data Analysis
T. Leonard

A Course in Large Sample Theory
T.S. Ferguson

Data Driven Statistical Methods
P. Sprent

Decision Analysis — A Bayesian Approach
J.Q. Smith

Elementary Applications of Probability Theory, Second Edition
H.C. Tuckwell

Elements of Simulation
B.J.T. Morgan

Epidemiology — Study Design and Data Analysis, Second Edition
M. Woodward

Essential Statistics, Fourth Edition
D.A.G. Rees

Extending the Linear Model with R: Generalized Linear, Mixed Effects and Nonparametric Regression Models
J.J. Faraway

A First Course in Linear Model Theory
N. Ravishanker and D.K. Dey

Generalized Additive Models: An Introduction with R
S. Wood

Interpreting Data — A First Course in Statistics
A.J.B. Anderson

An Introduction to Generalized Linear Models, Third Edition
A.J. Dobson and A.G. Barnett

Introduction to Multivariate Analysis
C. Chatfield and A.J. Collins

Introduction to Optimization Methods and Their Applications in Statistics
B.S. Everitt

Introduction to Probability with R
K. Baclawski

Introduction to Randomized Controlled Clinical Trials, Second Edition
J.N.S. Matthews

Introduction to Statistical Methods for Clinical Trials
Thomas D. Cook and David L. DeMets

Large Sample Methods in Statistics
P.K. Sen and J. da Motta Singer

Linear Models with R
J.J. Faraway

Markov Chain Monte Carlo — Stochastic Simulation for Bayesian Inference, Second Edition
D. Gamerman and H.F. Lopes

Mathematical Statistics
K. Knight

Modeling and Analysis of Stochastic Systems
V. Kulkarni

Modelling Binary Data, Second Edition
D. Collett

Modelling Survival Data in Medical Research, Second Edition
D. Collett

Multivariate Analysis of Variance and Repeated Measures — A Practical Approach for Behavioural Scientists
D.J. Hand and C.C. Taylor

Multivariate Statistics — A Practical Approach
B. Flury and H. Riedwyl

Pólya Urn Models
H. Mahmoud

Practical Data Analysis for Designed Experiments
B.S. Yandell

Practical Longitudinal Data Analysis
D.J. Hand and M. Crowder

Practical Statistics for Medical Research
D.G. Altman

A Primer on Linear Models
J.F. Monahan

Probability — Methods and Measurement
A. O'Hagan

Problem Solving — A Statistician's Guide, Second Edition
C. Chatfield

Randomization, Bootstrap and Monte Carlo Methods in Biology, Third Edition
B.F.J. Manly

Readings in Decision Analysis
S. French

Sampling Methodologies with Applications
P.S.R.S. Rao

Statistical Analysis of Reliability Data
M.J. Crowder, A.C. Kimber, T.J. Sweeting, and R.L. Smith

Statistical Methods for Spatial Data Analysis
O. Schabenberger and C.A. Gotway

Statistical Methods for SPC and TQM
D. Bissell

Statistical Methods in Agriculture and Experimental Biology, Second Edition
R. Mead, R.N. Curnow, and A.M. Hasted

Statistical Process Control — Theory and Practice, Third Edition
G.B. Wetherill and D.W. Brown

Statistical Theory, Fourth Edition
B.W. Lindgren

Statistics for Accountants
S. Letchford

Statistics for Epidemiology
N.P. Jewell

Statistics for Technology — A Course in Applied Statistics, Third Edition
C. Chatfield

Statistics in Engineering — A Practical Approach
A.V. Metcalfe

Statistics in Research and Development, Second Edition
R. Caulcutt

Survival Analysis Using S—Analysis of Time-to-Event Data
M. Tableman and J.S. Kim

The Theory of Linear Models
B. Jørgensen

Time Series Analysis
H. Madsen

Texts in Statistical Science

Pólya Urn Models

Hosam M. Mahmoud
The George Washington University
District of Columbia, U.S.A.

CRC Press is an imprint of the
Taylor & Francis Group, an **informa** business
A CHAPMAN & HALL BOOK

Chapman & Hall/CRC
Taylor & Francis Group
6000 Broken Sound Parkway NW, Suite 300
Boca Raton, FL 33487-2742

First issued in paperback 2022

ISBN 13: 978-1-03-247779-4 (pbk)
ISBN 13: 978-1-4200-5983-0 (hbk)

DOI: 10.1201/9781420059847

Publisher's Note
The publisher has gone to great lengths to ensure the quality of this reprint but points out that some imperfections in the original copies may be apparent.

Library of Congress Cataloging-in-Publication Data

Mahmoud, Hosam M. (Hosam Mahmoud), 1954–
Polya urn models / Hosam Mahmoud.
p. cm. -- (CRC texts in statistical science series ; 76)
Includes bibliographical references and index.
ISBN 978-1-4200-5983-0 (alk. paper)
1. Probabilities. 2. Distribution (Probability theory). I. Title.

QA273.M317 2008
519.2--dc22 2008013347

Visit the Taylor & Francis Web site at
http://www.taylorandfrancis.com

and the CRC Press Web site at
http://www.crcpress.com

Dedication

To Ma and Mona,

with much love and admiration

Contents

Preface

Putting and withdrawing balls in urns is a visual mechanism to conceptualize the principles of discrete probability. This book is intended to communicate probability ideas to intermediate students and self-studies through a modern and yet evolving theory. It deals with the subject of discrete probability via several modern developments in urn theory and its numerous applications. The book means to collect and present in a simple structure some recent results in Pólya urn theory. It is hoped that the book can help those interested in modeling dynamically evolving systems, where particles come and go according to governing rules. Scientists, engineers, mathematicians, and economists may find facets of modeling pertinent to developments in their fields.

One main point in the book is that much of the endeavor in the field of (discrete) probability has roots in urn models, or can be cast in the framework of urn models. Some classical problems, initially not posed as urn questions, turn out to have a very natural interpretation or representation in terms of urns. Some of these classical problems originated in gambling and such areas.

Chapter 1 is about basic discrete distributions and a couple of modern tools such as stochastic processes and exchangeability. To conform with the aim and scope of the book we switch to the language of urns as quickly as possible. Realization of distributions is substantiated with urn arguments, and examples for all concepts are in terms of urns. Chapter 2 presents some classical probability problems. Some of these problems were not originally given as urn problems but most of them can be recast as such. Chapter 3 is about dichromatic Pólya urns as a basic discrete structure, growing in discrete time. Chapter 4 considers an equivalent view in continuous time obtained by embedding the discrete Pólya urn scheme in Poisson processes, an operation called poissonization, from which we obtain the Pólya process. Chapter 5 gives heuristical arguments to connect the Pólya process to the discrete urn scheme (depoissonization). Chapter 6 is concerned with several extentions and generalizations (multicolor schemes, random additions). Chapter 7 presents an analytic view, where functional equations for moment generating functions can be obtained and solved. Chapter 8 is about applications to random trees, the kinds that appear in computer science as data structures or models for analysis of algorithms. Chapter 9 is on applications in bioscience (evolution, phylogeny, competitive exclusion, contagion, and clinical trial), where other types of Pólya-like urns appear. Chapter 10 is for urns evolving by multiple ball drawing and applications.

All the chapters have exercises at the end. The quality of the exercises vary considerably from easy to challenging, and for this reason we added solutions at the end of the book, where it was sometimes opportune for putting a little extra material or remarks that can shed more light on the entire scope.

Hosam M. Mahmoud
Washington, 2008

Acknowledgments

Countless names are connected to any field of science, from founding fathers and dedicated researchers to students and practitioners who perpetuate the interest. It is not possible to mention all the names associated with research in discrete applied probability.

However, three names deserve a special mention: I am indebted to Philippe Flajolet, Svante Janson, and Robert Smythe for their continued patronage. Over the years I learned a great deal from personal contact, discussion, and reading their scholarly work on Pólya urns.

Many friends, family members, and colleagues contributed to the birth of this work. The book came to existence with the encouragement of my friend Dr. Reza Modarres and my publisher Mr. Robert Stern. To Talaat Abdin, Rafik Aguech, Srinivasan Balaji, Michael Drmota, Kareem, Maï and Sherif Hafez, Benjamin Kedem, Moshira Mahmoud, Mostafa Nabih, Helmut Prodinger and Iman Youssef I say thank you all for believing in me and for your continued support. Special thanks are due to my wife Fari, for years of endurance and unfailing love.

1

Urn Models and Rudiments of Discrete Probability

The terminology of urns and urn models generally refers to a system of one or more urns containing objects of various types. One customarily calls the objects *balls* and distinguishes the object types by colors. Thus, in the usual setting there are colored balls in one or more urns. The system then evolves in time, subject to rules of throwing balls into the urns or drawing balls under predesignated replacement schemes. The rule, for example, might be to draw a ball *at random* (all balls being equally likely) and put it back in the urn together with another ball of the same color. The equally likely assumption in drawing a ball is usually thought of as an effect of purposeful randomization. It can be approximated reasonably well by shaking the urn well before a draw to induce good mixing of its contents.

Drawing balls from an urn may sometimes be referred to as *sampling*. Sampling can be carried out in different manners. Among the most common ways of sampling are *sampling with replacement* and *sampling without replacement*.

Urn schemes have a myriad of applications and serve as lucid models for a variety of common situations in science, economics, business, industry, and technology. To name a few, there are applications in evolution of species in biology, in particle systems in chemistry, and in the formation of social and economic networks in sociology. We shall have an opportunity to discuss some of these applications in later chapters.

Pólya urn models comprise an urn of colored balls with replacement schemes. Balls are sampled at random from the urn and, depending on the color of the ball withdrawn, balls of various colors are replaced in the urn.

Several standard urn schemes correspond to some basic probability distributions and classical problems. In this chapter we make these connections to basic discrete distributions, and in the next chapter we make the connection to several classical probability problems, just to show that a good proportion of interesting work in probability can be cast within the framework of drawing balls from urns. Another purpose is to set the notation for the material in later chapters and establish some basic formulas for probability distributions that will appear throughout. The exercise set at the end of this and the next chapter reinforce some probability notions from an urn point of view.

1.1 Probability Axioms

The set of all outcomes of a random experiment is the *sample space* and the points of this set are the *sample points*. The sample space is often denoted by Ω. This book is mostly concerned with discrete probability. A *discrete space* is a finite or an infinitely countable one. Subsets of a discrete sample space are called *events*, and to every event A we assign a probability $\mathbf{P}(A)$. Much of the theory of probability rests on very few axioms concerning the assignment of probabilities to events. Several variants of the probability axioms may be used. They rely on rather natural hypotheses, such as probability must be nonnegative, and countable additivity—the probability assigned to countable parts of a whole add up to the probability of the whole.

If the sample space is uncountable, it is often not possible to find nontrivial set functions (probability measures) that apply to or measure all the sets of the space, while the countable additivity still holds. In this case, we require the measure to at least be applicable to a large class of subsets, hoping the class is wide enough and of broad relevance to the sets that arise in practice. Such a collection of subsets of Ω is called a *sigma field* (or *sigma algebra*), and should enjoy nice closure properties. The usual definition of a sigma field $\mathcal{F}$ over Ω is that it is a collection of subsets of Ω satisfying:

- *Nonemptiness:* $\Omega \in \mathcal{F}$.
- *Closure under complements:* $A \in \mathcal{F}$ implies that A^c, the complement of A, is also in $\mathcal{F}$.
- *Closure under countable union:* If $A_1, A_2, \dots$ are in $\mathcal{F}$, then their union $\cup_{i=1}^{\infty} A_i$ is also in $\mathcal{F}$.

In the broader framework, it is customary to set up probability problems on a probability triple $(\Omega, \mathcal{F}, \mathbf{P})$, also called the *probability space*, where Ω is an arbitrary sample space, $\mathcal{F}$ is a suitable sigma field containing the events of interest, and $\mathbf{P}$ is any probability measure consistent with the axiomatic choices, but now only applicable to the measurable sets (events), of the sigma field, namely:

- *Axiom 1:* $\mathbf{P}(A) \ge 0$, for every $A \in \mathcal{F}$.
- *Axiom 2:* $\mathbf{P}(\Omega) = 1$.
- *Axiom 3:* If $A_1, A_2, \dots$ are mutually exclusive events in $\mathcal{F}$ then

$$\mathbf{P}\left(\bigcup_{i=1}^{\infty} A_i\right) = \sum_{i=1}^{\infty} \mathbf{P}(A_i).$$

These axioms are known as *Kolmogorov's axioms*, and have been around as the building blocks of probability theory since the 1930s.

1.2 Random Variables

Often we are interested in numerical values associated with the outcome of an experiment. For example, in a sequence of four coin tosses, the sample points are sequences of length four of H (heads) and T (tails). The number of heads in a sample point is a numerical value associated with the sample point. The sample point $HHTH$ has three heads. The number of heads in this experiment is an example of random variables.

To be able to measure probabilities of the numerical values falling in certain reasonable subsets on the real line, one should technically insist that the inverse image in Ω is a set to which the probability measure applies. That is, the inverse image must be a measurable set in the given probability space. The most common sigma field on the real line is the one generated by the open sets of the form (a, b) (the *Borel sigma field* on the real line). A function

$$X : \Omega \to \mathbb{R}$$

is measurable if the inverse image of a set from the Borel sigma field is an event.

A *random variable* is a measurable function X defined on Ω, the sample space of an experiment, that maps points of Ω into the set of real numbers. it is customary to write X for $X(\omega)$, for $\omega \in \Omega$.

1.3 Independence of Discrete Random Variables

The notion of independence of events is that knowledge of the outcome of one event does not convey information on the other. Knowing the outcome of one does not influence our a priori assessment of the chances of occurrence of the other. For instance one would tend to think that knowing any information about the price of silk in China on a given day does not tell us anything that was not known a priori about the number of accidents on Elm street in Los Angeles on the same day. If that number of accidents has a certain chance (probability) of being low, our view of it is that it remains in the low range regardless of whether the price of silk increases or not in China.

To formalize this, we use the following notation. The *conditional probability* of an event A given that event B has occurred is denoted by $\mathbf{P}(A \mid B)$ and is defined by the relation[1]

$$\mathbf{P}(A \mid B) = \frac{\mathbf{P}(A \cap B)}{\mathbf{P}(B)}, \qquad \text{if } \mathbf{P}(B) > 0$$

The event A is said to be *independent* of the event B if

$$\mathbf{P}(A \mid B) = \mathbf{P}(A);$$

[1] To avoid requiring $\mathbf{P}(B) > 0$, one can take $\mathbf{P}(A \cap B) = \mathbf{P}(A \mid B)\mathbf{P}(B)$ as a more general definition.

the rationale being that if A is independent of B, the occurrence of B does not enhance in any way our knowledge about how likely A is going to occur. By the definition of conditional probability, if A is independent of B, with $\mathbf{P}(A) > 0$, then

$$\mathbf{P}(B \mid A) = \frac{\mathbf{P}(B \cap A)}{\mathbf{P}(A)} = \frac{\mathbf{P}(A \mid B)\,\mathbf{P}(B)}{\mathbf{P}(A)} = \frac{\mathbf{P}(A)\,\mathbf{P}(B)}{\mathbf{P}(A)} = \mathbf{P}(B),$$

so B is also independent of A. Thus the events A and B are independent if

$$\mathbf{P}(A \cap B) = \mathbf{P}(A)\,\mathbf{P}(B),$$

otherwise the two events are said to be *dependent*. Independence renders the calculation of probabilities easier as it turns the probability of joint events into a product of individual probabilities. The concept extends straightforwardly to multiple events and infinite sequences of events. The events $A_1, \ldots, A_n$ are *totally independent*, if

$$\mathbf{P}\big(A_{i_1} \cap A_{i_2} \cap \ldots \cap A_{i_j}\big) = \mathbf{P}(A_{i_1})\,\mathbf{P}(A_{i_2}) \ldots \mathbf{P}(A_{i_j}),$$

for every subset $\{i_1, i_2, \ldots, i_j\} \subseteq \{1, \ldots, n\}$. We consider an infinite sequence of events to be independent if every finite collection is.

For discrete random variables, we say that X is independent of Y if

$$\mathbf{P}(X = x \mid Y = y) = \mathbf{P}(X = x),$$

for every feasible x and y, which is a variant of the definition of independence.

Example 1.1

Suppose we have two urns, one has m white balls numbered $1, \ldots, m$, the other has n blue balls numbered $1, \ldots, n$. Suppose we sample one ball from each urn and let the label of the white ball be W and that of the blue ball be B. It is reasonable to assume that under good mixing of the content of each urn that if the information about W is released, the recipient will still hold his a priori uniform assumption on the outcome of the blue balls:

$$\mathbf{P}(B = b \mid W = w) = \mathbf{P}(B = b) = \frac{1}{n}.$$

By contrast, the sum $S = W + B$ of the two outcomes is dependent, as one might expect, on W. If W is high, one would shift his a priori view on the distribution of S toward a view that anticipates higher values of S. In technical terms, for $w + 1 \le s \le w + n$,

$$\begin{aligned}
\mathbf{P}(S = s \mid W = w) &= \mathbf{P}(W + B = s \mid W = w) \\
&= \mathbf{P}(B = s - w \mid W = w) \\
&= \mathbf{P}(B = s - w), \qquad \text{by independence} \\
&= \frac{1}{n},
\end{aligned}$$

whereas the unconditional probability (for $2 \le s \le m + n$) is

$$\mathbf{P}(S = s) = \sum_{\substack{w+b=s \\ 1\le w\le m,\ 1\le b\le n}} \mathbf{P}(W = w, B = b) = \sum_{\substack{w+b=s \\ 1\le w\le m,\ 1\le b\le n}} \mathbf{P}(W = w)\, \mathbf{P}(B = b),$$

by the independence of W and B. Subsequently, for $m \ge 2$,

$$\mathbf{P}(S = 2) = \sum_{\substack{w+b=2 \\ 1\le w\le m,\ 1\le b\le n}} \frac{1}{m} \times \frac{1}{n} = \frac{1}{mn} \ne \frac{1}{n}.$$

The variables W and S are dependent. ◇

1.4 Realization of Basic Discrete Distributions via Urns

A random variable assuming a finite or an infinitely countable set of values is said to be a *discrete random variable*. For example, if $X \in \{x_1, x_2, \ldots, x_K\}$, for some integer K, then X is a discrete random variable belonging in a finite set. If $Y \in \{y_1, y_2, y_3, \ldots\}$, it is a discrete random variable with an infinite but countable range of values. The *range* is the admissible set of values.

A random variable X with range $\mathcal{X}$ has chances of *hitting* (a casual way to say *assuming*) a value $x \in \mathcal{X}$. This chance is measured by the probability $\mathbf{P}(X = x)$, and a list, description or specification by any other means of all such probabilities is called the *probability distribution* of X.

Of importance as measures of centrality is the *mean* or *expectation*:

$$\mathbf{E}[X] = \sum_{x\in\mathcal{X}} x\mathbf{P}(X = x).$$

The expectation of X is often denoted by μ_X or simply μ when no ambiguity is present. Two random variables with rather different ranges and rather different degrees of variablity within may have the same mean. A second discriminatory parameter is needed to aid in differentiating the degrees of randomness. A popular such parameter is the *variance*, which is a weighted accumulation of all squared differences from the mean:

$$\mathbf{Var}[X] = \sum_{x\in\mathcal{X}} (x - \mu)^2 \mathbf{P}(X = x).$$

The variance of X is often denoted by σ_X^2 or simply σ^2, when no ambiguity is present. Going further, one defines the rth moment of X as

$$\mathbf{E}[X^r] = \sum_{x\in\mathcal{X}} x^r \mathbf{P}(X = x), \qquad \text{for } r \ge 0.$$

Most common distributions are characterized by their sequence of moments. Therefore, a generating function of the moments when available in closed form is valuable in specifying and deriving numerous important facts about a distribution. Such a function needs a dummy variable, say t, the powers of which act as placeholders for the moments. The *moment-generating function* of X is

$$\phi_X(t) = \mathbf{E}[e^{Xt}] = 1 + \mathbf{E}[X]\,\frac{t}{1!} + \mathbf{E}[X^2]\,\frac{t^2}{2!} + \mathbf{E}[X^3]\,\frac{t^3}{3!} + \cdots,$$

when the series converges. We next take the grand tour of basic discrete distributions, and show how each one emerges naturally from some urn realization.

1.4.1 The Discrete Uniform Distribution

The discrete uniform distribution is one of the simplest and most common distributions. It underlies, for example, our notion of fair dice used in a plethora of commercial games. Suppose $\{a_1, \ldots, a_n\}$ is a set of numbers that are equally likely outcomes of an experiment, that is, we have a discrete uniform random variable U associated with the experiment, with the probability distribution

$$\mathbf{P}(U = a_k) = \frac{1}{n}, \qquad \text{for } k = 1, \ldots, n.$$

The most common set of outcomes is a consecutive set of integers $\{a, a + 1, \ldots, b\}$. We refer to this specialized form of the discrete uniform random variable as the Uniform$[a \,..\, b]$. In many applications $a = 1$ and $b = n$.

Suppose we have an urn with n balls labeled with the numbers $a_1, \ldots, a_n$. One drawing from the urn produces a discrete uniform random variable on the set $\{a_1, \ldots, a_n\}$.

1.4.2 The Bernoulli Distribution

The Bernoulli random variable has an extremely simple distribution. It is an indicator that measures the success of some event by assuming the value 1 if the event occurs, otherwise the indicator assumes the value 0. Its importance stems from its role as a building block for other significant random variables such as the binomial and the geometric.

If the event being indicated by a Bernoulli random variable has probability of success p, we shall use the informative notation Ber(p). The failure probability is $1 - p$, and is usually denoted by q. Thus, if X is a Ber(p) random variable, it has the distribution

$$X = \begin{cases} 1, & \text{with probability } p; \\ 0, & \text{with probability } q. \end{cases}$$

Imagine an urn containing w white balls and b blue balls. One ball is sampled from the urn at random. There is probability $p := w/(w + b)$ that the ball is white, and probability $q = 1 - p = b/(w + b)$ that the ball is blue.

The indicator of the event of withdrawing a white ball is therefore a Ber($\frac{w}{w+b}$) random variable.

1.4.3 The Binomial Distribution

A binomial random variable Bin(n, p) counts the number of successes (realizations of a certain event) in n independent identically distributed trials, within each the event in question has probability p of success, and probability $q = 1 - p$ of failure. Hence,

$$\text{Bin}(n, p) = X_1 + X_2 + \cdots + X_n,$$

and each $X_i, i = 1, \ldots, n$, is an independent Ber(p) random variable.

The sample space, or set of outcomes, can be mapped into strings of length n of the form

$$FFFSSFFSFS\ldots FFS$$

of successes (S) and failures (F). A *specific* string with k successes among the total number of experiments has probability $p^k q^{n-k}$. There are $\binom{n}{k}$ strings with k successes (appearing anywhere) among the letters in the n-long string. The probability of k successes is therefore the collective probability (sum) of these mutually exclusive strings (by Axiom 3 of probability measures). Hence, the binomial probability law of a random variable Y with distribution like Bin(n, p) is given by

$$\mathbf{P}(Y = k) = p^k q^{n-k} \binom{n}{k}, \qquad \text{for } k = 0, 1, \ldots, n.$$

Suppose we have an urn containing w white balls and b blue balls. Balls are sampled from the urn at random and with replacement (we always put the ball back in the urn). The experiment is carried out n times. In an individual drawing there is probability $p := w/(w + b)$ that the ball is white, and probability $q = 1 - p = b/(w + b)$ that the ball is blue. The indicator of the event of withdrawing a white ball in an individual experiment is therefore a Ber($\frac{w}{w+b}$) random variable. Let Y be the number of times a white ball is withdrawn. Then Y is a sum of n independent Ber(p) random variables, or

$$Y = \text{Bin}\Big(n, \frac{w}{w + b}\Big).$$

1.4.4 The Geometric Distribution

An experiment is performed until a certain observable event occurs. Each time the experiment is carried out, with probability of success p for the event, in a manner independent of all previously conducted experiments. The experiment is repeated as many times as necessary (possibly indefinitely) until the event is observed. A geometric random variable Geo(p) is the number of attempts until the event occurs.

The sample space here can be mapped into strings of varying length of successes (S) and failures (F) of the forms

$$S,$$

$$F\,S,$$

$$F\,F\,S,$$

$$F\,F\,F\,S,$$

$$\vdots$$

A string starting with $k-1$ failures followed by a success has probability $q^{k-1}p$. Hence, the geometric probability law of a random variable Y with distribution like Geo(p) is given by

$$\mathbf{P}(Y = k) = q^{k-1}p, \qquad \text{for } k \ge 1.$$

Suppose, we have an urn containing w white balls and b blue balls. Balls are sampled from the urn at random and with replacement until a white ball is taken out. The individual drawing has probability $p := w/(w+b)$ that the ball is white, and probability $q = 1 - p = b/(w+b)$ that the ball is blue. The indicator of the event of withdrawing a white ball in an individual experiment is therefore a Ber($\frac{w}{w+b}$) random variable. Let Y be the number of draws until a white ball appears for the first time. Then,

$$Y = \text{Geo}\left(\frac{w}{w+b}\right).$$

1.4.5 The Negative Binomial Distribution

The Geo(p) random variable is associated with performing an experiment until the first success is achieved. The negative binomial NB(r, p) is a generalization to achieve the rth success. The random variable $Y = \text{NB}(r, p)$ is the number of attempts until the indicated event is observed for the rth time.

The sample space of the negative binomial distribution can be mapped into strings of varying length of successes (S) and failures (F), and within each string there are r successes, the last of which occurs at the end of the string. For example, with $r = 4$ one outcome in the sample space is

$$F\,F\,S\,F\,F\,F\,F\,S\,S\,F\,S.$$

For this string Y realizes the value 11. A string in the event $\{Y = k\}$ has a certain form: It is of length $k \ge r$, and ends in a success (the rth) and has exactly $r-1$ successes among the first $k-1$ trials. An individual string from such an event has probability $p^r q^{k-r}$, and the positions of the first $r-1$ successes can be arbitrarily chosen among the first $k-1$ trials in $\binom{k-1}{r-1}$ ways. Hence, the negative binomial probability for Y with distribution like $\text{NB}(r, p)$ is given by

$$\mathbf{P}(Y = k) = \binom{k-1}{r-1} p^r q^{k-r}, \qquad \text{for } k \geq r.$$

It is not immediately obvious from the formal expression why the word "negative" got in the title of the distribution. This becomes clear from an equivalent expression involving negative signs:

$$\mathbf{P}(Y = k) = (-1)^{r-1} \binom{-(k-r+1)}{r-1} p^r q^{k-r}, \qquad \text{for } k \geq r,$$

when we impose the combinatorial interpretation that $\binom{-x}{s}$, for real x and nonnegative integer s, is the formal expanded definition of the binomial coefficients when continued to the real line:

$$\binom{-x}{s} = \frac{-x(-x-1)\ldots(-x-s+1)}{s!}.$$

An urn realization of $\text{NB}(r, p)$ for some rational $p = w/(w+b)$ is attained by the same mechanism that was set up for the $\text{Geo}(p)$ random variable, the one difference being that the drawing is continued until r white balls have been observed in the sample.

1.4.6 The Hypergeometric Distribution

The hypergeometric distribution comes up in conjunction with the uncertainty in sampling without replacement from two types of objects in a population of n_1 type 1 objects and n_2 type 2 objects, with $n_1 + n_2 = N$. The two types of objects can be thought of as balls of two colors in an urn, of which $w = n_1$ are white, and $b = n_2$ are blue. A sample of size $n \leq N$ is taken out without replacement. The number W of white balls in the sample has a hypergeometric distribution, given by

$$\mathbf{P}(W = k) = \frac{\binom{w}{k}\binom{b}{n-k}}{\binom{N}{n}};$$

the formula is for $\max(0, n-b) \leq k \leq \min(w, n)$, otherwise we interpret the probability as 0.

There are two equivalent views for collecting a sample of size n without replacement. One view is to think of the collection as a sequential process, where at each step a ball is chosen at random from the urn and set aside then the operation is repeated until n balls are taken out of the urn. The other view is to think of the balls as distinguishable (for example, in addition to color, they may carry labels such as $1, \ldots, N$). We imagine that a scoop is used to urn and lift n balls at once in such a way that all $\binom{N}{n}$ sets of size n y likely loads in the scoop.

The Negative Hypergeometric Distribution

ıestion that arises in sampling without replacement from two types is number of draws until a particular type attains a certain level. In terms ı urns, the question is as follows. An urn contains w white balls and b blue balls. Balls are to be drawn at random without replacement, until k white balls have been collected, with $0 \le k \le w$. *What is the distribution of X, the number of draws?* The name of the distribution bears on the notion of waiting until a kth event occur (kth white sample), just like the negative binomial distribution in connection to the geometric. For X to be x, with $k \le x \le b + k$, a favorable sequence must have $k-1$ white balls and $x-k$ blue balls appearing among the first $x-1$ draws, and the xth draw must be of a white ball. When the white ball is drawn at the xth selection, it is taken (conditionally independently from the past realization) from an urn containing $w-k+1$ white balls, and $b-x+k$ blue balls, and the probability of picking a white ball is $(w-k+1)/(w+b-x+1)$. There are $\binom{x-1}{k-1}$ equally likely favorable sequences obtained by fixing the $k-1$ positions for white balls among the first $x-1$ draws. The probability of the particular sequence achieving all the blue balls appearing before all the white balls is

$$\frac{b}{w+b} \times \frac{b-1}{w+b-1} \times \cdots \times \frac{b-x+k+1}{w+b-x+k+1} \times \frac{w}{w+b-x+k}$$
$$\times \frac{w-1}{w+b-x+k-1} \times \cdots \times \frac{w-k+2}{w+b-x+2} \times \frac{w-k+1}{w+b-x+1}.$$

The required probabilities can be written compactly in terms of falling factorials

$$\mathbf{P}(X = x) = \frac{(w)_k (b)_{x-k}}{(w+b)_x} \binom{x-1}{k-1},$$

where $(y)_s$ is *Pochhammer's symbol* for the falling factorial $y(y-1)\dots(y-s+1)$.

1.4.8 The Poisson Distribution

The Poisson distribution comes about as a limit for certain classes of binomial distributions. It is a limit for the probability that the binomial random variable $\text{Bin}(n, p_n)$ assumes the fixed value k, when both $n \to \infty$, and $p_n n$ converges to a limit, say λ. As the sequence of probabilities involved is diminishing, for large n, the success chance of the independent $\text{Ber}(p_n)$ underlying the binomial in an individual trial becomes slim. Therefore, some call it the distribution of rare events. Like all limits, the limiting probability can then be used as a good approximation for the probabilities of $\text{Bin}(n, p_n)$ for large values of n.

The limit distribution can be attained in a formal way. Let X_n, for $n \ge 1$, denote the binomial sequence $\mathrm{Bin}(n, p_n)$. Then, for any *fixed* k,

$$\mathbf{P}(X_n = k) = \binom{n}{k} p_n^k (1 - p_n)^{n-k} = \frac{n!}{k!\,(n-k)!}\, p_n^k (1 - p_n)^n (1 - p_n)^{-k}.$$

As we know from calculus, when $p_n \to 0$, $(1 - p_n)^r$ approaches 1, if r is fixed. But if $r = r(n) = n$, and $np_n \to \lambda$, the limit of $(1 - p_n)^n$ is $e^{-\lambda}$. So, for each fixed nonnegative integer k,

$$\begin{aligned}
\lim_{\substack{n\to\infty \\ p_n\to 0 \\ np_n\to\lambda}} \mathbf{P}(X_n = k) &= \lim_{\substack{n\to\infty \\ p_n\to 0 \\ np_n\to\lambda}} \frac{n(n-1)\ldots(n-k+1)}{k!}\, p_n^k \\
&\quad \times \lim_{\substack{n\to\infty \\ p_n\to 0 \\ np_n\to\lambda}} (1 - p_n)^n \lim_{\substack{n\to\infty \\ p_n\to 0 \\ np_n\to\lambda}} (1 - p_n)^{-k} \\
&= \lim_{\substack{n\to\infty \\ p_n\to 0 \\ np_n\to\lambda}} \frac{1}{k!} \left[\left(n^k + O\left(n^{k-1}\right)\right) p_n^k\right] \times e^{-\lambda} \times 1 \\
&= \frac{1}{k!} \lim_{\substack{n\to\infty \\ p_n\to 0 \\ np_n\to\lambda}} \left[\left(1 + O\left(\frac{1}{n}\right)\right) (np_n)^k\right] e^{-\lambda} \\
&= \frac{\lambda^k e^{-\lambda}}{k!}.
\end{aligned}$$

As an illustration of how the limit probability can be realized via urns, suppose we have a sequence of urns, the nth of which contains n balls, among which 3 are white, and the rest are blue for $n \ge 3$. (The single ball in urn 1 and the two balls in urn 2 can be of any color.) From the nth urn we pick n balls with replacement, the probability of a white ball in any individual trial is $\frac{3}{n}$, for $n \ge 3$. The number of white balls observed in the sample by the end of the drawing is $W_n = \mathrm{Bin}(n, \frac{3}{n})$. Accordingly,

$$\lim_{n\to\infty} \mathbf{P}(W_n = k) = \lim_{n\to\infty} \binom{n}{k} \frac{3^k (n-3)^{(n-k)}}{n^n} = \frac{3^k e^{-3}}{k!}, \qquad \text{for } k = 0, 1, 2, \ldots .$$

We shall denote a Poisson random variable with parameter λ by the notation $\mathrm{Poi}(\lambda)$.

1.4.9 The Multinomial Distributions

The multinomial distribution is a generalization of the binomial distribution. As in the binomial distribution, an experiment is performed independently n times, but instead of the dichotomy of outcomes into success and failure, a multinomial distribution arises by considering the outcome to be one of a number of possibilities, say r. For instance, we may be tossing n balls into r urns, and the probability of the ith urn to attract the ball is p_i, for $i = 1, \ldots, r$ (with $\sum_{i=1}^{r} p_i = 1$, of course). The multinomial distribution is a joint distribution concerned with the shares of the urns. Let X_i be the share of balls of the ith urn after tossing all n balls. The multinomial distribution is a probability law for the joint shares, that is, a statement about $\mathbf{P}(X_1 = x_1, \dots, X_r = x_r)$, for any real numbers $x_1, \dots, x_r$. These probabilities are 0 for any infeasible choice of shares, such as any set of number $x_1, \dots, x_r$ with the sum not totaling to n. Let us consider only nontrivial probabilities for feasible shares $x_1, \dots, x_r$, which are nonnegative integers satisfying

$$x_1 + \cdots + x_r = n. \tag{1.1}$$

A particular outcome can be described by a sequence of hits. That is, it can be specified by a string of length r made up of the indexes of the randomly chosen urns. A sequence that has the index 1 occuring x_1 times, the index 2 occuring x_2, times, and so on, occurs with probability $p_1^{x_1} p_2^{x_2} \dots p_r^{x_r}$. Such a string is favorable to the event we are considering in the multinomial law. So, the multinomial probability in question can be obtained by adding the probability of all such outcomes (Axiom 3 of probability measures). The question is how many sequences there are satisfying Eq. (1.1). The count of such sequences is called the *multinomial coefficient*, and is denoted by $\binom{n}{x_1, x_2, \dots, x_r}$. We can realize this number by counting all permutations of $\{1, \dots, n\}$ in two different ways. We can count by first creating r groups of numbers, of sizes $x_1, x_2, \dots, x_r$. We can then choose the numbers for these groups in $\binom{n}{x_1, x_2, \dots, x_r}$ ways, which is the number we would like to determine. Subsequently, we permute the members of each group, with each permutation so constructed being distinct. There are $\binom{n}{x_1, x_2, \dots, x_r} x_1!\, x_2! \dots x_r!$ such permutations, and this number must also be the same as the usual count by $n!$. Equating the two numbers we see that

$$\binom{n}{x_1, x_2, \dots, x_r} = \frac{n!}{x_1!\, x_2! \dots x_r!}, \qquad \text{for feasible } x_1, \dots, x_r.$$

The multinomial law we are looking for is therefore

$$\mathbf{P}(X_1 = x_1, \dots, X_r = x_r) = \binom{n}{x_1, x_2, \dots, x_r} p_1^{x_1} p_2^{x_2} \dots p_r^{x_r}.$$

The case $r = 2$ is the usual binomial distribution, in which we chose to write $\binom{n}{k, n-k}$ in the more familiar notation $\binom{n}{k}$.

1.5 The Poisson Process

A stochastic process is a collection of random variables $\{X_\alpha : \alpha \in \mathcal{I}\}$, with $\mathcal{I}$ being an indexing set, usually viewed as time. Typical cases are discrete processes $X_1, X_2, X_3, \dots$ evolving in discrete time, or time chopped up into unit ticks, and $X(t)$, where t is the real time. An example of the continuous-time process is a *counting process* that records the number of occurrences of a certain event by time t in the variable $X(t)$. Being a count, it must be a nondecreasing integer-valued process that starts at value 0 at time 0, and by time t it is a nonnegative integer.

The Poisson process is perhaps one of the most fundamental of counting processes. It can be formulated in a number of equivalent ways, of which we choose one standard way for the presentation. We define the *Poisson process with intensity* $\lambda > 0$ as a counting process satisfying:

1. *Stationarity:* The distribution of the number of events in a time interval depends only on the length of that interval, and not its location.
2. *Independent increments:* The number of events in disjoint intervals are independent.
3. *Linear rate:* The probability that exactly one event arrives in an interval of length h is asymptotically (as $h \to 0$) equivalent to λh, and the probability of two or more arrivals within the interval is asymptotically negligible.

The intensity is also called the *rate*.

Let $X(t)$ be the integer value of the Poisson process at time t, which is the number of events that have arrived by time t (including any jumps at time t). The points of jump are sometimes referred to as the *epochs* of the process. For $b > a$, $X(b) - X(a)$ is the increment in the process, or the number of arrivals within the interval $(a, b]$. Let $a < b$, and $c < d$. Stationarity means that the distribution of the number of event arrivals in $(a, b]$ is identical to the distribution of the number of arrivals in $(c, d]$, whenever $b-a = d-c$. In other words, stationarity means that the distribution within intervals of the same length remains the same throughout, and does not change with time, or the distribution of the increments is *homogeneous*. The condition of independent increments means that if the two intervals $(a, b]$ and $(c, d]$ are disjoint, the increments $X(b) - X(a)$ and $X(d) - X(c)$ are independent. The condition of linear rates says, as $h \to 0$,

$$\mathbf{P}(X(t) = 1) = \lambda h + o(h),$$

where the little o notation indicates any function $y(h)$ that is negligible with respect to h, that is, $y(h)/h \to 0$, as $h \to 0$. For very short intervals, the essential character of this probability is linearly proportional to the length of the interval. The condition of linear rates assigns probability tending to 0, for $\mathbf{P}(X(t) \ge 2)$, as $h \to 0$.

The definition of the Poisson process induces an exponential probability for the occurrence of no events in an interval:

$$Q(t) := \mathbf{P}(X(t) = 0) = e^{-\lambda t}. \tag{1.2}$$

To prove such an assertion we can use the properties of a Poisson process to write a simple differential equation in the following manner:

$$\begin{aligned} Q(t+h) &= \mathbf{P}(X(t)=0,\ X(t+h)-X(t)=0) \\ &= \mathbf{P}(X(t)=0)\,\mathbf{P}(X(t+h)-X(t)=0), \\ &\qquad\qquad\qquad\qquad \text{by independent increments} \\ &= Q(t)\,\mathbf{P}(X(h)=0), \qquad \text{by stationarity} \\ &= Q(t)(1-\mathbf{P}(X(h)=1)-\mathbf{P}(X(h)\ge 2)) \\ &= Q(t)(1-\lambda h+o(h)), \qquad \text{by linear rate.} \end{aligned}$$

Hence, as $h \to 0$,

$$\frac{Q(t+h)-Q(t)}{h} = -\lambda Q(t) + o(1) \to -\lambda Q(t).$$

The derivative $Q'(t)$, exists and satisfies the first-order differential equation

$$Q'(t) = -\lambda Q(t),$$

which is to be solved under the initial condition $Q(0) = 1$ (a count must start at 0). As stated, the solution is $e^{-\lambda t}$.

A variable V with distribution function

$$F(v) := \mathbf{P}(V \le v) = \begin{cases} 1 - e^{-v/a}, & \text{if } v > 0; \\ 0, & \text{otherwise,} \end{cases}$$

is an exponential random variable $\text{Exp}(a)$ with mean a, and probability density

$$f(v) = F'(v) = \begin{cases} \frac{1}{a} e^{-v/a}, & \text{if } v > 0; \\ 0, & \text{otherwise.} \end{cases}$$

The relation in Eq. (1.2) tells us something about the first arrival. The first arrival time exceeds t, if and only if $X(t) = 0$. If we let Y_1 be the time of the first arrival, then

$$\mathbf{P}(Y_1 > t) = \mathbf{P}(X(t) = 0) = e^{-\lambda t}.$$

Thus, $Y_1 \stackrel{\mathcal{D}}{=} \text{Exp}(1/\lambda)$. In fact, together with the properties of the process, Eq. (1.2) tells us that all interarrival times are independent exponentially distributed random variables, a case we argue briefly next. For instance, if Y_2 is the second interarrival time (the time it takes for the count to increase from 1 to 2), then

$$\mathbf{P}(Y_2 > t) = \int_{-\infty}^{\infty} \mathbf{P}(Y_2 > t \mid Y_1 = s)\, f_{Y_1}(s)\, ds,$$

where $f_{Y_1}(s)$ is the density of Y_1, which we already know. The conditional probability, for $s \ge 0$, in this integral can be computed from

$$\begin{aligned}
\mathbf{P}(Y_2 > t \mid Y_1 = s) &= \mathbf{P}(X(s+t) - X(s) = 0 \mid Y_1 = s) \\
&= \mathbf{P}(X(s+t) - X(s) = 0), \quad \text{by independent increments} \\
&= \mathbf{P}(X(t) = 0), \qquad \text{by stationarity} \\
&= e^{-\lambda t}.
\end{aligned}$$

Hence

$$\begin{aligned}
\mathbf{P}(Y_2 > t) &= \int_0^\infty \mathbf{P}(Y_2 > t \mid Y_1 = s)\lambda e^{-\lambda s}\, ds \\
&= \int_0^\infty \lambda e^{-\lambda t} e^{-\lambda s}\, ds \\
&= e^{-\lambda t}.
\end{aligned}$$

The interarrival time Y_2 is distributed like $\text{Exp}(1/\lambda)$, too. Because it turned out that

$$\mathbf{P}(Y_2 > t) = \mathbf{P}(Y_2 > t \mid Y_1 = s),$$

we conclude that Y_2 is independent of Y_1. The argument can be repeated for higher indexes, to show that generally Y_k, the interarrival time between the $(k-1)$st and kth events, for $k \ge 1$, is an $\text{Exp}(1/\lambda)$ random variable that is independent of $Y_1, \ldots, Y_{k-1}$.

A main result from which the Poisson process derives its name is the distribution of the number of arrivals by time t. It is a Poisson random variable with mean λt, as we argue next. Let W_n be the waiting time until the nth event occurs. By the definition of interarrival times,

$$W_n \overset{\text{a.s.}}{=} Y_1 + Y_2 + \cdots + Y_n.$$

A sum of n independent $\text{Exp}(a)$ random variables is distributed like $\text{Gamma}(n, a)$, a gamma random variable with parameters n and a. Generally the $\text{Gamma}(\alpha, \beta)$ random variable has the density

$$f(x) = \begin{cases} \dfrac{1}{\Gamma(\alpha)\beta^\alpha} x^{\alpha-1} e^{-x/\beta}, & \text{if } x > 0; \\ 0, & \text{elsewhere.} \end{cases}$$

Thus, $W_n \overset{\mathcal{D}}{=} \text{Gamma}(n, 1/\lambda)$, with density

$$\frac{\lambda^n}{(n-1)!} x^{n-1} e^{-\lambda x}.$$

In the Poisson process with intensity λ, the distribution of $X(t)$, the number of events by time t, follows from a duality principle:

$$X(t) \ge n \iff W_n \le t.$$

Whence,

$$\begin{aligned} \mathbf{P}(X(t) = n) &= \mathbf{P}(X(t) \ge n) - \mathbf{P}(X(t) \ge n+1) \\ &= \mathbf{P}(W_n \le t) - \mathbf{P}(W_{n+1} \le t) \\ &= \int_0^t \lambda e^{-\lambda x} \frac{(\lambda x)^{n-1}}{(n-1)!}\, dx - \int_0^t \lambda e^{-\lambda x} \frac{(\lambda x)^n}{n!}\, dx. \end{aligned}$$

The last integration on the right-hand side has an inductive definition—if we integrate it by parts, we have

$$\int_0^t \lambda e^{-\lambda x} \frac{(\lambda x)^n}{n!}\, dx = -\int_0^t \frac{(\lambda x)^n}{n!}\, de^{-\lambda x} = \int_0^t \lambda e^{-\lambda x} \frac{(\lambda x)^{n-1}}{(n-1)!}\, dx - \frac{(\lambda t)^n}{n!} e^{-\lambda t}.$$

If we insert this integration back in the distribution, we affect a cancelation of the remaining integrals, and only the simple formula

$$\mathbf{P}(X(t) = n) = \frac{(\lambda t)^n}{n!} e^{-\lambda t}$$

is left. Indeed,

$$X(t) \overset{\mathcal{D}}{=} \mathrm{Poi}(\lambda t).$$

We shall have occasion to consider the depositing of balls in or sampling of balls from urns, where a ball added or removed takes its turn according to an associated Poisson process. That is, the ball is picked or deposited at the epochs of a Poisson process (at independent exponentially distributed interarrival times).

1.6 Markov Chains

Markovian systems are memoryless systems characterized by a property that the future depends only on the present, without regard to the particular past that led to it. One would encounter such a Markovian system in some board games of strategy, such as Tic-Tac-Toe where the best move available at a position is only determined by its configuration, in complete oblivion to the sequence of moves that led to that position. By this token, chess is not Markovian, even though it comes close. The rules of castling and en passant impose exceptions: There are positions where there would be a mate-in-one (immediate win, hence best move) depending on whether the last move was an en passant or not, and there are positions where castling, if admissible, is the best move, but that depends on whether the king had moved or not. If his majesty had moved and returned to his initial position, some move other than the inadmissible castling is best possible, and so on.

Suppose $X_0, X_1, \ldots$ are random variables, each assuming values in a finite set of cardinality K. We can map that set in 1–1 manner onto $1, \ldots, K$, or rename its members with labels from the latter counting set. The more general theory deals with the possibility of an infinite countable set of values ($K = \infty$). We will restrict our attention to finite outcomes, as usually happens in urn schemes. One could think of the variable index as the discrete time. One says that the system is in state i at time n, if $X_n = i$. These random variables are said to be a *finite Markov chain*, if the future state depends only on the current state. If we are given the full history since the beginning, the next state at time $n+1$ is determined only by the state at time n. That is to say, the system is a (homogeneous finite-state) Markov chain if

$$\begin{aligned} p_{ij} &:= \mathbf{P}(X_{n+1} = j \mid X_0 = x_0, X_1 = x_1, \ldots, X_{n-1} = x_{n-1}, X_n = i) \\ &= \mathbf{P}(X_{n+1} = j \mid X_n = i), \end{aligned}$$

which is simply the probability of making the transition to state j, if the system is in state i, and it depends only on the selection of the future state j and the present state i, regardless of how we got to it. The values p_{ij}, for $i = 1, \ldots, K$, and $j = 1, \ldots, K$, define the *transition matrix* $\mathbf{M}$ as

$$\mathbf{M} = \begin{pmatrix} p_{1,1} & p_{1,2} & \cdots & p_{1,K} \\ p_{2,1} & p_{2,2} & \cdots & p_{2,K} \\ \vdots & \vdots & \ddots & \vdots \\ p_{K,1} & p_{K,2} & \cdots & p_{K,K} \end{pmatrix}.$$

If we sum across the ith row of the matrix ($i = 1, \ldots, K$), we exhaust the states we can go to from state i, and we must have

$$\sum_{j=1}^{K} p_{ij} = 1, \qquad \text{for each row } i = 1, \ldots, K.$$

Of interest is the state probability row vector

$$\boldsymbol{\pi}^{(n)} = \begin{pmatrix} \pi_1^{(n)} & \pi_2^{(n)} & \cdots & \pi_K^{(n)} \end{pmatrix},$$

where the component $\pi_i^{(n)}$ is the a priori probability of getting in state i after n transitions from the beginning moment of time.

A vehicle that aids in computing $\boldsymbol{\pi}^{(n)}$ is the n-step transition matrix $\mathbf{M}^{(n)}$, the entry $p_{ij}^{(n)}$ of which is the probability of going to step j from state i in n steps. For instance, $\mathbf{M}^{(1)} = \mathbf{M}$, and

$$\mathbf{P}(X_1 = j) = \sum_{i=1}^{K} P(X_1 = j \mid X_0 = i)\, \mathbf{P}(X_0 = i) = \sum_{i=1}^{K} p_{ij}\pi_i^{(0)},$$

which is the jth component in the row vector $\pi^{(0)}\mathbf{M}$. That is,

$$\pi^{(1)} = \pi^{(0)}\mathbf{M}^{(1)}. \quad (1.3)$$

1.6.1 Classification of Markov Chains

A chain is *irreducible* if it is possible to go from any state to every other state. This means there is positive probability to go from state i to state j for every $i, j \in \{1, \ldots, K\}$. In an irreducible chain it is possible to make a transition from i to j in *some* number of steps, say n, and $p_{ij}^{(n)}$ is positive. For some n, the entry at row i and column j in $\mathbf{M}^{(n)}$ is positive for an appropriate choice of n. The value of n that makes $p_{ij}^{(n)} > 0$ may be different from n' that makes $p_{k,\ell}^{(n')} > 0$.

There are some natural and simple examples where a chain is not irreducible, as for example systems that have transient states in which we can get early in the development, and leave them without the possibility of ever coming back. An instance of this is a Markov chain with the transition matrix

$$\mathbf{M} = \begin{pmatrix} \frac{1}{2} & \frac{1}{2} & 0 \\ \frac{1}{2} & \frac{1}{2} & 0 \\ \frac{1}{2} & \frac{1}{2} & 0 \end{pmatrix}.$$

This chain never gets in state 3, unless it starts in it, in which case it leaves it in one transition and never comes back to it.

There are reducible Markov chains that are essentially several chains within, such as one on four states numbered $1, \ldots, 4$, that gets locked in between the states 1 and 2, if we start in either one, or locked in between the states 3 and 4 if we start in either one. Such a chain may have a transition matrix like

$$\mathbf{M} = \begin{pmatrix} \frac{1}{3} & \frac{2}{3} & 0 & 0 \\ \frac{4}{7} & \frac{3}{7} & 0 & 0 \\ 0 & 0 & \frac{1}{2} & \frac{1}{2} \\ 0 & 0 & \frac{1}{2} & \frac{1}{2} \end{pmatrix}.$$

An irreducible chain is periodic with period $r \geq 2$ if the value of n that makes $p_{ij}^{(n)} > 0$ is *always* a multiple of r, for all possible pairs (i, j). Basically, in a periodic chain starting from a state we can come back to it (with positive probability) in r transitions, and r is the same for all the states. A chain that is not periodic is called *aperiodic*. An aperiodic irreducible Markov chain is called *ergodic*.

1.6.2 Chapman-Kolmogorov Equations

To see a pattern, let us work out the two-transition matrix for a small chain. Suppose a two-state Markov chain started in the state $X_0 = 1$ with some probability $\pi_1^{(0)}$, and started in the state $X_0 = 2$ with probability $\pi_2^{(0)} = 1 - \pi_1^{(0)}$. The probability of being in state 2 after two steps is

$$\begin{aligned}
\mathbf{P}(X_2 = 2) &= \mathbf{P}(X_2 = 2 \mid X_1 = 1)\,\mathbf{P}(X_1 = 1) + \mathbf{P}(X_2 = 2 \mid X_1 = 2)\,\mathbf{P}(X_1 = 2)\\
&= p_{12}\left(\mathbf{P}(X_1 = 1 \mid X_0 = 1)\pi_1^{(0)} + \mathbf{P}(X_1 = 1 \mid X_0 = 2)\pi_2^{(0)}\right)\\
&\qquad + p_{22}\left(\mathbf{P}(X_1 = 2 \mid X_0 = 1)\,\pi_1^{(0)} + \mathbf{P}(X_1 = 2 \mid X_0 = 2)\,\pi_2^{(0)}\right)\\
&= p_{12}\left(p_{11}\pi_1^{(0)} + p_{21}\pi_2^{(0)}\right) + p_{22}\left(p_{12}\pi_1^{(0)} + + p_{22}\pi_2^{(0)}\right),
\end{aligned}$$

which is the second component of $\boldsymbol{\pi}^{(0)}\mathbf{M}^2$. Likewise, one finds that $\mathbf{P}(X_2 = 1)$ is the first component of $\boldsymbol{\pi}^{(0)}\mathbf{M}^2$. Thus, $\boldsymbol{\pi}^{(2)} = \boldsymbol{\pi}^{(0)}\mathbf{M}^2$. Note also that this expression can be written as $\boldsymbol{\pi}^{(1)}\mathbf{M}$. The two forms are manifestations of the Chapman-Kolmogorov equations, which say in words that the transition probability from state i to state j in n steps, can be obtained by first visiting state $k \in \{1, \ldots, K\}$ in r steps, with $r \in \{1, \ldots, n-1\}$, then followed by $n - r$ transitions starting in state k and ending in state j. Since visiting the various k states are mutually exclusive events, the required n-transition probability is assembled from summing over k.

THEOREM 1.1
The entries of the n-step transition matrix in a K-state Markov chain are given by

$$p_{ij}^{(n)} = \sum_{k=1}^{K} p_{ik}^{(r)} p_{kj}^{(n-r)},$$

for every $0 < r < n$.

PROOF

$$\begin{aligned}
p_{ij}^{(n)} &= \mathbf{P}(X_n = j \mid X_0 = i)\\
&= \sum_{k=1}^{K} \mathbf{P}(X_n = j,\, X_r = k \mid X_0 = i)\\
&= \sum_{k=1}^{K} \mathbf{P}(X_n = j \mid X_r = k,\, X_0 = i)\,\mathbf{P}(X_r = k \mid X_0 = i)\\
&= \sum_{k=1}^{K} \mathbf{P}(X_n = j \mid X_r = k)\,\mathbf{P}(X_r = k \mid X_0 = i),
\end{aligned}$$

which is the same as the statement. ∎

COROLLARY 1.1
For any $n \geq 2$, *and* $0 < r < n$, *we have*

$$\mathbf{M}^{(n)} = \mathbf{M}^{(r)}\mathbf{M}^{(n-r)} = \mathbf{M}^n.$$

PROOF We can prove this corollary by induction on n. For the basis $n = 2$, the only feasible r is 1. For the one-step transition matrix, $\mathbf{M}^{(1)} = \mathbf{M}$. So, by Chapman-Kolmogorov equations

$$p_{ij}^{(2)} = \sum_{k=1}^{K} p_{ik}^{(1)} p_{kj}^{(1)} = \sum_{k=1}^{K} p_{ik} p_{kj},$$

for all $1 \le i, j \le K$. That is, $\mathbf{M}^{(2)} = \mathbf{M}^2$. If the corollary is true up to $n-1$, with $n \ge 3$, then $\mathbf{M}^{(r)} = \mathbf{M}^r$, and $\mathbf{M}^{(n-r)} = \mathbf{M}^{n-r}$. Again, by Chapman-Kolmogorov equations

$$p_{ij}^{(n)} = \sum_{k=1}^{K} p_{ik}^{(r)} p_{kj}^{(n-r)} = \sum_{k=1}^{K} m_{ik}(r) m'_{kj}(n-r),$$

where $m_{ik}(r)$ is the (i, j) entry in $\mathbf{M}^{(r)} = \mathbf{M}^r$, and $m'_{kj}(n-r)$ is the (i, j) entry in $\mathbf{M}^{(n-r)} = \mathbf{M}^{n-r}$, by the induction hypothesis. ■

COROLLARY 1.2

$$\boldsymbol{\pi}^{(n)} = \boldsymbol{\pi}^{(0)} \mathbf{M}^n.$$

PROOF We can prove this corollary by induction on n. From the state probability vector $\boldsymbol{\pi}^{(n-1)}$ we can get to $\boldsymbol{\pi}^{(n)}$ in one transition via multiplication by $\mathbf{M}$ as in Eq. (1.3). The case $\boldsymbol{\pi}^{(1)} = \boldsymbol{\pi}^{(0)}\mathbf{M}$ provides a basis for the induction. If we assume the relation holds for some $n \ge 0$, we obtain

$$\boldsymbol{\pi}^{(n)} = \boldsymbol{\pi}^{(n-1)}\mathbf{M} = \boldsymbol{\pi}^{(0)}\mathbf{M}^{n-1}\mathbf{M} = \boldsymbol{\pi}^{(0)}\mathbf{M}^{n-1}. \quad ■$$

1.6.3 Limit and Stationary Distributions

A well-behaved Markov chain can be predictable. One could compute with high accuracy the probability of being in various states, if the chain runs long enough. For instance, from Corollary 1.2 if $\mathbf{M}^n$ converges to a limit $\mathbf{M}_\infty$, one would expect $\pi_i^{(n)}$ to stabilize near a limit, obtained at the ith component of the vector $\boldsymbol{\pi}^{(0)}\mathbf{M}_\infty$. In other words, one would see that

$$\mathbf{P}(X_n = i) \to q_i,$$

where the numbers q_i are not dependent on n. Of course, these limiting probabilities must add up to 1. If X is a random variable with range $\{1, \ldots, K\}$, and with the distribution

$$\mathbf{P}(X = q_i), \qquad \text{for } i = 1, \ldots, K,$$

one claims X as a limiting distribution for X_n, and writes

$$X_n \xrightarrow{\mathcal{D}} X, \qquad \text{as } n \to \infty.$$

An ergodic chain has such limiting behavior.

While a limit may not always exist, as we shall see shortly in an instance of the Ehrenfest urn scheme, stationary distributions always exist. A *stationary distribution* is a probability distribution on the states of a Markov chain, such that, if we start in a state according to its probability in that distribution, the transition to all the other states is again given by the stationary distribution. There can be more than one stationary distribution. Any of the stationary distributions can be represented by a row vector of probabilities

$$\pi = (\pi_1 \quad \pi_2 \quad \ldots \quad \pi_K),$$

where π_i is the stationary probability of being at state i. By its definition, if we are in the stationary distribution on the states, a one-step transition puts us back into the stationary distribution:

$$\pi = \pi \mathbf{M},$$

and of course

$$\sum_{i=1}^{K} \pi_i = 1.$$

Example 1.2

We provide an example from the Ehrenfest urn scheme, which will be discussed thoroughly in chapter 3. An Ehrenfest urn starts out with one white and one blue ball. Balls are sampled from the urn. Whenever a ball of a certain color is picked, we discard that ball and replace in the urn a ball of the opposite color. The urn always returns to the initial state after an even number of draws, and is always out of that state after an odd number of draws. When it is out of the initial state, it can be in one of two equally probable states: A state with two white balls or a state with two blue balls (see Figure 1.1). In the figure the bullets represent blue balls, the small circles represent the white balls, the large circles are the states of the Markov chain, with one large double circle marking the starting state. The numbers by the arrows are the probabilities of transition.

Let W_n be the number of white balls in the urn after n draws. Thus,

$$W_n = \begin{cases} 1, & \text{if } n \text{ is even;} \\ 2\text{Ber}\left(\dfrac{1}{2}\right), & \text{if } n \text{ is odd.} \end{cases}$$

There are two subsequences of random variables with two different distributions, and there cannot possibly be any convergence in distribution, even though there is convergence on average (the average is always 1).

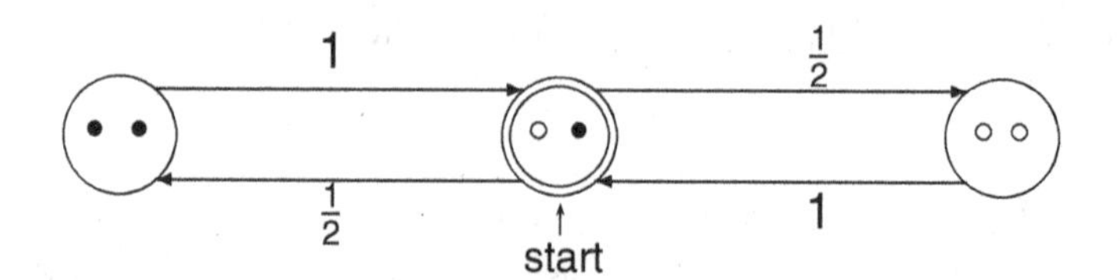

FIGURE 1.1
The Markov chain for an Ehrenfest urn.

A simple randomization mechanism, however, stabilizes the process in the long run. As Figure 1.1 illustrates, this process is a Markov chain. The states can be represented with the number of white balls in the urn. We have a Markov chain with three states, which can be labeled 0, 1, and 2, and with transition matrix

$$\mathbf{M} = \begin{pmatrix} 0 & 1 & 0 \\ \frac{1}{2} & 0 & \frac{1}{2} \\ 0 & 1 & 0 \end{pmatrix}.$$

Such a chain has a stationary distribution given by the row vector $\pi = (\pi_0 \;\; \pi_1 \;\; \pi_2)$ that solves the equations

$$\pi \mathbf{M} = \pi,$$

and $\pi_0 + \pi_1 + \pi_2 = 1$. Namely, we have

$$\pi \mathbf{M} = (\, \pi_0 \quad \pi_1 \quad \pi_2 \,) \begin{pmatrix} 0 & 1 & 0 \\ \frac{1}{2} & 0 & \frac{1}{2} \\ 0 & 1 & 0 \end{pmatrix} = \left(\frac{1}{2}\pi_1 \quad \pi_0 + \pi_2 \quad \frac{1}{2}\pi_1 \right).$$

Equating this to $(\, \pi_0 \quad \pi_1 \quad \pi_2 \,)$ we readily see that the stationary distribution is a binomial distribution on the states: $\pi = (\frac{1}{4}, \frac{1}{2}, \frac{1}{4})$.

Moreover, this Markov chain does not have a limit distribution, in view of the periodicity

$$\begin{pmatrix} 0 & 1 & 0 \\ \frac{1}{2} & 0 & \frac{1}{2} \\ 0 & 1 & 0 \end{pmatrix} = \mathbf{M} = \mathbf{M}^3 = \mathbf{M}^5 = \ldots,$$

and

$$\begin{pmatrix} \frac{1}{2} & 0 & \frac{1}{2} \\ 0 & 1 & 0 \\ \frac{1}{2} & 0 & \frac{1}{2} \end{pmatrix} = \mathbf{M}^2 = \mathbf{M}^4 = \mathbf{M}^6 = \ldots,$$

that is, the chain is periodic, with period 2, and $\mathbf{M}^{(n)}$ alternates between two values without reaching a limit. If we are in any of the three states, it can take as few as two transitions to come back to it. ◇

1.7 Exchangeability

Independence is an important vehicle in probability. The probability of joint independent events decomposes into a product of probabilities of simpler individual events. When there is no independence, more complicated probability structures appear, and we need additional tools. Exchangeability is an important concept of modern probability that formalizes the notion of symmetry and provides tools for it. An example of symmetry arises, when X and Y are the independent outcomes of rolling two fair dice. Then we have the symmetry

$$\mathbf{P}(X = x, Y = y) = \mathbf{P}(Y = x, X = y),$$

for any $x, y \in \{1, \dots, 6\}$, and of course both are equal to $\mathbf{P}(X = x) \times \mathbf{P}(Y = y) = \frac{1}{36}$. In fact, the equality holds for any real numbers x and y—if either is infeasible as dice outcome the two probabilities are 0.

Exchangeability goes one step beyond independence. Formally, the random variables $X_1, X_2, \dots, X_k$ are exchangeable, if

$$(X_1, \dots, X_k) \overset{\mathcal{D}}{=} (X_{i_1}, \dots, X_{i_k}),$$

for any arbitrary permutation $(i_1, \dots, i_k)$ of the indexes in $\{1, \dots, k\}$. That is to say,

$$\mathbf{P}(X_1 \le x_1, \dots, X_k \le x_k) = \mathbf{P}(X_{i_1} \le x_1, \dots, X_{i_k} \le x_k),$$

or equivalently,

$$\mathbf{P}(X_1 \le x_1, \dots, X_k \le x_k) = \mathbf{P}(X_1 \le x_{i_1}, \dots, X_k \le x_{i_k}),$$

for any arbitrary permutation $(i_1, \dots, i_k)$ of the indexes in $\{1, \dots, k\}$.

For events $A_1, \dots, A_k$, we say they are exchangeable, when their indicators $\mathbf{1}_{A_1}, \dots, \mathbf{1}_{A_k}$ are exchangeable random variables (the indicator $\mathbf{1}_{\mathcal{E}}$ of the event $\mathcal{E}$ is a random variable that assumes the value 1, when $\mathcal{E}$ occurs, and assumes the value 0 otherwise). An infinite sequence of random variables (events) is said to be *infinitely exchangeable*, or just exchangeable, if every finite collection of its variables (events) is exchangeable.

Evidently, independent random variables are exchangeable, but there are exchangeable random variables that are not independent. For instance, Let $X_1, \dots, X_k$ be independent and identically distributed random variables, and Z be another independent random variable, and define

$$Y_i = X_i + Z, \qquad \text{for } i = 1, \dots, k.$$

The random variables Y_i are dependent, at the same time they are exchangeable (we leave the proof as an exercise).

Our purpose here is to touch on the subject to acquaint a student at the intermediate level with the topic, and also show that it is a property that arises naturally in Pólya urn structures, and can be a useful vehicle in proofs.

This chapter is an opportunity to introduce an important exchangeability theorem, due to de Finetti (1937). First, we get a feel for exchangeability through an urn example.

Example 1.3
Consider the white-blue Pólya-Eggenberger urn scheme in which balls are sampled with replacements, and whenever a ball color appears in the sample, we add an extra ball of the same color in the urn, supposing we started with one ball of each color. To see the mechanism of exchangeability, let $\hat{W}_i$ be the indicator of the event of picking a white ball in the ith draw, for all $i \ge 1$. That is, $\hat{W}_i$ assumes the value 1 if the ball in the ith draw is white, and assumes the value 0, if it is blue. We can quickly compute

$$\mathbf{P}(\hat{W}_1 = 1, \hat{W}_2 = 1, \hat{W}_3 = 0) = \frac{1}{2} \times \frac{2}{3} \times \frac{1}{4} = \frac{1}{12}.$$

Likewise,

$$\mathbf{P}(\hat{W}_1 = 1, \hat{W}_2 = 0, \hat{W}_3 = 1) = \frac{1}{2} \times \frac{1}{3} \times \frac{2}{4} = \frac{1}{12},$$

and

$$\mathbf{P}(\hat{W}_1 = 0, \hat{W}_2 = 1, \hat{W}_3 = 1) = \frac{1}{2} \times \frac{1}{3} \times \frac{2}{4} = \frac{1}{12}.$$

All sequences with only one blue sample have the same probability. The reason for this equality, irrespective of the sequence of events, is an inherent exchangeability. The three possibilities with only one white ball in the first three picks also have equal probabilities:

$$\begin{aligned}\mathbf{P}(\hat{W}_1 = 1, \hat{W}_2 = 0, \hat{W}_3 = 0) &= \mathbf{P}(\hat{W}_1 = 0, \hat{W}_2 = 1, \hat{W}_3 = 0)\\ &= \mathbf{P}(\hat{W}_1 = 0, \hat{W}_2 = 0, \hat{W}_3 = 1)\\ &= \frac{1}{12}.\end{aligned}$$

Just to complete the list, we add

$$\mathbf{P}(\hat{W}_1 = 1, \hat{W}_2 = 1, \hat{W}_3 = 1) = \frac{1}{2} \times \frac{2}{3} \times \frac{3}{4} = \frac{1}{4},$$

and

$$\mathbf{P}(\hat{W}_1 = 0, \hat{W}_2 = 0, \hat{W}_3 = 0) = \frac{1}{2} \times \frac{2}{3} \times \frac{3}{4} = \frac{1}{4}.$$

The complete list shows that

$$\mathbf{P}(\hat{W}_1 = \hat{w}_1, \hat{W}_2 = \hat{w}_2, \hat{W}_3 = \hat{w}_3) = \mathbf{P}(\hat{W}_{i_1} = \hat{w}_1, \hat{W}_{i_2} = \hat{w}_2, \hat{W}_{i_3} = \hat{w}_3), \quad (1.4)$$

for any permutation (i_1, i_2, i_3) of the set $\{1, 2, 3\}$, and any feasible sequence $\hat{w}_1, \hat{w}_2, \hat{w}_3$, where feasible means $\hat{w}_i \in \{0, 1\}$. If one or more of the numbers

$\hat{w}_1$, $\hat{w}_2$, and $\hat{w}_3$ is not feasible, both events $\{\hat{W}_1 = \hat{w}_1, \hat{W}_2 = \hat{w}_2, \hat{W}_3 = \hat{w}_3\}$ and $\{\hat{W}_{i_1} = \hat{w}_1, \hat{W}_{i_2} = \hat{w}_2, \hat{W}_{i_3} = \hat{w}_3\}$ become impossible, with probability 0, and we can also assert the form in Eq. (1.4) in this case, too. For any sequence of real numbers the form in Eq. (1.4) is valid. Indeed, $\hat{W}_1$, $\hat{W}_2$, and $\hat{W}_3$ are exchangeable. ◇

The argument in Example 1.3 extends to any initial sequence of indicators $\hat{W}_1, \dots, \hat{W}_n$, for $n \ge 2$, and in fact for any subset of indicators, as we argue next. For transparency of the argument, let us first consider an instance with a longer drawing sequence. Suppose we drew 9 balls according to the rules, and white balls were observed in the sampled balls on draws 4, 7, and 8 (blue balls appear in all other draws). The probability of this event is

$$\frac{1}{2} \times \frac{2}{3} \times \frac{3}{4} \times \frac{1}{5} \times \frac{4}{6} \times \frac{5}{7} \times \frac{2}{8} \times \frac{3}{9} \times \frac{6}{10} = \frac{3!\,6!}{10!} = \frac{1}{840}.$$

Notice how we get an increasing product of consecutive integers in the numerator until the first draw of a white ball (4th draw). When this pattern is broken, a second pattern of increasing consecutive integers is started. That first white ball drawing does not change the number of blue balls, and the trend of increasing numbers continues in the first pattern with the next blue ball drawing, picking up right after the first white ball drawing. The number of white balls remains unchanged (at two) until the second drawing of a white ball (7th draw), and that is when the second pattern picks up its next increasing integer, etc.; two factorials form in the numerator. In the denominator a pattern of increasing consecutive integers starts at 2 and remains uninterrupted, as we consistently add a ball after each draw.

More generally, take the case of $\hat{W}_1, \dots, \hat{W}_n$ for arbitrarily large but fixed $n \ge 2$. Suppose there is a total of k among the indicators that are 1, the rest are 0, and suppose the ones occur at positions

$$1 \le i_1 < i_2 < \cdots < i_k \le n.$$

The probability of this event is

$$\begin{aligned}
\mathbf{P}(&\hat{W}_1 = 0, \hat{W}_2 = 0, \dots, \hat{W}_{i_1-1} = 0, \hat{W}_{i_1} = 1, \hat{W}_{i_1+1} = 0, \dots, \hat{W}_{i_k-1} = 0, \\
&\hat{W}_{i_k} = 1, \hat{W}_{i_k+1} = 0, \dots, \hat{W}_n = 0) \\
&= \frac{1}{2} \times \frac{2}{3} \times \cdots \times \frac{i_1 - 1}{i_1} \times \frac{1}{i_1 + 1} \times \frac{i_1}{i_1 + 2} \times \frac{i_1 + 1}{i_1 + 3} \times \cdots \times \frac{i_2 - 2}{i_2} \\
&\quad \times \frac{2}{i_2 + 1} \times \frac{i_2 - 1}{i_2 + 2} \times \cdots \times \frac{i_k - k}{i_k} \times \frac{k}{i_k + 1} \times \frac{i_k - k + 1}{i_k + 2} \\
&\quad \times \cdots \times \frac{n - k}{n + 1} \\
&= \frac{k!\,(n-k)!}{(n+1)!}.
\end{aligned}$$

On the other hand

$$\begin{aligned}\mathbf{P}(\hat{W}_1 = 1, \ldots, \hat{W}_k = 1, \hat{W}_{k+1} = 0, \ldots, \hat{W}_n = 0) \\ &= \frac{1}{2} \times \frac{2}{3} \times \frac{3}{4} \times \cdots \times \frac{k}{k+1} \times \frac{1}{k+2} \times \frac{2}{k+3} \times \cdots \times \frac{n-k}{n+1} \\ &= \frac{k!\,(n-k)!}{(n+1)!},\end{aligned}$$

too. Note how the final expression does not include $i_1, \ldots, i_k$. The probability of having k white ball draws in n draws does not depend on where in the sequence the white balls are drawn. What only matters is that there are k of them, and all sequences with the same number of white ball drawings have the same probability. The sequence $\hat{W}_1, \hat{W}_2, \ldots$ is infinitely exchangeable.

As a corollary, if we let W_n be the number of white balls in the urn after n draws, then $W_n = k + 1$ if there are k white samples in the first n draws. These white samples can take place anywhere among the first n equally likely (exchangeable) events of drawing a white ball in $\binom{n}{k}$ ways and

$$\mathbf{P}(W_n = k + 1) = \binom{n}{k} \times \frac{k!\,(n-k)!}{(n+1)!} = \frac{1}{n+1}, \qquad \text{for } k = 0, 1, \ldots, n. \tag{1.5}$$

1.7.1 Mixtures of Distributions

Central to the ideas in de Finetti's exchangeability theorem is the notion of mixing. We illustrate mixing by some urn examples. Suppose we have two urns A and B, with different compositions: A has $\frac{1}{3}$ of its balls being white, the rest are blue, and B has $\frac{3}{4}$ of its balls being white, the rest are blue. If we sample with replacement from A, it takes Geo($\frac{1}{3}$) attempts to get a white ball, but if we sample from B it takes Geo($\frac{3}{4}$) attempts to get a white ball. Let F_A be the distribution function of the geometric random variable Geo($\frac{1}{3}$), and let F_B be the distribution function of Geo($\frac{3}{4}$). Suppose we decided from which urn we sample by flipping a fair coin. If the coin turns up heads (event $\mathcal{H}$) we sample from A, otherwise the coin turns up tails (event $\mathcal{T}$) and we sample from B. How long does it take to get a white ball? Let this waiting time be denoted by W. Then

$$\begin{aligned}P(W \le k) &= \mathbf{P}(W \le k \,|\, \mathcal{H})\, \mathbf{P}(\mathcal{H}) + \mathbf{P}(W \le k \,|\, \mathcal{T})\, \mathbf{P}(\mathcal{H}) \\ &= \frac{1}{2} F_A(k) + \frac{1}{2} F_B(k).\end{aligned}$$

We say that the two geometric distributions have been *mixed* with equal probability, or that W has a *mixture distribution* of two equally likely geometric distributions. The two mixing probabilities are those of a Bernoulli random variable Ber($\frac{1}{2}$), and we say the Ber($\frac{1}{2}$) is a *mixer*.

One can do the same with any two (or more) distribution functions. For instance if $F_1(x), F_2(x), \ldots$ are distribution functions of any kind, and M is a discrete mixing random variable with distribution $\mathbf{P}(M = k) = p_k$, then one calls the distribution function

$$F(x) = \sum_{k=1}^{\infty} a_k F_k(x)$$

a mixed distribution function, or calls its underlying random variable a mixture of the random variables with distribution functions $F_1(x), F_2(x), \ldots,$ with mixer M.

The choice of the distribution can even be determined by picking a value from a continuum. For example, we may have a choice from normal distributions $\mathcal{N}(x, x^2)$, where the value of x is obtained from some continuous distribution with density $f_X(x)$. In this case, the mixed random variable $Y = \mathcal{N}(X, X^2)$ has the mixed distribution

$$F(y) = \int_{-\infty}^{\infty} \mathbf{P}(Y \le y \mid X = x) f_X(x)\, dx = \int_{-\infty}^{\infty} \Phi^*(x) f_X(x)\, dx,$$

with X being the normal mixer (here $\Phi^*(x)$ is the distribution function of $\mathcal{N}(x, x^2)$).

1.7.2 De Finetti's Theorem for Indicators

We discuss in this section an important form of de Finetti's theorem specialized for indicators. We first illustrate by an example the underpinning idea.

Example 1.4

A trial may result in a success event B with probability $p \in (0, 1)$. Suppose, a series of n independent success-failure trials are performed, and let B_i be the event that success occurs in the ith trial. Let Y_n be the number of successes. As usual, Y_n has a binomial distribution, given by

$$\mathbf{P}(Y_n = k) = \binom{n}{k} p^k (1 - p)^{n-k}.$$

Suppose next that p is not fixed, but rather is a random variable Θ that has a *prior distribution* on the interval $(0, 1)$, such as the Uniform (0,1), say. The notion of randomized parameters is a core idea in Baysian statistics, where one leaves the door open to learn from statistical data, by updating the prior distribution to a posterior distribution, after having obtained some sample that may sway our prior opinion one way or the other. Let Θ have density function $f(\theta)$, supported on $(0, 1)$. In other words, we assume that θ is first obtained as a sample from the prior distribution, then is used as the success probability to generate an independent sequence of Bernoulli(θ) random variables. That is, while the events B_i are dependent, they are *conditionally independent*, given

$\Theta = \theta$. Despite their dependence, we demonstrate by a quick calculation that the events B_i are exchangeable: For any subset of indexes $\{i_1, \dots, i_k\} \subseteq \{1, \dots, n\}$, we have

$$\begin{aligned}
\mathbf{P}(B_{i_1} \cap \dots \cap B_{i_k}) &= \int_0^1 \mathbf{P}(B_{i_1} \cap \dots \cap B_{i_k} \mid \Theta = \theta) f(\theta)\, d\theta \\
&= \int_0^1 \mathbf{P}(B_{i_1} \mid \Theta = \theta) \cdots \mathbf{P}(B_{i_k} \mid \Theta = \theta) f(\theta)\, d\theta \\
&= \int_0^1 \theta^k f(\theta)\, d\theta \\
&= \mathbf{P}(B_1 \cap \dots \cap B_k).
\end{aligned}$$

We note that the probability expression is free of the choice $\{i_1, \dots, i_k\} \subseteq \{1, \dots, n\}$. All collections of size k among $B_1, \dots, B_n$ have the same probability of joint occurrence. The events $B_1, B_2, \dots, B_n$ are exchangeable, for each $n \ge 2$. ◇

In Example 1.4 let $\mathbf{1}_{B_i}$ be the indicator of the event B_i. The number of successes can be represented as a sum of exchangeable random variables:

$$Y_n = \mathbf{1}_{B_1} + \mathbf{1}_{B_2} + \cdots + \mathbf{1}_{B_n}.$$

So $Y_n = k$, if any k of the indicators are 1, and the rest are 0. That is,

$$\begin{aligned}
\mathbf{P}(Y_n = k) &= \int_0^1 \mathbf{P}(Y_n = k \mid \Theta = \theta) f(\theta)\, d\theta \\
&= \int_0^1 \binom{n}{k} \mathbf{P}\big(\mathbf{1}_{B_1} = 1, \dots, \mathbf{1}_{B_k} = 1, \\
&\qquad \mathbf{1}_{B_{k+1}} = 0, \dots, \mathbf{1}_{B_n} = 0 \mid \Theta = \theta\big) f(\theta)\, d\theta \\
&= \int_0^1 \binom{n}{k} \theta^k (1-\theta)^{n-k} f(\theta)\, d\theta,
\end{aligned}$$

within the range $k = 0, 1, \dots, n$.

A key idea here is that we could represent Y_n as a mixture of binomial random variables (each is a sum of conditionally independent Bernoulli random variables). This illustrates a more general principle that a sum of infinitely exchangeable random variables can be obtained as a mixture of conditionally independent random variables, and their probabilities can be obtained by integration over *some* distribution function.

De Finetti's theorem, presented next, asserts that the partial sum in every infinitely exchangeable sequence of Bernoulli random variables can be represented as a mixture of binomial random variables.

THEOREM 1.2

(De Finetti, 1937). Let $X_1, X_2, \ldots$ be an infinitely exchangeable sequence of indicators. There is a distribution function $F(x)$ such that

$$\mathbf{P}(X_1 = 1, \ldots, X_k = 1, X_{k+1} = 0, \ldots, X_n = 0) = \int_0^1 x^k (1-x)^{n-k} \, dF(x),$$

for each n and $0 \le k \le n$.

PROOF The proof we present here is due to Heath and Sudderth (1976). As the given sequence is exchangeable, all $\binom{m}{r}$ realizations of $X_1, \ldots, X_m$, for any $m \ge 1$, that attain the same total sum S_m have the same probability. We consider $m \ge n$. Conditioning on the sum being $r \ge k$, there are m-long sequences that favor the event $X_1 = 1, \ldots, X_k = 1, X_{k+1} = 0, \ldots, X_n = 0$. Every one of these equally likely sequences has the right assignments up to n, and any arbitrary assignments of ones and zeros to the *tail* variables $X_{r+1}, \ldots, X_m$, that are consistent with the sum being r. Of the total of r ones, there are k ones among $X_1, \ldots, X_n$. So, $r - k$ ones should appear in the tail, giving

$$\begin{aligned}
&\mathbf{P}(X_1 = 1, \ldots, X_k = 1, X_{k+1} = 0, \ldots, X_n = 0) \\
&\quad = \sum_{r=k}^{m} \mathbf{P}(X_1 = 1, \ldots, X_k = 1, X_{k+1} = 0, \ldots, X_n = 0 \mid S_m = r) \times \mathbf{P}(S_m = r) \\
&\quad = \sum_{r=k}^{m} \frac{\binom{m-n}{r-k}}{\binom{m}{r}} \mathbf{P}(S_m = r) \\
&\quad = \sum_{r=k}^{m} \frac{(m-n)!}{(r-k)!\,(m-n-r+k)!} \times \frac{r!\,(m-r)!}{m!} \mathbf{P}(S_m = r) \\
&\quad = \sum_{r=k}^{m} \frac{(m-n)!}{m!} \times \frac{r!}{(r-k)!} \times \frac{(m-r)!}{(m-n-r+k)!} \mathbf{P}(S_m = r) \\
&\quad = \sum_{r=k}^{m} \frac{(r)_k (m-r)_{n-k}}{(m)_n} \mathbf{P}(S_m = r).
\end{aligned}$$

The result is expressed in terms of Pochhammer's symbol for falling factorials: $(y)_j = y(y-1)\ldots(y-j+1)$, for real y and integer j. Set $\theta = r/m$, and write the discrete probability of the sum S_m as the amount of jump in its distribution function F_m at the point j. The given sequence is infinitely exchangeable, and we can take m as large as we please to express the probability as

$$\begin{aligned}\mathbf{P}(X_1 = 1, \ldots, X_k = 1, X_{k+1} = 0, \ldots, X_n = 0) \\ = \sum_{\theta=k/m}^{1} \frac{(\theta m)_k (m - m\theta)_{n-k}}{(m)_n}\, dF_m(\theta);\end{aligned}$$

in the sum θ ascends from k/m to 1 on ticks that are $\frac{1}{m}$ apart, and are the jump points of F_m.

Suppose the sequence $\{F_m\}_{m=1}^{\infty}$ has a subsequence $\{F_{m_j}\}_{j=1}^{\infty}$ converging to a limit distribution function F, it will follow that

$$\begin{aligned}\mathbf{P}(X_1 &= 1, \ldots, X_k = 1, X_{k+1} = 0, \ldots, X_n = 0) \\ &= \lim_{j\to\infty} \sum_{\theta=k/m_j}^{1} \frac{(\theta m_j)_k (m_j - m_j\theta)_{n-k}}{(m_j)_n}\, dF_{m_j}(\theta) \\ &\to \int_0^1 \theta^k (1-\theta)^{n-k}\, dF(\theta).\end{aligned}$$

The existence and uniqueness of such a limit is guaranteed by a technical selection theorem known as Helly's theorem. The idea is like that of the existence of an at least one accumulation point in every sequence of real numbers (see Feller, 1971, Vol. II, p. 261). ■

Example 1.5

Let us revisit the white-blue Pólya-Eggenberger urn scheme of Example 1.3 in which balls are sampled with replacement, and whenever a certain ball color appears in the sample, we add an extra ball of the same color in the urn. We started with one ball of each color. Let W_n be the number of white balls after n draws, and $\hat{W}_n$ be the indicator of the event of drawing a white ball in the nth sampling, and we have the representation

$$W_n = 1 + \hat{W}_1 + \cdots + \hat{W}_n.$$

We showed in Eq. (1.5) by direct calculation that W_n has the distribution of a Uniform $[1\,..\,n+1]$ random variable.

Another way to get this result is via de Finetti's theorem, which is an opportunity to drill its concept. According to the theorem, there is a distribution function $F_V(x)$, corresponding to a mixing random variable V, such that

$$\mathbf{P}(\hat{W}_1 = 1, \ldots, \hat{W}_k = 1, \hat{W}_{k+1} = 0, \ldots, \hat{W}_n = 0) = \int_0^1 \theta^k (1-\theta)^{n-k} dF_V(\theta),$$

for all $0 \le k \le n$. To characterize V, we apply this result at $k = n$, for each $n \ge 1$, and get

$$\frac{1}{2} \times \frac{2}{3} \times \cdots \times \frac{n}{n+1} = \mathbf{P}(\hat{W}_1 = 1, \ldots, \hat{W}_n = 1) = \int_0^1 \theta^n dF_V(\theta) = \frac{1}{n+1}.$$

However, we also know that the moments of U, a standard Uniform(0, 1) random variable, are

$$\mathbf{E}[U^n] = \int_0^1 u^n \, du = \frac{1}{n+1}.$$

The moments of U and V are the same, and the uniform is one of the distributions uniquely characterized by its moments. Hence, the mixer is uniform:

$$V \overset{\mathcal{D}}{=} \text{Uniform}(0, 1),$$

with distribution function $F_V(\theta) = \theta$, for $\theta \in (0, 1)$. We can proceed to find the distribution of the number of white balls from the relation

$$\begin{aligned}
\mathbf{P}(W_n = k+1) &= \mathbf{P}\left(\sum_{i=1}^{n} \hat{W}_i = k\right) \\
&= \sum_{\substack{\hat{w}_1 + \cdots + \hat{w}_k = k \\ \hat{w}_i \in \{0,1\},\ i=1,\ldots,k}} \mathbf{P}(\hat{W}_1 = \hat{w}_1, \ldots, \hat{W}_n = \hat{w}_n) \\
&= \sum_{\substack{\hat{w}_1 + \cdots + \hat{w}_k = k \\ \hat{w}_i \in \{0,1\},\ i=1,\ldots,k}} \mathbf{P}(\hat{W}_1 = 1, \ldots, \hat{W}_k = 1, \hat{W}_{k+1} = 0, \ldots, \hat{W}_n = 0) \\
&= \binom{n}{k} \mathbf{P}(\hat{W}_1 = 1, \ldots, \hat{W}_k = 1, \hat{W}_{k+1} = 0, \ldots, \hat{W}_n = 0) \\
&= \binom{n}{k} \int_0^1 \theta^k (1-\theta)^{n-k} \, d\theta \\
&= \binom{n}{k} \beta(k+1, n-k+1) \\
&= \binom{n}{k} \times \frac{k!\,(n-k)!}{(n+1)!} \\
&= \frac{1}{n+1},
\end{aligned}$$

for $k = 0, \ldots, n$, and we have rederived the uniform distribution for W_n via de Finetti's theorem. ◇

Exercises

1.1 People take the liberty in using statements like "choose an integer at random," usually meaning consider the integers to be equally likely. Show that this statement is truly void of meaning by demonstrating that there does not exist a probability measure that assigns equal probability to all the singletons on a discrete probability space $(\Omega, \mathcal{P}(\Omega))$, where Ω is infinite but countable, and $\mathcal{P}(\Omega)$ is its power set (set of all subsets of Ω).

1.2 Let Ω be a sample space and let $\mathcal{F}$ be a sigma-field defined on it. Suppose P_1 and P_2 are probability measures defined on $(\Omega, \mathcal{F})$. Choose an arbitrary $\alpha \in [0, 1]$. Prove that the set function $\mathbf{P}(A) = \alpha\mathbf{P}_1(A) + (1 - \alpha)\mathbf{P}_2(A)$, for any measurable set A, is also a probability measure on $(\Omega, \mathcal{F})$.

1.3 What is the minimum number of points a sample space must contain in order that there exist n independent events $A_1, \ldots, A_n$, none of which has probability 0 or 1?

1.4 An urn contains five white balls and three blue balls. Three balls are taken out (one at a time) and each draw is at random and without replacement.

(a) What is the probability that all three balls are white?

(b) What is the probability that two balls in the sample are white and one is blue?

(c) If the sequence of balls withdrawn is also recorded, what is the probability that the first two balls in the sample are white and the third is blue?

1.5 Urn A contains three white and six blue balls; urn B contains four white and five blue balls. A ball is sampled at random from urn A, and is replaced in urn B. A ball is now sampled from urn B.

(a) What is the conditional probability that the second ball sampled is blue, given the first was blue?

(b) What is the probability the second ball sampled is blue?

1.6 For tennis practice, a coach pools a large number of balls in a big bin for a champion he is training. Commercially available bins have a capacity for 100 balls. On a particular day, there are 91 balls in the bin. Of these balls 40 are new, and the rest are used. The coach thinks that the champion's performance is very slightly improved when he plays with brand new balls. The coach chooses six balls at random from the bin for the first set. After the set, the balls are set aside, and six other balls are taken from the bin at random for the second set. What is the probability that the balls used in the first set are all new? What is the probability that the balls used in the second set are all new?

1.7 An urn contains 9 white balls and 11 blue balls. A common commercial fair die is thrown, and the number showing on the upturned face determines the number of balls to be taken out randomly from the urn. What is the probability that there are more white balls than blue in the sample?

1.8 (Bernstein, 1946). An urn contains four balls labeled 000, 110, 101, and 011. A ball is drawn at random (all balls being equally likely). The event A_i, for $i = 1, 2, 3$, is the event that 1 is observed in the ith position. For instance, if the ball labeled 101 is the one withdrawn, the events A_1 and A_3 occur, but A_2 does not. Are the events A_1, A_2, A_3 pairwise independent? Are they totally independent?

1.9 Three urns A, B, and C are initially empty. A ball is tossed at random into one of the three urns, hitting any of them with the same probability. In discrete steps the ball is taken from whichever urn it is in and tossed into one of the other two urns with equal probability. What is the probability that the ball is in urn A after n moves? If the ball is initially deposited in urn A instead of tossing it randomly in one of the three urns, does the probability of the event that the ball is back in urn A after n moves converge to a limit?

1.10 An urn contains n white and n blue balls. Balls are drawn at random in stages until one color is depleted. The number of draws until this event happens is a waiting time. What is the distribution of this waiting time?

1.11 An urn receives n balls of random colors. Each ball deposited is painted at random: With probability p the paint is white, and with probability $1 - p$ it is blue. A sample of size k is then taken out with replacement. What is the average number of white balls appearing in the sample? Argue that the distribution of the number of white balls in the sample is the compound binomial $\text{Bin}(k, \frac{1}{n}\text{Bin}(n, p))$.

1.12 (Woodbury, 1949). An urn contains w white balls and b blue balls. A ball is sampled at random from the urn. If the ball withdrawn is white, it is replaced in the urn and there is no change in the urn composition, but if the ball withdrawn is blue, a white ball is removed from the urn, and the blue ball is replaced in the urn together with an extra blue ball. This urn scheme was suggested as a model for infestation, where the white balls are healthy members of a community, and blue balls are infested. Eventually all the balls in the urn become blue, representing total prevalence of the infestation in the community. How long (in terms of ball draws) does it take on average to reach a state of total infestation?

1.13 An urn receives a colored ball at each epoch of a Poisson process with intensity λ. The ball color is determined randomly by flipping a fair coin. If the upturned coin face is heads, a white ball is added into the urn, if the upturned face is tails, a blue ball is added into the urn. What is the distribution of the number of white balls at time t?

1.14 An urn has one white and one blue ball. A ball is drawn at random and discarded. A ball of a color opposite to the color of the ball withdrawn is deposited in the urn, and a second ball drawing takes place. Let $\hat{W}_i$ be the indicator of drawing a white ball on draw $i \in \{1, 2\}$. Show that $\hat{W}_1$ and $\hat{W}_2$ are exchangeable.

1.15 In the previous exercise, show that there does not exist a distribution function $F(x)$, so that

$$\mathbf{P}(\hat{W}_1 = 1, \ldots, \hat{W}_k = 1, \hat{W}_{k+1} = 0, \ldots, \hat{W}_2 = 0) = \int_0^1 x^k (1-x)^{2-k} dF(x),$$

for $k = 0, 1, 2$. How do you reconcile the conclusion with de Finetti's theorem?

1.16 (Heath and Sudderth, 1976). An urn contains w white and b blue balls. The balls are sampled at random and without replacement until all the urn contents are taken out. Let X_i be 1 if the ball in the ith draw is white, otherwise X_i is 0. Show that the random variables $X_1, \ldots, X_{w+b}$ are exchangeable. Are they independent?

2

Some Classical Urn Problems

A number of classical problems, attributed to some famous mathematicians belonging to the early school of probability theory, have connections to Pólya urns and other types of urns. We review a few of these problems in this chapter. In later chapters we shall revisit these problems and recast most of them within the Pólya urn framework.

2.1 Ballot Problems

These types of problems are concerned with the progress of a vote. The specific variation presented below was first proposed and solved in 1878 by W. Whitworth, who included the solution in the fourth edition of his book *Choice and Chance* in 1886. In 1887, J. Bertrand presented another solution, and asked for a direct proof. The problem was solved in the same year by D. André, who introduced the reflection principle.

Suppose two candidates A and B are running against each other in an election, and A wins (or is tied) by receiving $m \ge n$ votes, where n is the number of votes received by B. During the count the media makes projections about winning. Such a projection is reinforced with every vote received, if A remains ahead throughout. The classical question is: *What is the probability that A stays ahead of B throughout the voting*? If $m = n$, the probability of the event in question is 0, because by the time the last vote is received there is an equal number of votes for each, and A is not ahead after the $(m + n)$th vote is cast.

We can view the progress of votes as white and blue balls (ballots) being deposited in an urn (the ballot box). White balls represent A, and there is a total of m of them, and blue balls represent B, and there is a total of n of them. To fix our probability measure, we assume that all $(m+n)!$ sequences of ballot deposits are equally likely. Let the notation $A \triangleright B$ stand for the event that A is always ahead of B. This event corresponds to a certain proportion of the sequences of possible deposits. The desired probability $\mathbf{P}(A \triangleright B)$ depends on both m and n; we can also denote it by a notation that reflects this dependence, such as Q_{mn}.

We first write a recurrence for the probability, by conditioning on who receives the last vote. Let L be the last vote cast in the race. In our urn correspondence L is either a white ball (W) or a blue ball (B). Therefore,

$$Q_{mn} = \mathbf{P}(A \triangleright B \mid L = W)\,\mathbf{P}(L = W) + \mathbf{P}(A \triangleright B \mid L = B)\,\mathbf{P}(L = B).$$

The event $L = W$ occurs by reserving any of the m white balls to be last in the sequence. The rest of $m + n - 1$ balls can be permuted in $(m + n - 1)!$ ways. Thus,

$$\mathbf{P}(L = W) = m \times \frac{(m+n-1)!}{(m+n)!} = \frac{m}{m+n}.$$

Likewise,

$$\mathbf{P}(L = B) = n \times \frac{(m+n-1)!}{(m+n)!} = \frac{n}{m+n}.$$

For $A \triangleright B \mid L = W$ to happen, A has to stay ahead through the first $m + n - 1$ deposits, too, which is recursively the same problem, but on $m - 1$ votes for A, and n votes for B. And so, the probability of this conditional event is $Q_{m-1,n}$. Likewise, the probability $\mathbf{P}(A \triangleright B \mid L = B)$ is $Q_{m,n-1}$. Thus,

$$Q_{mn} = \frac{m}{m+n} Q_{m-1,n} + \frac{n}{m+n} Q_{m,n-1}.$$

We shall show by induction on $m + n$ that

$$Q_{mn} = \frac{m-n}{m+n},$$

for $m \ge n$. The expression is true for $m = n$, yielding 0 probability as already discussed; we assume in the sequel that $m > n$. For a basis, we take $m + n = 1$, necessitating that $m = 1 > n = 0$. In this boundary case $A \triangleright B$ is a sure event, with probability one; we readily check $(m - n)/(m + n) = (1 - 0)/(1 + 0) = 1$.

Assuming the proposed formula is true for $m + n - 1 \ge 1$, we have

$$\begin{aligned}
Q_{mn} &= \frac{m}{m+n} \times \frac{(m-1)-n}{(m-1)+n} + \frac{n}{m+n} \times \frac{m-(n-1)}{m+(n-1)} \\
&= \frac{m^2 - n^2 + n - m}{(m+n)(m+n-1)} \\
&= \frac{(m-n)(m+n) - (m-n)}{(m+n)(m+n-1)} \\
&= \frac{m-n}{m+n}.
\end{aligned}$$

Take note of the recursive side that required the pair $m - 1$ and n. As $m > n$, it follows that $m - 1 \ge n$, and if $m - 1$ is still greater than n, this invokes further recursion, but in the case of equality $m - 1 = n$, we interpret $Q_{m-1,n}$ to be 0, as discussed. The induction is complete.

2.2 Occupancy Problems

The occupancy problem was first studied by de Moivre in 1713, who considered it a problem in the *Doctrine of Measurement of Chance*. There are n balls that will be thrown into k urns. Each ball is independently deposited in a uniformly chosen urn. *What is the probability that no urn will be empty*?

If we denote by $U_1, U_2, \ldots, U_n$ the sequence of urns chosen as the balls are deposited in succession, all k^n sequences of numbers from the set $\{1, \ldots, k\}$ are equally likely for $U_1, U_2, \ldots, U_n$. This is what got to be known in physics long after de Moivre by the name the Maxwell-Boltzmann statistic, a model for gases of indistinguishable particles in distinguishable energy states. The probability that a particular urn, say the ith, is not chosen at an individual throw is $1 - 1/k$. Let A_i be the event that at the end of tossing all the balls urn i is empty. After n independent ball throws,

$$\mathbf{P}(A_i) = \left(1 - \frac{1}{k}\right)^n.$$

We extend the argument, for any collection of urns indexed $1 \leq i_1 < i_2 < \ldots < i_j \leq k$. The probability that a particular throw hits any of the j specified urns is j/k. The probability that they are all empty after one ball tossing is $1 - j/k$, and in n ball throws

$$\mathbf{P}(A_{i_1} \cap A_{i_2} \cap \ldots \cap A_{i_j}) = \left(1 - \frac{j}{k}\right)^n.$$

The event that no urn is empty is the complement of the event $A = A_1 \cup A_2 \cup \ldots \cup A_k$. We can find the probability of A via the principle of inclusion–exclusion:

$$\begin{aligned}
\mathbf{P}(A^c) &= 1 - \mathbf{P}(A_1 \cup A_2 \cup \ldots \cup A_k) \\
&= 1 - \Bigg(\sum_{i=1}^{k} \mathbf{P}(A_i) - \sum_{1 \leq i < j \leq k}^{k} \mathbf{P}(A_i \cap A_j) \\
&\qquad + \cdots + \sum_{1 \leq i_1 < i_2 < \ldots < i_j \leq k} (-1)^{j-1} \mathbf{P}(A_{i_1} \cap A_{i_2} \cap \ldots \cap A_{i_j}) \\
&\qquad + \cdots + (-1)^{k-1} \mathbf{P}(A_1 \cap A_2 \cap \ldots \cap A_k)\Bigg) \\
&= 1 - k\mathbf{P}(A_1) + \binom{k}{2} \mathbf{P}(A_1 \cap A_2) + \cdots + (-1)^j \binom{k}{j} \mathbf{P}(A_1 \cap A_2 \cap \ldots \cap A_j) \\
&\qquad + \cdots + (-1)^k \mathbf{P}(A_1 \cap A_2 \cap \ldots \cap A_k) \\
&= \sum_{j=0}^{k} (-1)^j \binom{k}{j} \left(1 - \frac{j}{k}\right)^n.
\end{aligned}$$

This kind of probability formula is remarkable, as it involves a finite series of terms with alternating signs. But of course, probability is always nonnegative. Somehow, the positive terms dominate, a fact that is not a priori obvious from the formula itself.

2.3 Coupon Collector's Problems

A class of urn problems is associated with coupon collection. There are n different coupons (an indefinite supply of balls of n different colors in an urn from which we sample without replacement) that come with a product. The coupons are obtained at random, each purchase generates a coupon numbered from 1 to n. When a coupon is collected the buyer keeps it, and the company replaces the product in the market. *How soon does the collector get all n different coupons*?

A second urn view is to visualize the n coupon types as n balls of different colors placed in an urn from which we draw with replacement. In terms of the latter urn analogy, the balls are sampled at random, with replacement at each stage, and the sampling is continued until all n colors are observed. How long does it take to produce such a sample?

A version of the problem was popularized in the 1930s by the Dixie Cup Company. The Dixie Cup Company produced ice cream cups with a cartoon cover that had hidden on the underside a likeable picture. At first the company used some animals pictures. Later on, pictures of Hollywood stars and other themes were used. For instance, the 1952 theme was a collection of 24 baseball players. Baseball fans, mostly young boys, were among the most avid of collectors. The fans were enticed to buy more ice cream to complete the collection, believing its value will appreciate over time. Today, in 2008 a complete set of baseball players' pictures obtained cheaply in 1952 is worth about 100 dollars.

The solution to this problem is related to the probability in the occupancy problem. Let Y be the number of draws required to secure a set of balls of n different colors. The event $Y \le y$ requires that all n coupon types appear before y draws. This is just like the occupancy event that if we drop y balls at random in n urns, each urn will get at least one hit and none of the urns will be empty. Therefore,

$$\mathbf{P}(Y \le y) = \sum_{j=0}^{n} (-1)^j \binom{n}{j} \left(1 - \frac{j}{n}\right)^y,$$

for $y = n, n+1, n+2, \ldots$. As remarked at the end of Section 2.2, it is remarkable that such a probability has negative terms (but the terms collectively come out nonnegative). Computing the average waiting time from the formula

$$\mathbf{E}[Y] = \sum_{y=n}^{\infty} y\,\mathbf{P}(Y = y)$$

is a daunting task. Exercise 2.6 offers an indirect, but algebraically easier, route for it.

2.4 The Gambler's Ruin

The gambler's ruin is a classical problem directly related to urns. The problem is associated with a number of famous scientists and mathematicians including a few Bernoullis, de Moivre, Feller, Fermat, Huygens, Laplace, Lagrange, Pascal, and several others. There are correspondences between Fermat and Pascal about it that are mentioned in Huygens (1657).

Two gamblers A and B keep going at betting until one of them goes bankrupt. Gambler A has in his possession m dollars, and gambler B has n. They bet repeatedly on the outcome of a game of flipping a coin: A bets on heads, and B bets on tails. The probability of a head turning up in an individual flip is p, and the probability of a tail is $q = 1 - p$. After a flip, the loser in that individual game gives a dollar to the winner. *What is the probability that B goes bankrupt (and A wins all the money)?*

In the language of urns, we have two urns A and B that start out with m balls in A and n in B. A coin is flipped repeatedly. If the result of the flip is heads (H), which occurs with probability p, we transfer a ball from urn B to A, otherwise the outcome is tails (T), and we transfer a ball from A to B. The game continues until it is no longer possible to proceed because one urn has run out of balls. What is the probability that at stoppage urn A has all the balls and urn B is empty? This urn scheme is similar to a Pólya urn scheme, called the Ehrenfest urn, which we shall discuss in Section 3.5. That urn scheme corresponds to two gas chambers, with a randomly chosen particle switching sides at each stage. The only difference between the gambler's ruin urn scheme and the Ehrenfest urn scheme is the model of randomness. The switching probabilities in the gambler's ruin problem remain constant throughout, as they are basically determined by the intrinsic nature of a coin independent of urn contents; the Ehrenfest scheme has switching probabilities that are adaptive in time and change according to the stochastic path.

Let $\mathcal{B}$ be the event that B is ruined in the gambler's ruin problem. We write the probability of $\mathcal{B}$ in terms of the given parameters as $Q_{m,n}$. We can write a recurrence for $Q_{m,n}$, by conditioning on the outcome of the first coin flip. Let the first flip be F, which is in the set $\{H, T\}$. We can then say

$$Q_{mn} = \mathbf{P}(\mathcal{B}) = \mathbf{P}(\mathcal{B} \mid F = H)\,\mathbf{P}(F = H) + \mathbf{P}(\mathcal{B} \mid F = T)\,\mathbf{P}(F = T).$$

The two conditional events have a recursive structure similar to the original problem. If the first game is won by A, the two urns progress to a state in which there are $m+1$ balls in urn A and $n-1$ balls in urn B. Similarly, we can think of $\mathcal{B} \mid F = T$, as a gambler's ruin problem, but only starting with parameters $m-1$ and $n+1$. We can write

$$Q_{mn} = Q_{m+1,n-1}\,p + Q_{m-1,n+1}\,q.$$

To turn this into a recurrence equation solvable by iterative algebra, we write the left-hand side as

$$Q_{mn} = Q_{mn}(p+q) = pQ_{mn} + qQ_{mn}.$$

So, we can reorganize the recurrence in the form

$$p\,Q_{m+1,n-1} - pQ_{mn} = qQ_{mn} - qQ_{m-1,n+1},$$

or

$$\begin{aligned} Q_{m+1,n-1} - Q_{mn} &= \frac{q}{p}(Q_{mn} - Q_{m-1,n+1}) \\ &= \left(\frac{q}{p}\right)^2 (Q_{m-1,n+1} - Q_{m-2,n+2}) \\ &\ \ \vdots \\ &= \left(\frac{q}{p}\right)^m (Q_{1,m+n-1} - Q_{0,m+n}). \end{aligned}$$

If urn A starts with 0 balls, it represents an already ruined player; it is a sure event that A loses all his money and impossible for B to be ruined, that is, $Q_{0,m+n} = 0$. We can iterate

$$Q_{m+1,n-1} = Q_{mn} + \left(\frac{q}{p}\right)^m Q_{1,m+n-1}$$

to get

$$\begin{aligned} Q_{m+1,n-1} &= Q_{m-1,n+1} + \left(\frac{q}{p}\right)^{m-1} Q_{1,(m-1)+(n+1)-1} + \left(\frac{q}{p}\right)^m Q_{1,m+n-1} \\ &\ \ \vdots \\ &= Q_{0,m+n} + \left[1 + \frac{q}{p} + \left(\frac{q}{p}\right)^2 + \cdots + \left(\frac{q}{p}\right)^m\right] Q_{1,m+n-1} \\ &= \left[1 + \frac{q}{p} + \cdots + \left(\frac{q}{p}\right)^m\right] Q_{1,m+n-1}. \end{aligned}$$

Therefore,

$$Q_{mn} = \begin{cases} \dfrac{1-(q/p)^m}{1-(q/p)}\, Q_{1,m+n-1}, & \text{if } \dfrac{q}{p} \neq 1; \\ m\,Q_{1,m+n-1}, & \text{if } p = q = \dfrac{1}{2}. \end{cases}$$

We got $Q_{0,m+n}$ from the boundary condition that A is ruined. We can use the other end (A wins, and B is ruined) to determine

$$Q_{m+n,0} = 1 = \begin{cases} \dfrac{1-(q/p)^{m+n}}{1-(q/p)} Q_{1,m+n-1}, & \text{if } \dfrac{q}{p} \neq 1; \\ (m+n)\, Q_{1,m+n-1}, & \text{if } p = q = \dfrac{1}{2}. \end{cases}$$

Hence,

$$Q_{1,m+n-1} = \begin{cases} \dfrac{1-(q/p)}{1-(q/p)^{m+n}}, & \text{if } \dfrac{q}{p} \neq 1; \\ \dfrac{1}{m+n}, & \text{if } p = q = \dfrac{1}{2}. \end{cases}$$

The ruin probability of player B follows:

$$Q_{mn} = \begin{cases} \dfrac{1-(q/p)^{m}}{1-(q/p)^{m+n}}, & \text{if } \dfrac{q}{p} \neq 1; \\ \dfrac{m}{m+n}, & \text{if } p = q = \dfrac{1}{2}. \end{cases}$$

2.5 Banach's Matchbox Problem

The famous Polish mathematician Stefan Banach (1892–1945) was a heavy smoker. In a speech in honor of Banach, Hugo Steinhaus (1887–1972), another noted Polish mathematician, concocted this problem as a humorous reference to Banach's smoking habits.

A pipe smoker has two matchboxes. He puts one box (an urn) in each of the two pockets of his pants (we refer to them as left and right pockets). Each complete new pack has n matches (balls) in it. Whenever the smoker needs a match, he reaches into a randomly selected pocket and takes a match from the box. *When the smoker finds the chosen box empty for the first time, how many matches are there in the other box?* We call that random number Y.

Switching to the language of urns, the balls are taken out until an empty urn is discovered. Conditioned on the event that the left urn is emptied first, the event $Y = k$, for $k = 0, , \ldots n$, is fulfilled, if $n+1$ attempts (successes) are made on the left urn and $n-k$ balls are taken out of the right urn, and the last of the $(n+1)+(n-k)$ attempts fails to find a ball in the left urn. Thus, Y is distributed like the negative binomial random variable $\text{NB}(n+1, \frac{1}{2})$, which awaits its $(n+1)$st success. Note that the support of a negative binomial distribution is unbounded, and $\text{NB}(n+1, \frac{1}{2})$ can assume very large values (with small probability), but the relevant portion of that distribution to the smoker's problem is for $2n+1-k = n+1, \ldots, 2n+1$ attempts. The event $Y = k$ can occur if either of the two symmetrical events that the left or the right urn is first discovered to be empty. The unconditional probability of this event is obtained by doubling the probability that $\text{NB}(n+1, \frac{1}{2})$ equals $2n+1-k$:

$$\mathbf{P}(Y = k) = \binom{2n-k}{n} \times \frac{1}{2^{2n-k}}.$$

Exercises

2.1 (Hudde and Huygens, 1657). Each of two rich players, A and B has an urn. Player A's urn has two white balls and one blue ball. Player B's urn has an unknown number of white and blue balls, but we know the odds ratio $p : q$ (the probability of drawing a white ball relative to the probability of drawing a blue ball). The two players alternate turns in betting on their drawings with replacement, until one of them draws a white ball and that is when the game is terminated and the player who draws it takes all the stakes. When a player draws a blue ball he adds one Ducat[1] to the stakes. If the starting stake is nill, and A draws first, what should the odds ratio be in B's urn for the game to be fair?

2.2 (Bernoulli, 1713). An urn has n white balls and n blue balls. The white balls are numbered $1, 2, \ldots, n$, and so are the blue balls. A sample of size k balls is taken out without replacement. What is the distribution of the number of white-blue pairs with the same numbers remaining in the urn? (Hint:

$$\sum_{j=0}^{\infty} \binom{i}{j} \binom{i-j}{s-j} = 2^s \binom{i}{s}.)$$

2.3 The roulette wheel has 38 slots; 18 are red, 18 are black, and 2 are green. The wheel is spun. It stops at one of the slots, presumably with equal probability. A player bets one chip on red 50 times. The lucky player wins 30 times and loses 20 times. Find the probability that the lucky player's net gain was always positive.

2.4 (Takács, 1997). An urn contains n balls numbered $x_1, \ldots, x_n$, with $x_i \geq 0$, for $i = 1, \ldots, n$, and

$$x = \sum_{i=1}^{n} x_i \leq n,$$

with each x_i being a nonnegative integer. The balls are taken out one at a time without replacement. Let Y_i be the label on the ball that appears

[1] A gold coin that was used as a trade currency throughout Europe before World War I.

in the ith draw. Show that

$$\mathbf{P}\left(\bigcap_{i=1}^{n}\{Y_1 + Y_2 + \cdots + Y_i < i\}\right) = 1 - \frac{x}{n}.$$

(Hint: Note that the result is true for $x = n$ (why?). Assume $x < n$ and do an induction on n, with the assumption that the result holds while one ball is omitted as an induction hypothesis).)

2.5 (Dvortzky and Motzkin, 1947). Deduce the probability in the ballot problem (see Section 2.1) from Exercise 2.4. (Hint: Let Y_j be the number of votes for B between the jth and $(j+1)$st votes for A. The natural interpretation for Y_m is the number of votes for B after the last for A.)

2.6 In the Dixie Cup competition (offering n coupons), what is the average of Y, the total waiting time to collect all coupon types? (Hint: Let X_i be the waiting time between the states of collecting $i-1$ and i coupons—if the collector has assembled a set of any $i-1$ coupons, with $1 \le i \le n-1$, the number of ball draws to get a new ball color (coupon type) is X_i. You can get the average from the representation

$$Y = X_1 + X_2 + \cdots + X_n.)$$

2.7 (Feller, 1968). Generalize Banach's matchbox problem to the case of selecting from the two pockets with different probability. This can have interpretation in games of skill, such as tennis or fencing where two players compete to score a certain odd number of points (three sets in the case of tennis, and 45 points in the case of fencing). The players may have probability p and $1 - p$ of winning an individual point, and the score for each point is independent of other points. What is the probability distribution of the score of the loser when the match is finished?

2.8 It is possible in the gambler's ruin game that the challenge does not come to an end. For example, if the two gamblers A and B start with 10 dollars each, and **A** is the event that A wins an individual game, and **B** is the event that B wins an individual game, sequences like **ABABAB**..., and **AABBAABBAABB**..., etc., represent non-ending games. What is the probability that the gambler's ruin game does not reach an and?

3

Pólya Urn Models

Pólya urns are one model of urns that involve only one urn and general methods of replacement under tenable conditions. Historically, urn schemes are very old. The book by Johnson and Kotz (1977) does an admirable job of presenting a historical perspective. They trace urns all the way back to the *Old Testament*. There are mentions of urns in the post-Renaissance era, in works of Huygens, de Moivre, Laplace, Bernoulli, and other noted mathematicians and scientists. The form we know today as Pólya urn and its numerous derivatives seem to have first appeared in work of Markov's around 1905–1907. A special form appeared in 1907 in the work of Paul Ehrenfest and Tatyana Ehrenfest, and was proposed as a model for the mixing of gases. The follow up of Eggenberger and Pólya (1923), with an alternative model, popularized the subject. The Pólya-Eggenberger urn was intended to model contagion. Epidemics and other such spreading phenomena have a branching nature within a population. Thus, steadily Pólya urn schemes acquired importance in all branching phenomena and in many phenomena with an underlying random tree structure.

A *Pólya urn* is an urn containing balls of up to k different colors. The urn evolves in discrete time steps. At each step, we shake the urn well and a ball is sampled uniformly at random (all balls being equally likely). The color of the ball withdrawn is observed, and the ball is returned to the urn. If at any step the color of the ball withdrawn is $i, i = 1, \dots, k$, then A_{ij} balls of color j are placed in the urn, $j = 1, \dots, k$, where A_{ij} follows a discrete distribution on a set of integers. Generally speaking, the entries A_{ij} can be deterministic or random, positive or negative.

It is customary to represent the urn scheme by a square ball addition matrix, or *schema*:

$$\mathbf{A} = \begin{pmatrix} A_{1,1} & A_{1,2} & \dots & A_{1,k} \\ A_{2,1} & A_{2,2} & \dots & A_{2,k} \\ \vdots & \vdots & \ddots & \vdots \\ A_{k,1} & A_{k,2} & \cdots & A_{k,k} \end{pmatrix},$$

the rows of which are indexed by the color of the ball picked, and the columns are indexed by the color of the balls added. We tend to use interchangeably the terms schema, ball addition matrix, or stochastic replacement matrix, and other similar terms for deterministic schemes, and reserve the use of the term "generator" for a random ball addition matrix.

The primary interest lies in the long-term composition of the urn and in the stochastic path leading to it. The number of balls of each color and the number of times a ball of a particular color is drawn are examples of important parameters.

3.1 Tenability

Toward an asymptotic theory, we need our urn to be tenable and withstand the test of time. A tenable urn is one from which we can perpetuate the drawing according to a given scheme on every possible stochastic path. Plainly put, in a tenable urn scheme it is always possible to draw balls and follow the replacement rules; we never get "stuck" in following the rules. As we shall see, the tenability of an urn scheme is a combination of what schema is given as stochastic replacement rules, and the initial conditions.

For example, the instance

$$\begin{pmatrix} 2 & 1 \\ 1 & 2 \end{pmatrix},$$

of Bernard Friedman's urn is tenable, under whichever nonempty initial state it starts in. (In fact an urn is tenable, if all the entries A_{ij} are nonnegative, under any nonempty starting conditions.) By contrast, an urn scheme of white and blue balls with the schema

$$\begin{pmatrix} -1 & -X \\ 3 & 4 \end{pmatrix},$$

with X being a Ber($\frac{4}{7}$) random variable, may or may not be tenable, depending on the initial conditions. This urn is not tenable if it starts out with at least as many white balls as blue. For instance, if the urn starts out with three white balls and two blue balls, it is not possible to perpetuate drawing according to the rules along some stochastic paths; if the event that a white ball is drawn and $X = 1$ persists three times, which is one possible stochastic path, on the third draw there are no blue balls to be taken out and the urn cannot progress. On the other hand, if initially the number of white balls is less than the number of blue balls, the urn is obviously tenable. Even on the most resistant path to growth, when $X = 1$ always persists whenever there is a chance to pick white balls, the number of white balls ultimately cycles in the set $\{0, 1, 2, 3\}$, and the number of blue balls is $\frac{1}{4}n + O(1)$, as $n \to \infty$. See Exercise 3.2 for a combinatorial flavor of this statement.

Example 3.1

A relevant example of an untenable urn is the urn that underlies the ballot problem (cf. Section 2.1). We reverse the view, and think of the ballots as balls of two colors, say m white (votes for A) and n blue (votes for B) in an urn. These balls are taken out one at a time (without replacement) to register a vote. The schema is

$$\begin{pmatrix} -1 & 0 \\ 0 & -1 \end{pmatrix}.$$

The drawing continues so long as there are balls in the urn. This urn is depleted after $m+n$ ball drawings. In fact, such a schema is untenable under any starting conditions. ◇

We illustrate the conditions of tenability on deterministic two-color schemes. Suppose we have an urn of white balls and blue balls. The urn is to progress under the replacement scheme

$$\mathbf{A} = \begin{pmatrix} a & b \\ c & d \end{pmatrix}.$$

In the sequel, for two-color schemes we consider the two colors to be white and blue, the author's favorite colors (the author grew up by the Azur (the Mediterranean Sea)). We shall always take white to preceed blue, in the sense that in the corresponding 2×2 replacement matrix the first row represents the actions taken upon withdrawing a white ball, and the first column to comprise the number of white balls added when a ball of either color is withdrawn.

Let W_n and B_n be the number of white and blue balls after n draws. Thus, the initial numbers of white and blue balls are W_0 and B_0. If all four numbers in the replacement matrix are nonnegative, we clearly have a tenable urn, for we simply always add balls of certain colors or leave them the same. Tenability becomes a critical issue if some of the numbers in the matrix are negative.

Generally, a 2×2 schema with two negative entries in the same column cannot be tenable. Both negative entries reduce a certain color whenever any ball of any color is picked. It is not hard to see then that no matter how the urn starts, the number of balls of a particular color are reduced until it is no longer possible to take out balls of that color because there are not enough of them to meet the demand.

We categorize the cases according to the number of negative entries in the schema. We shall use a notation of pluses, zeros, and minuses to indicate the positions of positive, zero, and negative entries in the schema. For instance,

$$\begin{pmatrix} + & - \\ 0 & + \end{pmatrix}$$

will indicate a schema where $a > 0$, $b < 0$, $c = 0$, and $d > 0$. We shall use the notation $\oplus$ to stand for an entry that is $+$ or 0.

3.1.1 Four Negative Numbers

Evidently, $\begin{pmatrix} - & - \\ - & - \end{pmatrix}$ cannot be tenable under any initial conditions, because there are two negative entries in each column.

3.1.2 Three Negative Numbers

Any matrix with three negative entries cannot be tenable, because two of them must be in the same column. For example, $\begin{pmatrix} - & - \\ - & + \end{pmatrix}$ is not tenable.

3.1.3 Two Negative Numbers

If an urn scheme is tenable, and its schema has two negative entries, they cannot be in the same column. This excludes

$$\begin{pmatrix} - & \oplus \\ - & \oplus \end{pmatrix} \qquad \begin{pmatrix} \oplus & - \\ \oplus & - \end{pmatrix}$$

from the class of tenable Pólya urn schemes. Also, the configuration

$$\begin{pmatrix} \oplus & - \\ - & \oplus \end{pmatrix}$$

is not tenable, because if a color is present in the initial urn, a sufficient number of draws from that color depletes the other color.

This leaves the cases

$$\underset{\text{(i)}}{\begin{pmatrix} - & - \\ \oplus & \oplus \end{pmatrix}} \qquad \underset{\text{(ii)}}{\begin{pmatrix} \oplus & \oplus \\ - & - \end{pmatrix}} \qquad \underset{\text{(iii)}}{\begin{pmatrix} - & \oplus \\ \oplus & - \end{pmatrix}}.$$

The cases (i) and (ii) are symmetric through the renaming of the colors. So, we consider (i) for (i) and (ii). Suppose we have the case $\begin{pmatrix} - & - \\ \oplus & \oplus \end{pmatrix}$, where a and b are both negative integers and c and d are nonnegative integers. Consider the "most critical path" which depletes the urn whenever possible. We call a state *critical*, if the urn has 0 white balls. Starting with W_0 white balls and a sufficient number B_0 of blue balls, always draw a white ball, whenever possible. Then W_0 must be a multiple of $|a|$. After $W_0/|a|$ draws, the urn is cleared of white balls, entering its first critical state. For the urn to rebound, there must be blue balls at this stage. The number of blue balls has been reduced to $B_0 - W_0|b|/|a|$. Thus,

$$B_0 - W_0\frac{b}{a} > 0 \tag{3.1}$$

is one of the necessary tenability conditions. One blue ball is forcibly in the next draw. The urn then progresses into a state where there are c white balls and $B_0 - W_0b/a + d$ blue balls. Depletion of white balls continues on this critical path, taking the urn into its second critical state, where there are 0 white balls (so, c must also be a multiple of $|a|$), and $B_0 - W_0b/a + d - cb/a$ blue balls. For tenability, this number of blue balls must be positive. A blue ball drawing is the only choice, putting the urn in a state, with c white balls and $B_0 - W_0b/a + 2d - cb/a$ blue balls, and so on. In the ith critical state there are 0 white balls, and $B_0 - W_0b/a + (i-1)d - (i-1)cb/a$ blue balls. For the urn to be tenable, it has to be recursively tenable from the ith critical state for each $i \ge 1$, with a positive number of blue balls (as there are no white balls). That is,

$$B_0 - W_0\frac{b}{a} + (i-1)d - (i-1)c\frac{b}{a} > 0, \qquad \text{for all } i \ge 1,$$

or

$$B_0 - W_0\frac{b}{a} > (i-1)\left(c\frac{b}{a} - d\right), \qquad \text{for all } i \ge 1.$$

Because i is arbitrarily large, and the left-hand side is positive (cf. Eq. (3.1)) this can only happen if $cb/a \le d$.

Summarizing all the conditions in determinant form, case (i) is tenable only if

- W_0 and c are both multiples of $|a|$.
- $\mathbf{det(A)} \le 0$.
- $\mathbf{det}\begin{pmatrix} a & b \\ W_0 & B_0 \end{pmatrix} < 0$.

Likewise, case (ii) is tenable by an obvious symmetrical change in the conditions.

Case (iii) requires less stringent conditions. It suffices to have

- W_0 and c are both multiples of $|a|$.
- B_0 and b are both multiples of $|d|$.
- Both b and c are positive.

The argument is similar and simpler than that for the case of two negative entries.

3.1.4 One Negative Number

With only one negative entry in the schema, the possible configurations are

$$\underset{\text{(i)}}{\begin{pmatrix} - & \oplus \\ \oplus & \oplus \end{pmatrix}} \qquad \underset{\text{(ii)}}{\begin{pmatrix} \oplus & - \\ \oplus & \oplus \end{pmatrix}} \qquad \underset{\text{(iii)}}{\begin{pmatrix} \oplus & \oplus \\ - & \oplus \end{pmatrix}} \qquad \underset{\text{(iv)}}{\begin{pmatrix} \oplus & \oplus \\ \oplus & - \end{pmatrix}}.$$

The cases (i) and (iv) are tenable under suitable easily met conditions. For (i), we only need W_0 and c to be both multiples of $|a|$, and if $b = 0$, B_0 must be positive; case (iv) is symmetrical to (i). For the symmetrical cases (ii) and (iii), stringent conditions are needed. For instance, (ii) can only be tenable if it has the form

$$\begin{pmatrix} \oplus & - \\ 0 & \oplus \end{pmatrix},$$

and we start with only a positive number of blue balls in the urn (no white). We leave the detailed argument of this instance as an exercise.

3.2 The Pólya-Eggenberger Urn

The earliest studies of Pólya urns focused on 2×2 schemata. One of the very first studies is Eggenberger and Pólya (1923), but it is reported that the model had been considered in Markov (1917) and Tchuprov (1922). The urn scheme was introduced as a model for contagion, and its underlying distributions are discussed further in Pólya (1931).

Eggenberger and Pólya (1923) is concerned with the the fixed schema

$$\begin{pmatrix} s & 0 \\ 0 & s \end{pmatrix}, \tag{3.2}$$

where one adds to the urn s (a positive integer) balls of the same color as the ball withdrawn. This urn is commonly known as the Pólya-Eggenberger urn (sometimes referred to in casual writing as Pólya's urn). Much of the rest of the ensuing theory is generalization in many different directions. It was natural in the first approach to the problem to seek discrete distributions underlying the process in exact form (in the style of nineteenth-century research). Indeed, the discrete distribution found in the Pólya-Eggenberger urn defined a fundamentally new distribution.

In what follows τ_n will refer to the total number of balls in an urn after n draws, and W_n and B_n will denote the number of white and blue balls, respectively, after n draws.

THEOREM 3.1
(Eggenberger and Pólya, 1923). Let $\tilde{W}_n$ be the number of white ball draws in the Pólya-Eggenberger urn after n draws. Then,

$$\mathbf{P}(\tilde{W}_n = k) = \frac{W_0(W_0+s)\dots(W_0+(k-1)s)B_0(B_0+s)\dots(B_0+(n-k-1)s)}{\tau_0(\tau_0+s)\dots(\tau_0+(n-1)s)}\binom{n}{k}.$$

PROOF In a string of n draws achieving k white ball drawings, there has to be $n-k$ draws of blue balls. Suppose $1 \le i_1 < i_2 < \dots < i_k \le n$ are the time indexes of the white ball draws. The probability of this particular string is

$$\frac{B_0}{\tau_0} \times \frac{B_0+s}{\tau_1} \times \frac{B_0+2s}{\tau_2} \times \dots \times \frac{B_0+(i_1-2)s}{\tau_{i_1-2}} \times \frac{W_0}{\tau_{i_1-1}}$$
$$\times \frac{B_0+(i_1-1)s}{\tau_{i_1}} \times \dots \times \frac{B_0+(i_2-3)s}{\tau_{i_2-2}}$$
$$\times \frac{W_0+s}{\tau_{i_2-1}} \times \frac{B_0+(i_2-2)s}{\tau_{i_2}} \times \dots \times \frac{B_0+(n-k-1)s}{\tau_{n-1}}.$$

Note that this expression does not depend on the indexes. The time indexes can be chosen in $\binom{n}{k}$ ways. ■

We presented a standard proof that bears an idea similar to the derivation of the binomial law on n independent trials. The difference is that in the binomial case the trials are identical, but here the probabilities are adaptive in time. Another proof is possible via de Finetti's theorem (see Theorem 1.2, and Exercise 3.4).

Calculations involving combinatorial reductions of the forms found in Riordan (1968) yield the moments.

COROLLARY 3.1
(Eggenberger and Pólya, 1923). Let W_n be the number of white balls in the Pólya-Eggenberger urn after n draws. Then,

$$\mathbf{E}[W_n] = \frac{W_0}{\tau_0} sn + W_0.$$
$$\mathbf{Var}[W_n] = \frac{W_0 B_0 s^2 n(sn+\tau_0)}{\tau_0^2(\tau_0+s)}.$$

PROOF Let $\tilde{W}_n$ be the number of times a white ball is drawn in n drawings from the urn, so that $W_n = s\tilde{W}_n + W_0$, and $\mathbf{E}[W_n] = s\mathbf{E}[\tilde{W}_n] + W_0$. We start with the classical definition

$$\mathbf{E}[\tilde{W}_n] = \sum_{k=0}^{\infty} k\,\mathbf{P}(\tilde{W}_n = k).$$

The probability $\mathbf{P}(\tilde{W}_n = k)$ in Theorem 3.1 can be written in terms of rising factorials as

$$\mathbf{P}(\tilde{W}_n = k) = s^k \frac{W_0}{s}\Big(\frac{W_0}{s}+1\Big)\dots\Big(\frac{W_0}{s}+(k-1)\Big)$$
$$\times \frac{s^{n-k}\frac{B_0}{s}\Big(\frac{B_0}{s}+1\Big)\dots\Big(\frac{B_0}{s}+(n-k-1)\Big)}{s^n\frac{\tau_0}{s}\Big(\frac{\tau_0}{s}+1\Big)\dots\Big(\frac{\tau_0}{s}+(n-1)\Big)}\binom{n}{k} \qquad (3.3)$$
$$= \binom{n}{k}\frac{\Big\langle\frac{W_0}{s}\Big\rangle_k\Big\langle\frac{B_0}{s}\Big\rangle_{n-k}}{\Big\langle\frac{\tau_0}{s}\Big\rangle_n}.$$

We use the combinatorial identity

$$k\binom{n}{k} = n\binom{n-1}{k-1}$$

to simplify the expectation:

$$\mathbf{E}[\tilde{W}_n] = n\sum_{k=1}^{n}\binom{n-1}{k-1}\frac{\Big\langle\frac{W_0}{s}\Big\rangle_k\Big\langle\frac{B_0}{s}\Big\rangle_{n-k}}{\Big\langle\frac{\tau_0}{s}\Big\rangle_n}$$
$$= n\frac{\frac{W_0}{s}}{\frac{\tau_0}{s}}\sum_{k=1}^{n}\binom{n-1}{k-1}\frac{\Big\langle\frac{W_0+s}{s}\Big\rangle_{k-1}\Big\langle\frac{B_0}{s}\Big\rangle_{n-k}}{\Big\langle\frac{\tau_0+s}{s}\Big\rangle_{n-1}}$$
$$= \frac{W_0 n}{\tau_0}\sum_{j=0}^{n-1}\binom{n-1}{j}\frac{\Big\langle\frac{W_0+s}{s}\Big\rangle_j\Big\langle\frac{B_0}{s}\Big\rangle_{(n-1)-j}}{\Big\langle\frac{\tau_0+s}{s}\Big\rangle_{n-1}}.$$

The last sum is 1, as it is the sum of all the probabilities for the number of times white balls appear in the sample in $n-1$ draws from a Pólya urn following the same schema but starting with $W_0 + s$ white balls and B_0 blue balls.

The proof for the variance is similar,[1] and starts with the second moment

$$\sum_{k=0}^{\infty} k^2\,\mathbf{P}(\tilde{W}_n = k).$$

We omit the details. ■

[1] In modern times the standard practice is to employ a computer algebra system like Maple or Mathematica. Even then it requires quite a bit of "human guidance" as Kirschenhofer and Prodinger (1998) puts it. We shall even see much more involved variance calculations in later chapters in applications in random trees and random circuits, and they were also done with the aid of computer algebra.

The limiting distribution involves $\beta(\alpha_1, \alpha_2)$, the beta random variable with parameters α_1, and α_2.

THEOREM 3.2

(Eggenberger and Pólya, 1923). Let $\tilde{W}_n$ be the number of white ball drawings in the Pólya-Eggenberger urn after n draws. Then,

$$\frac{\tilde{W}_n}{n} \xrightarrow{\mathcal{D}} \beta\left(\frac{W_0}{s}, \frac{B_0}{s}\right).$$

PROOF If either W_0 or B_0 is 0, we have a degenerate urn, progressing on only one color, with no randomness. In this case Theorem 3.1 remains valid through the appropriate boundary interpretation. The present theorem also remains valid through the appropriate interpretation of a beta distribution, when one of its parameters is 0.

Assume both W_0 and B_0 to be greater than 0. The proof is based on the appropriate passage to the limit. Rewrite the exact distribution of Theorem 3.1 (as given in the form in Eq. (3.3)) as

$$\mathbf{P}(\tilde{W}_n = k) = \frac{\Gamma\left(k + \frac{W_0}{s}\right)\Gamma\left(n - k + \frac{B_0}{s}\right)}{\Gamma\left(\frac{W_0}{s}\right)\Gamma\left(\frac{B_0}{s}\right)\Gamma\left(n + \frac{\tau_0}{s}\right)/\Gamma\left(\frac{\tau_0}{s}\right)} \times \frac{n!}{k!\,(n-k)!}.$$

So, for $x \in [0, 1]$, the distribution function of the white ball drawings is

$$\mathbf{P}(\tilde{W}_n \le nx) = \sum_{k=0}^{\lfloor nx \rfloor} \frac{\Gamma\left(k + \frac{W_0}{s}\right)\Gamma\left(n - k + \frac{B_0}{s}\right)\Gamma\left(\frac{\tau_0}{s}\right)}{\Gamma\left(\frac{W_0}{s}\right)\Gamma\left(\frac{B_0}{s}\right)\Gamma\left(n + \frac{\tau_0}{s}\right)} \times \frac{\Gamma(n+1)}{\Gamma(k+1)\,\Gamma(n-k+1)}.$$

Stirling's approximation to the ratio of gamma functions is a classical formula, given by

$$\frac{\Gamma(x+r)}{\Gamma(x+s)} = x^{r-s} + O(x^{r-s-1}), \qquad \text{as } x \to \infty.$$

Apply Stirling's approximation to the Gamma functions, and proceed to the limit, as $n \to \infty$, with

$$\mathbf{P}\left(\frac{\tilde{W}_n}{n} \le x\right) \to \frac{\Gamma\left(\frac{\tau_0}{s}\right)}{\Gamma\left(\frac{W_0}{s}\right)\Gamma\left(\frac{B_0}{s}\right)} \int_0^x u^{W_0/s-1}(1-u)^{B_0/s-1}\,du;$$

the right-hand side is the distribution function of the $\beta(\frac{W_0}{s}, \frac{B_0}{s})$ random variable. ■

It is curious that the limiting properties of a Pólya-Eggenberger urn depend critically on the initial conditions.

3.3 Bernard Friedman's Urn

Bernstein (1940) and Savkevich (1940) generalize the basic model in Display (3.2). Friedman (1949) extends it to one where one adds s balls of the same color, and a balls of the antithetical color, with the symmetric ball replacement matrix:

$$\begin{pmatrix} s & a \\ a & s \end{pmatrix}. \tag{3.4}$$

Assume as before the two colors to be white and blue, and the respective number of balls of these colors after n draws are W_n and B_n. For mathematical convenience, as well as æsthetics, Friedman (1949) (and most ensuing classical studies) stayed with the case of constant row sum (say K). We shall generally refer to these urn schemes as *balanced*. The reason why balance is convenient is that it gives a steady linear nonrandom rate of increase in the total number of balls in view of the relation $\tau_n := W_n + B_n = \tau_0 + Kn$.

Friedman (1949) develops functional equations for the number of white balls in an urn following the scheme in Display (3.4). Of course, the case $s = a$ is *degenerate*, where $W_n = W_0 + sn$. This degenerate case is of no interest, as there is no randomness in it.

THEOREM 3.3

(Friedman, 1949). Let W_n be the number of white balls in a nondegenerate Bernard Friedman's urn after n draws. The moment generating function $\phi_n(t) = \mathbf{E}[e^{W_n t}]$ satisfies the difference-differential equation

$$\phi_{n+1}(t) = e^{at}\left[\phi_n(t) + \frac{e^{(s-a)t} - 1}{\tau_n}\phi'_n(t)\right].$$

PROOF Let $\mathbf{1}_n^W$ and $\mathbf{1}_n^B \equiv 1 - \mathbf{1}_n^W$ be respectively the indicators of the events of drawing a white or a blue ball at the nth step. The number of white balls after $n+1$ draws is what it was after n steps, plus the addition (possibly negative) incurred by the ball sampled at step $n+1$:

$$W_{n+1} = W_n + s\mathbf{1}_{n+1}^W + a\mathbf{1}_{n+1}^B = W_n + (s-a)\mathbf{1}_{n+1}^W + a.$$

Then

$$\mathbf{E}\left[e^{W_{n+1}t} \mid W_n\right] = e^{(W_n+a)t}\,\mathbf{E}\left[e^{(s-a)\mathbf{1}_n^W t} \mid W_n\right]. \tag{3.5}$$

Further, we have the conditional expectation

$$\begin{aligned}\mathbf{E}\left[e^{(s-a)\mathbf{1}_n^W t} \mid W_n\right] &= \mathbf{E}\left[e^{(s-a)\mathbf{1}_n^W t} \mid W_n, \mathbf{1}_n^W = 0\right] \mathbf{P}\big(\mathbf{1}_n^W = 0 \mid W_n\big) \\ &\quad + \mathbf{E}\left[e^{(s-a)\mathbf{1}_n^W t} \mid W_n, \mathbf{1}_n^W = 1\right] \mathbf{P}\left(\mathbf{1}_n^W = 1 \mid W_n\right) \\ &= \left[1 - \frac{W_n}{\tau_n}\right] e^0 + \frac{W_n}{\tau_n} e^{(s-a)t}.\end{aligned}$$

Plug this into Eq. (3.5) to get

$$\begin{aligned}\mathbf{E}\left[e^{W_{n+1}t} \mid W_n\right] &= e^{(W_n+a)t}\left(1 - \frac{W_n}{\tau_n} + \frac{W_n}{\tau_n} e^{(s-a)t}\right) \\ &= e^{at}\left(e^{W_n t} + \frac{1}{\tau_n}(e^{(s-a)t} - 1) W_n e^{W_n t}\right).\end{aligned}$$

The result follows by taking expectations. ∎

The functional equation in Theorem 3.3 is not particularly easy to solve for any arbitrary combination of values of s and a. Nevertheless, explicit solutions are available for special cases, such as $a = 0$ (Pólya-Eggenberger's urn), and $s = 0$. Friedman (1949) suggested the transformation

$$\psi_n(t) = (1 - e^{-t(s-a)})^{\delta+\alpha n}\, \phi_n(t), \tag{3.6}$$

with

$$\delta := \frac{\tau_0}{s-a}, \qquad \alpha := \frac{s+a}{s-a}.$$

This gives a slightly simpler recurrences:

$$\psi_{n+1}(t) = \frac{e^{st}}{\tau_n}(1 - e^{-t(s-a)})^{\alpha+1}\, \psi_n'(t). \tag{3.7}$$

We shall discuss a solvable instance of this functional equation in one of the applications (See Subsection 8.2.1 on recursive trees). The solution in this special case should give us general hints on how to approach the functional equation of Theorem 3.3.

Twentieth century research paid attention to simplifying results by focusing on the essential elements or "asymptotics." Freedman (1965) developed an asymptotic theory for Bernard Friedman's urn.

THEOREM 3.4

(Freedman(1965)). Let W_n be the number of white balls in a nondegenerate Bernard Friedman's urn after n draws. Let $\rho = (s-a)/(s+a)$. If $\rho < \frac{1}{2}$, then

$$\frac{W_n - \frac{1}{2}(s+a)n}{\sqrt{n}} \xrightarrow{\mathcal{D}} \mathcal{N}\left(0, \frac{(s-a)^2}{4(1-2\rho)}\right).$$

A proof will be given when we present results on the more general Bagchi-Pal urn. Freedman (1965) gives an expository section on Bernard Friedman's urn, where the results of Friedman (1949) are mostly presented in terms of the difference $W_n - B_n$: For $\rho < 1/2$ the limiting distribution is normal under the $\sqrt{n}$ scale, and it is interesting to note that in the case $\rho = \frac{1}{2}$, one needs a different scale factor to attain a Gaussian limit distribution:

$$\frac{W_n - B_n}{\sqrt{n \ln n}} \xrightarrow{\mathcal{D}} \mathcal{N}(0, (s-a)^2).$$

For $\rho > 1/2$ the behavior is radically different:

$$\frac{W_n - B_n}{n^\rho} \xrightarrow{\mathcal{D}} \beta\left(\frac{W_0}{s}, \frac{B_0}{s}\right).$$

It is curious that in the case $\rho \le \frac{1}{2}$ the effect of any initial condition is washed out asymptotically. The urn balances itself in the long run—each color constitutes half the urn content on average. Contrast this with the case $\rho > \frac{1}{2}$, where the asymptotic proportion of colors depends critically on the initial conditions, as we saw in the Pólya-Eggenberger urn where ρ is 1.

3.4 The Bagchi-Pal urn

For the next step in the historical development of Pólya urn theory, it was natural to think of breaking the perfect symmetry of Bernard Friedman's urn in Display (3.4). Bagchi and Pal (1985) considers the more general case

$$\begin{pmatrix} a & b \\ c & d \end{pmatrix},$$

for any four integers, so long as the choice is tenable consistently with the initial conditions, with the exception of a few intricate cases. Namely, they require a balance condition of constant row sum ($a + b = c + d =: K$), and (to guarantee tenability) $b > 0$, $c > 0$, and if $a < 0$, then a divides W_0 and c, and if $d < 0$, then d divides B_0 and b. We exclude degenerate cases: The case $b = c = 0$, which is the Pólya-Eggenberger urn, and the case $a = c$, such a case has no randomness. The case where one minor diagonal element is zero ($bc = 0$, $\max(b, c) > 0$), is also excluded. We first develop the mean and variance.

PROPOSITION 3.1

(Bagchi and Pal, 1985). Let W_n be the number of white balls after n draws from a nondegenerate Bagchi-Pal urn, where K balls are added at each step. Then,

$$\mathbf{E}[W_n] = \frac{c}{b+c} K n + o(n).$$

If $a - c < \frac{1}{2}K$,

$$\mathbf{Var}[W_n] = \frac{bcK(a-c)^2}{(b+c)^2(K-2(a-c))}\, n + o(n),$$

and if $a - c = \frac{1}{2}K$,

$$\mathbf{Var}[W_n] = \frac{bc}{K}\, n \ln n + O(1).$$

PROOF We have the recurrence

$$\mathbf{P}(W_{n+1} = W_n + a \mid W_n) = \frac{W_n}{\tau_n}; \tag{3.8}$$

$$\mathbf{P}(W_{n+1} = W_n + c \mid W_n) = 1 - \frac{W_n}{\tau_n}. \tag{3.9}$$

This gives conditionally

$$\mathbf{E}[W_{n+1} \mid W_n] = (W_n + a)\frac{W_n}{\tau_n} + (W_n + c)\left(1 - \frac{W_n}{\tau_n}\right) = \left(1 + \frac{a-c}{\tau_n}\right) W_n + c.$$

The unconditional expectation is therefore

$$\mathbf{E}[W_{n+1}] = \left(1 + \frac{a-c}{\tau_n}\right) \mathbf{E}[W_n] + c,$$

for $n \ge 0$. The setting

$$Y_n = W_n - \frac{c}{b+c}\tau_n \tag{3.10}$$

puts the equation into an iteratable form:

$$\mathbf{E}[Y_{n+1}] = \left(1 + \frac{a-c}{\tau_n}\right) \mathbf{E}[Y_n].$$

The solution is obtained by unwinding the recurrence all the way back to $n = 0$:

$$\begin{aligned}\mathbf{E}[Y_n] &= \left(W_0 - \frac{c}{b+c}\tau_0\right) \prod_{j=0}^{n-1} \left(1 + \frac{a-c}{\tau_j}\right) \\ &= \left(W_0 - \frac{c}{b+c}\tau_0\right) \frac{\Gamma\left(\frac{\tau_0}{K}\right) \Gamma\left(n + \frac{\tau_0 + a - c}{K}\right)}{\Gamma\left(\frac{\tau_0 + a - c}{K}\right) \Gamma\left(n + \frac{\tau_0}{K}\right)}.\end{aligned}$$

By the Stirling approximation of the Gamma functions, one finds $\mathbf{E}[Y_n] = O(n^{(a-c)/K})$. We have $a - c < K$, $\mathbf{E}[Y_n] = o(n)$, and

$$\mathbf{E}[W_n] = \frac{c}{b+c}\tau_n + \mathbf{E}[Y_n] = \frac{cK}{b+c}n + o(n);$$

the linear term is dominant.

We shall not develop the variance in detail, we shall take a brief look into its structure because the phase change in it is worthy of notice. To detect the phase change, write Eqs. (3.8) and (3.9) in the form

$$\mathbf{P}\big(W_{n+1}^2 = (W_n + a)^2 \mid W_n\big) = \frac{W_n}{\tau_n};$$

$$\mathbf{P}\big(W_{n+1}^2 = (W_n + c)^2 \mid W_n\big) = 1 - \frac{W_n}{\tau_n}.$$

With the transformation in Eq. (3.10), we have a recurrence:

$$\mathbf{E}[Y_{n+1}^2] = \left(1 + \frac{2(a-c)}{\tau_n}\right)\mathbf{E}[Y_n^2] + \frac{(b-c)(a-c)^2}{(b+c)\tau_n}\mathbf{E}[Y_n] + \frac{bc(a-c)^2}{(b+c)^2)}.$$

If $a - c < \frac{1}{2}K$, this recurrence asymptotically has a linear solution. By contrast, if $a - c = \frac{1}{2}K$, this recurrence simplifies to

$$\mathbf{E}[Y_{n+1}^2] = \frac{\tau_{n+1}}{\tau_n}\mathbf{E}[Y_n^2] + \frac{b^2 - c^2}{\tau_n}\mathbf{E}[Y_n] + bc,$$

the complementary solution of which is linear (in fact it is τ_n), but the recurrence in this case has a superlinear particular solution. ■

COROLLARY 3.2
In a nondegenerate Bagchi-Pal urn, if $a - c \le \frac{1}{2}K$, then

$$\frac{W_n}{n} \xrightarrow{P} \frac{cK}{b+c}.$$

PROOF By Chebychev's inequality, for any fixed $\varepsilon > 0$,

$$\mathbf{P}(|W_n - \mathbf{E}[W_n]| > \varepsilon) \le \frac{\mathbf{Var}[W_n]}{\varepsilon^2}.$$

Replace ε by $\varepsilon\mathbf{E}[W_n]$, to get

$$\mathbf{P}\left(\left|\frac{W_n}{\mathbf{E}[W_n]} - 1\right| > \varepsilon\right) = \frac{O(n\ln n)}{\varepsilon^2\mathbf{E}^2[W_n]} = O\left(\frac{\ln n}{n}\right) \to 0.$$

Therefore,

$$\frac{W_n}{\mathbf{E}[W_n]} \xrightarrow{P} 1.$$

We have shown that

$$\frac{\mathbf{E}[W_n]}{n} \to \frac{cK}{b+c},$$

and we can multiply the last two relations according to Slutsky's theorem (cf. Billingsley, 1995) to get

$$\frac{W_n}{n} \xrightarrow{P} \frac{cK}{b+c}.$$

We prove next the asymptotic normality under certain eigenvalue conditions by the method of recursive moments, a method that proved to be very successful in the studies of random structures and algorithms, and has been popularized in the recent work of Chern, Hwang, and Tsai (2002). ■

THEOREM 3.5

(Bagchi and Pal, 1985.) Let W_n be the number of white balls after n draws from a nondegenerate Bagchi-Pal urn, where K balls are added at each step. If $a - c < \frac{1}{2}K$, we get

$$W_n^* := \frac{W_n - \frac{cK}{b+c}n}{\sqrt{n}} \xrightarrow{\mathcal{D}} \mathcal{N}\left(0, \frac{bcK(a-c)^2}{(b+c)^2(K-2(a-c))}\right).$$

If $a - c = \frac{1}{2}K$,

$$W_n^* := \frac{W_n - \frac{cK}{b+c}n}{\sqrt{n\ln n}} \xrightarrow{\mathcal{D}} \mathcal{N}\left(0, \frac{bc}{K}\right).$$

PROOF *(sketch)* This theorem is proved by showing that the moments of W_n^* converge to those of the normal distribution specified, with the normal distribution being uniquely characterized by its moments.

Recall the definition of Y_n as $W_n - c\tau_n/(b+c)$. The shift quantity is deterministic, and Y_n and W_n have the same variance. Set

$$Z_n := \frac{W_n - \mathbf{E}[W_n]}{\sqrt{\mathbf{Var}[W_n]}} = \frac{Y_n - \mathbf{E}[Y_n]}{\sqrt{\mathbf{Var}[Y_n]}}.$$

Asymptotic normality of Z_n is equivalent to that of W_n^*.

Like in the mean and variance, we start the computation of any moment by conditioning. If a white ball is drawn, the number of white balls is increased by a, and

$$Y_{n+1} = W_{n+1} - \frac{c\tau_{n+1}}{b+c} = W_n + a - (\tau_n + K)\frac{c}{b+c}.$$

Likewise, if a blue ball is drawn, the number of white balls is increased by c, and

$$Y_{n+1} = W_{n+1} - \frac{c\tau_{n+1}}{b+c} = W_n + c - (\tau_n + K)\frac{c}{b+c}.$$

For $r \geq 1$,

$$\begin{aligned}\mathbf{E}[Y_{n+1}^r \mid Y_n] &= \left(W_n + a - (\tau_n + K)\frac{c}{b+c}\right)^r \frac{W_n}{\tau_n} \\ &\quad + \left(W_n + c - (\tau_n + K)\frac{c}{b+c}\right)^r \frac{B_n}{\tau_n} \\ &= \left(Y_n + a - \frac{Kc}{b+c}\right)^r \left(\frac{c}{b+c} + \frac{Y_n}{\tau_n}\right) \\ &\quad + \left(Y_n + c - \frac{Kc}{b+c}\right)^r \left(\frac{b}{b+c} - \frac{Y_n}{\tau_n}\right).\end{aligned}$$

We take expectations and write a recurrence for the rth moment:

$$\mathbf{E}[Y_{n+1}^r] - \left(1 + \frac{r(a-c)}{\tau_n}\right) \mathbf{E}[Y_n^r] = \sum_{j=1}^{r} \left(p_{r,r-j} + \frac{q_{r,r-j}}{\tau_n}\right) \mathbf{E}[Y_n^{r-j}], \qquad (3.11)$$

where $p_{r,r-j}$ and $q_{r,r-j}$ are coefficients that appear in the binomial theorem upon raising to power r, namely,

$$p_{r,r-j} = \binom{r}{j}\left(\frac{a-c}{b+c}\right)^j \frac{bc}{b+c}(b^{j-1} + (-1)^j c^{j-1}),$$

and

$$q_{r,r-j} = \binom{r}{j+1}\left(\frac{a-c}{b+c}\right)^{j+1}(b^{j+1} + (-1)^j c^{j+1}).$$

Note that $p_{r,r-1} = 0$, $q_{r,0} = 0$. The initial conditions are

$$\mathbf{E}[Y_0^r] \equiv Y_0^r = \left(W_0 - \frac{c\tau_0}{b+c}\right)^r,$$

for $r \geq 1$.

We first take up the case $a - c < \frac{1}{2}K$. Let σ^2 be the coefficient of linearity in the variance in Proposition 3.1. The strategy for the rest of the proof is to show that

$$\mathbf{E}\left[\frac{Z_n^r}{\sigma^r}\right] \to \begin{cases} (r-1)!!, & \text{for even } r; \\ 0, & \text{for odd } r, \end{cases}$$

as are the moments of a standard normal random variate—the double factorial notation stands for $(r-1)\times(r-3)\times\cdots\times 5\times 3\times 1$. We shall show by induction on even r that $\mathbf{E}[Y_n^r] = (r-1)!!\,\sigma^r n^{r/2} + o(n^{r/2})$. We have already demonstrated the basis for this induction in the second moment ($r = 2$) in Proposition 3.1. We are seeking a solution for Eq. (3.11). This comprises a homogeneous solution $H(n,r)$ for

$$\mathbf{E}\left[Y_{n+1}^r\right] - \left(1 + \frac{r(a-c)}{\tau_n}\right)\mathbf{E}\left[Y_n^r\right] = 0,$$

and a particular solution $Q(n, r)$. The homogeneous solution is obtained by iteration

$$H(n,r) = \left(1 + \frac{r(a-c)}{\tau_n}\right)\left(1 + \frac{r(a-c)}{\tau_{n-1}}\right)\cdots\left(1 + \frac{r(a-c)}{\tau_0}\right) \mathbf{E}\big[Y_0^r\big]$$
$$= \frac{\Gamma\left(\frac{\tau_0}{K}\right)}{\Gamma\left(\frac{\tau_0 + r(a-c)}{K}\right)} \times \frac{\Gamma\left(n+1+\dfrac{\tau_0 + r(a-c)}{K}\right)}{\Gamma\left(n+1+\dfrac{\tau_0}{K}\right)} \mathbf{E}\big[Y_0^r\big].$$

By the Stirling approximation to Gamma functions, we see that $H(n, r)$ is $o(n^{r/2})$.

We next show that $Q(n, r)$ is asymptotically equivalent to $(r-1)!!\,\sigma^r n^{r/2}$, constituting the dominant term in the solution (and $H(n, r)$ is asymptotically negligible). Use the substitution $Q(n,r) = g(n,r) + (r-1)!!\,\sigma^r(\tau_n/K)^{r/2}$ in Eq. (3.11), and write by the induction hypothesis

$$g(n+1,r) + (r-1)!!\,\sigma^r \left(\frac{\tau_n + K}{K}\right)^{r/2}$$
$$- \left(1 + \frac{r(a-c)}{\tau_n}\right)\left(g(n,r) + (r-1)!!\,\sigma^r\left(\frac{\tau_n}{K}\right)^{r/2}\right)$$
$$= \left(p_{r,r-1} + \frac{q_{r,r-1}}{\tau_n}\right) \times O(n^{(r-1)/2})$$
$$+ \left(p_{r,r-2} + \frac{q_{r,r-2}}{\tau_n}\right)(r-3)!!\,\sigma^{r-2}\left(\frac{\tau_n}{K}\right)^{r/2-1}$$
$$+ o(n^{r/2-1}) + O(n^{(r-3)/2}).$$

Recall the definitions σ^2, and of $p_{i,j}$ and $q_{i,j}$, and that $p_{r,r-1} = 0$, and write

$$g(n+1,r) + (r-1)!!\,\sigma^r \left[\left(\frac{\tau_n}{K}\right)^{r/2} + \frac{r}{2}\left(\frac{\tau_n}{K}\right)^{r/2-1} + O(n^{(r/2-2)})\right]$$
$$- \left(1 + \frac{r(a-c)}{\tau_n}\right)\left(g(n,r) + (r-1)!!\,\sigma^r\left(\frac{\tau_n}{K}\right)^{r/2}\right)$$
$$= (r-3)!!\binom{r}{2}\frac{\sigma^2}{K}(K - 2(a-c))\,\sigma^{r-2}\left(\frac{\tau_n}{K}\right)^{r/2-1} + o(n^{r/2-1}).$$

Since $(r-1)!! = (r-3)!!\,(r-1)$, this reduces to

$$g(n+1,r) = \left(1 + \frac{r(a-c)}{\tau_n}\right) g(n,r) + o(n^{r/2-1}).$$

And so,

$$g(n,r) = H(n,r)\left(g(n, j_0) + \sum_{j=j_0}^{n-1} \frac{o(n^{(r/2-1)})}{H(j+1,r)}\right),$$

for some suitable j_0. Again, by the Gamma function technique, if $(a-c) < \frac{1}{2}K$, we find

$$g(n, r) = o(n^{r/2}).$$

For odd r, the basis at $r = 1$ is the result on the mean in Proposition 3.1. We set $Q(n, r) = g(n, r)$ and proceed similarly. We omit a rather similar proof for the case $a - c = \frac{1}{2}K$. ■

The case where one minor diagonal element is zero ($bc = 0, \max(b, c) > 0$), has been excluded. It is handled via an elegant martingale technique in Gouet (1989) to show a strong law for W_n/n, and via the functional central limit theorem in Gouet (1993) to show convergence (of a normalized version) to a Gaussian law modulated by a multiplicative independent random variable.

3.5 The Ehrenfest Urn

The Ehrenfest scheme furnishes a different flavor from the types of Pólya urns we saw so far. In the preceeding sections we discussed Pólya urns that grow in time. The population of the Ehrenfest urn does not change in size over time, though it might experience internal fluctuation in the details of its constant population size.

The Ehrenfest urn scheme has the replacement matrix

$$\mathbf{A} = \begin{pmatrix} -1 & 1 \\ 1 & -1 \end{pmatrix}.$$

The matrix stands for an urn scheme in which whenever a ball is picked, it is replaced by a ball of the opposite color. The Ehrenfest urn has applications as a model for the mixing of particles in two connected gas chambers, A and B. There are a number of particles in each chamber. Initially there is a partition. The partition is removed at time 0. At each subsequent discrete stage in time, a particle from the population of the two chambers is chosen at random. The chosen particle switches sides. The situation can be likened to an urn containing two types of balls, white and blue. The white balls represent the particles in chamber A, the blue represent those in chamber B. A particle changing chambers is a ball in the urn acquiring the opposite color. Note that the population of balls is constant in size, but experiences changes in the profile of white and blue balls within. In other words, the total number of balls in the urn is fixed, say τ, but the number of balls of one particular color has a discrete distribution over nonnegative integers in the range $\{0, 1, \ldots, \tau\}$.

Generally speaking, there may be no convergence for the random variables associated with the urn. Example 1.2 presents an Ehrenfest urn with a total of $\tau = 2$ balls, starting in a state with one white and one blue ball. In this instance of an Ehrenfest urn scheme, the number of white balls after n draws does not converge to a limit. Nonetheless, a stationary binomial distribution exists. That is, the number of white balls follows a distribution, if the initial state is appropriately randomized.

THEOREM 3.6
Suppose the Ehrenfest urn starts out with $W_0 = \text{Bin}(\tau, \frac{1}{2})$ *white balls and* $B_0 = \tau - W_0$ *blue balls. Let* W_n *be the number of particles in the urn after* n *draws. Then*

$$W_n \xrightarrow{\mathcal{D}} \text{Bin}\left(\tau, \frac{1}{2}\right).$$

PROOF We can determine the binomial stationary distribution by a lengthy linear algebra calculation from the underling Markov chain transition matrix. We present an alternative approach. After n draws, there are W_n white and B_n blue balls, respectively, in the Ehrenfest urn. For W_n to be k after n draws, we must either have had $k+1$ white balls in the previous step, and draw a white ball, thus decreasing the white balls by one, or have had $k-1$ white balls in the previous step, and increase the white balls by drawing a blue ball:

$$\mathbf{P}(W_n = k) = \frac{k+1}{\tau}\,\mathbf{P}(W_{n-1} = k+1) + \frac{\tau - k + 1}{\tau}\,\mathbf{P}(W_{n-1} = k-1).$$

The grounds for justifying the existence of a limit $\mathbf{P}(W_n = k) \to p(k)$ are in its being a stationary distribution of an irreducible Markov chain. We find that the limit must satisfy

$$p(k) = \frac{k+1}{\tau}\,p(k+1) + \frac{\tau - k + 1}{\tau}\,p(k-1).$$

Reorganize as

$$\tau\big(p(k) - p(k-1)\big) = (k+1)\,p(k+1) - (k-1)\,p(k-1).$$

From this write down a system of equations for $k, k-1, \ldots, 0$ (with the natural interpretation that $p(-1) = 0$), and add them up to obtain

$$p(k+1) = \frac{\tau - k}{k+1}\,p(k).$$

Iteration gives

$$\begin{aligned}
\mathbf{P}(W_n = k) &= p(k) \\
&= \frac{(\tau - (k-1))(\tau - (k-2)) \ldots \tau}{k(k-1) \ldots 1}\,p(0) \\
&= \binom{\tau}{k}\mathbf{P}(W_0 = 0) \\
&= \frac{1}{2^\tau}\binom{\tau}{k},
\end{aligned}$$

for each $k = 0, \ldots, \tau$. ■

The focus so far was on the composition of an urn after a sequence of drawings, or the number of appearances of a certain color in the sequence. In conclusion of this chapter we discuss a problem of waiting times or *stopping times*, which is a path less often trodden when dealing with urns.

An Ehrenfest urn of white and blue balls starts out all white. The initial number of $2M$ white balls are in the urn at the beginning. *What is the average waiting time $Y(2M)$ until equilibrium occurs for the first time?* Equilibrium is defined as the state of having an equal number of balls of each color. In the model of mixing of gases, the question is *if initially all the gas is in one chamber, how soon on average do we get a split with half the amount of gas in each chamber?* The problem is discussed in Blom, Holst, and Sandell (1994), and our presentation below is modeled after theirs.

Let us describe the states of a Markov chain underlying the model. Let A_i be the state in which i blue balls are in the urn. For instance, the urn starts in the state A_0, with no blue balls, but moves into the state A_1 right after the first draw (in which one must pick a white ball); the white ball in the first draw is recolored blue and replaced in the urn, and so forth. Let X_i be the time it takes the urn to go from state A_i to state A_{i+1}. The waiting time $Y(2M)$ is

$$Y(2M) = X_0 + X_1 + \cdots + X_{M-1}. \tag{3.12}$$

If the urn is in state A_i (with i blue balls and $2M - i$ white balls in the urn), it can go right away into state A_{i+1} in one step, if a white ball is drawn, with probability $(2M-i)/(2M)$. Otherwise, there is a setback in ascending toward equilibrium, a blue ball is drawn (with probability $i/(2M)$), and the number of blue balls actually decreases, and the urn renters the state A_{i-1}. Afterward, one has to wait another amount of time distributed like X_{i-1} time just to get back into the state A_i, and an additional amount of time distributed like X_i to go up into the state A_{i+1}. Thus,

$$X_i \stackrel{\mathcal{D}}{=} \begin{cases} 1, & \text{with probability } \dfrac{2M-i}{2M}, \\[2ex] 1 + X_{i-1} + X_i, & \text{with probability } \dfrac{i}{2M}. \end{cases}$$

On average, we have

$$\mathbf{E}[X_i] = 1 \times \frac{2M-i}{2M} + (1 + \mathbf{E}[X_{i-1}] + \mathbf{E}[X_i]) \times \frac{i}{2M},$$

which we rearrange as

$$\mathbf{E}[X_i] = \frac{2M}{2M-i} + \frac{i}{2M-i}\mathbf{E}[X_{i-1}]. \tag{3.13}$$

This is not an easy recurrence, because the sequences $\mathbf{E}[X_{i-1}]$ and $2M-i$ are going in "opposite directions." One possibility is to seek an integral representation. We shall show by induction that

$$\mathbf{E}[X_i] = 2M \int_0^1 x^{2M-i-1}(2-x)^i \, dx := 2MI_i.$$

The formula is true at the basis $i = 0$, as

$$2M \int_0^1 x^{2M-0-1}(2-x)^0 \, dx = 2M \int_0^1 x^{2M-1} \, dx = 1,$$

which we know to be $\mathbf{E}[X_0]$, because it takes only one transition to move away from the state of all white.

Assuming the formula to be true for some $k \ge 0$, then, according to the recurrence in Eq. (3.13),

$$\begin{aligned} \mathbf{E}[X_{k+1}] &= \frac{2M + (k+1)\mathbf{E}[X_k]}{2M - (k+1)} \\ &= \frac{2M}{2M-k-1} + \frac{2M(k+1)}{2M-k-1} \int_0^1 x^{2M-k-1}(2-x)^k \, dx. \end{aligned}$$

However, when we integrate I_{k+1} by parts, we get

$$\begin{aligned} I_{k+1} &= \int_0^1 (2-x)^{k+1} \, d\frac{x^{2M-k-1}}{2M-k-1} \\ &= \frac{x^{2M-k-1}}{2M-k-1}(2-x)^{k+1}\Big|_{x=0}^1 - \frac{1}{2M-k-1} \int_0^1 x^{2M-k-1} \, d(2-x)^{k+1} \\ &= \frac{1}{2M-k-1} + \frac{k+1}{2M-k-1} \int_0^1 x^{2M-k-1}(2-x)^k \, dx \\ &= \frac{1 + (k+1)\mathbf{E}[X_k]/(2M)}{2M-k-1}. \end{aligned}$$

Hence $2MI_{k+1}$ coincides with $\mathbf{E}[X_{k+1}]$, completing the induction.

Now, the expectation of Eq. (3.12) yields

$$\begin{aligned} \mathbf{E}[Y(2M)] &= \mathbf{E}[X_0] + \mathbf{E}[X_1] + \cdots + \mathbf{E}[X_{M-1}] \\ &= 2M \sum_{i=0}^{M-1} \int_0^1 x^{2M-i-1}(2-x)^i \, dx \\ &= 2M \int_0^1 x^{2M-1} \sum_{i=0}^{M-1} \left(\frac{2-x}{x}\right)^i \, dx. \end{aligned}$$

Summing the geometric series results in

$$\mathbf{E}[Y(2M)] = 2M \int_0^1 x^{2M-1} \frac{\left(\frac{2-x}{x}\right)^M - 1}{\frac{2-x}{x} - 1}\, dx$$
$$= 2M \int_0^1 x^M \frac{(2-x)^M - x^M}{2-2x}\, dx.$$

Some computational cleaning up is facilitated by the substitution $x = 1-y$, which gives

$$\mathbf{E}[Y(2M)] = M \int_0^1 \frac{1}{y}(1-y)^M((1+y)^M - (1-y)^M)\, dy \qquad (3.14)$$
$$= M \int_0^1 \frac{1}{y}((1-y^2)^M - (1-y)^{2M})\, dy$$
$$= M \int_0^1 \frac{1}{y}(1-(1-y)^{2M})\, dy - M \int_0^1 \frac{1}{y}(1-(1-y^2)^M)\, dy.$$

Interpret the integrands as sums of geometric series in view of the identity

$$\sum_{k=0}^{n-1}(1-z)^k = \frac{1-(1-z)^n}{1-(1-z)}.$$

We arrive at

$$\mathbf{E}[Y(2M)] = M \int_0^1 \sum_{k=0}^{2M-1}(1-y)^k\, dy - M \int_0^1 y \sum_{j=0}^{M-1}(1-y^2)^j\, dy$$
$$= M \sum_{k=0}^{2M-1} \frac{1}{k+1} - \frac{M}{2} \sum_{j=0}^{M-1} \frac{1}{j+1}$$
$$= MH_{2M} - \frac{M}{2} H_M,$$

where H_r is the rth harmonic number $1 + \frac{1}{2} + \cdots + \frac{1}{r}$. The harmonic numbers have a well-known approximation

$$H_r = \ln r + \gamma + O\left(\frac{1}{r}\right), \qquad \text{as } r \to \infty,$$

where $\gamma = 0.5772\ldots$ is Euler's constant. For large M, the average time until equilibrium is

$$\mathbf{E}[Y(2M)] = M\left(\ln(2M) + \gamma + O\left(\frac{1}{M}\right)\right) - \frac{M}{2}\left(\ln M + \gamma + O\left(\frac{1}{M}\right)\right)$$
$$= \frac{1}{2} M \ln M + M\left(\ln 2 + \frac{\gamma}{2}\right) + O(1).$$

There was nothing special about starting with $2M$ balls, nor was there any particular reason to specify the state of equilibrium (the state A_M) as a target. We could have started with N white balls, and asked how long it takes to reach the state A_k, for arbitrary $k = 1, 2, \dots, N$, and the argument would need only minor adjustments in the parameters. We deal with one other instance in the exercises.

Exercises

3.1 Let us utilize a system of pluses, minuses, and $\oplus$ to indicate positive, negative, and nonnegative entries in the schema of a Pólya urn. When can an urn of the form

$$\begin{pmatrix} - & \oplus \\ \oplus & \oplus \end{pmatrix}$$

be tenable? Justify the answer.

3.2 An urn scheme of white and blue balls has the replacement matrix

$$\begin{pmatrix} -1 & -1 \\ 3 & 4 \end{pmatrix};$$

the top row corresponds to the replacements upon drawing a white ball, and the first column is for the replacements in the white balls. Let W_n and B_n be respectively the number of white and blue balls after n draws. Show that, if the draws follow a most critical stochastic path, always depleting the white balls as quickly as possible, then for $n \ge W_0$,

$$W_n = (4 - ((n - W_0) \,\mathbf{mod}\, 4))(1 - \mathbf{1}_{\{(n - W_0 \,\mathbf{mod}\, 4) = 0\}}),$$

and

$$B_n = B_0 - W_0 + \frac{1}{4}((n - W_0) - (n - W_0) \,\mathbf{mod}\, 4)$$
$$+(5 - (n - W_0) \,\mathbf{mod}\, 4)(1 - \mathbf{1}_{\{((n - W_0) \,\mathbf{mod}\, 4) = 0\}}).$$

3.3 A Pólya-Eggenberger scheme with the ball replacement matrix

$$\begin{pmatrix} 2 & 0 \\ 0 & 2 \end{pmatrix},$$

starts with four white balls and three blue balls.

(a) What is the probability that the second draw picks a white ball, given the first was blue?

(b) Show that irrespective of the draw number, the probability of picking a blue ball is $\frac{3}{7}$.

(c) Show that the probability that two consecutive drawings are the same is $\frac{13}{21}$.

(Hint: For (b) and (c) you may use induction or exchangeability.)

3.4 Prove Theorem 3.1 via exchangeability. (Hint: In the standard binomial-distribution-like proof presented we have shown that the probability is the same for any sequence of n draws with the same number k of white ball drawings, regardless of where they occur in the sequence).

3.5 Let $\hat{W}_n$ be the indicator of the event that a white ball appears in the nth sample in a Bernard Friedman's urn with the scheme $\begin{pmatrix} 0 & 1 \\ 1 & 0 \end{pmatrix}$. Show that the sequence $\{\hat{W}_n\}_{n=1}^{\infty}$ is not necessarily exchangeable.

3.6 (Knuth and McCarthy, 1991). Lipitor is prescribed for high cholesterol patients. In light cases, a daily dose of 10 mg is recommended. A patient presents a prescription to a pharmacy requesting 100 pills of the 10 mg dosage. On that day, the pharmacy has only 60 such pills, but they have plenty of 20 mg Lipitor pills in stock. The pharmacist advises the patient to take sixty 10 mg pills and twenty of the 20 mg pills, and each 20 mg pill is to be broken in two units, each to be taken on a separate day. The patient takes the recommended combination in a bottle. Every time she needs a pill, she chooses one at random from the bottle. If the pill is a 20 mg one, the patient breaks it into two halves and returns one half to the bottle, but if the pill is of the correct dosage (half of a previously broken one or a pill in the 10 mg denomination) she just takes it for the day.

(a) What Pólya schema represents the administering of these pills?

(b) What is the average number of 10 mg units left, when there are no more 20 mg pills in the bottle?

(Hint: Let the 10 mg pills be white balls and the 20 mg pills be blue balls in an urn. The number $\tilde{W}_1$ of white balls withdrawn until a blue one appears has a shifted negative hypergeometric distribution with parameters 20 and 60, then $\tilde{W}_2$, the number of white balls withdrawn until the second blue ball appears, is also a shifted negative hypergeometric random variable with parameters 19 and $61 - \tilde{W}_1$, and so on. The negative hypergeometric random variable that counts the draws until the first ball of the first type appears from within a population of m balls of the first type and n of the second has the average $(m+n+1)/(m+1)$.)

3.7 (Blom, Holst, and Sandell, 1994). An Ehrenfest urn of white and blue balls starts out all white. An initial number of M white balls are in the urn initially. What is the average waiting time until complete reversal of colors occurs for the first time (i.e., the average number of draws needed for all the balls to become blue for the first time)?

4

Poissonization

Associated with a Pólya urn growing in discrete time is a process obtained by embedding it in continuous time. It comprises a (changing) number of parallel Poisson processes. Poissonization has been used as a heuristic to understand discrete processes for some time. Early work on poissonization can be traced back to Kac (1949). In the context of Pólya urns, the embedding was introduced in Athreya and Karlin (1968) to model the growth of an urn in discrete time according to certain rules. Although it has been around for some time, no one until very recently (see Balaji and Mahmoud (2006)) had called it by the name the Pólya process. We think it is an appropriate name. We shall refer to the embedded process by that name. Of course, a Pólya urn growing in discrete time defines a stochastic process, too. However, to create a desirable distinction, we shall refer to a discrete Pólya system as we did before as a *scheme,* and reserve the title *Pólya process* to the poissonized process obtained by embedding the scheme in continuous time.

Even though it is meant as a mathematical transform for the discrete Pólya urn scheme, the inverse transform is fraught with difficulty, but it remains as one of the best heuristics to analyze the discrete schemes. It is also of interest to study the Pólya process in its own right because it models realistically a variety of real-world applications, too.

4.1 The Pólya Process

The Pólya process is a renewal process with rewards. It is derived from a Pólya urn scheme. It comprises a changing number of colored processes running in parallel. The various parallel processes are independent, but generally the numbers of processes of different colors can be dependent. For ease of introductory exposition, we take up first the two-color tenable Pólya process with deterministic schemata. We shall consider only tenable processes. The process grows out of a certain number of white and blue balls (thought to be contained in an urn). At time t, let the number of white balls be $W(t)$ and the number of blue balls be $B(t)$. Thus, initially we have $W(0)$ white balls, and $B(0)$ blue balls in the urn. Each ball generates a renewal after an independent Exp(1), an exponentially distributed random variable with parameter 1. We think of the renewal moment for a ball as the moment the ball is withdrawn from the urn.

We call a process evolving from a white ball a *white process*, and a process evolving from a blue ball a *blue process*. When a renewal occurs, the Pólya process emulates the behavior of a discrete process with the same schema. A certain number of balls is added, and that number depends on the color of the ball picked (the color of the process that induced the renewal). It is assumed that ball additions take place instantaneously at the renewal times. If a white process causes the renewal, we add a white balls and b blue balls to the urn, and if a blue process causes the renewal, we add c white balls and d blue balls to the urn.

As we did in the discrete schemes, it will organize the discussion of the embedded process to think of the ball addition scheme as a 2×2 matrix $\mathbf{A}$, the rows of which are indexed by the color of the process inducing the renewal, and the columns of which are labeled by the colors of balls added to the urn:

$$\mathbf{A} = \begin{pmatrix} a & b \\ c & d \end{pmatrix}.$$

To understand the behavior over time, it is helpful to think via an analogy from a Markovian race. At any stage, consider every ball in the urn as a runner in a race, whose running time is an independent Exp(1), an exponential random variable with parameter 1. The runners are grouped in teams wearing shirts of the color of the process they represent. When a runner from the white team wins the race, we say the white team wins the race, then a runners (balls) wearing white shirts and b runners wearing blue shirts enter the race (are added to the urn). Alternatively, when a runner from the blue team wins the race, we say the blue team wins the race, then c runners wearing white shirts and d runners wearing blue shirts enter the race. Every new runner is endowed with an independent Exp(1) clock.

Another race among all the existing runners is immediately restarted. The collective process enjoys a memoryless property as it is induced by individuals based on the exponential distribution—if a runner has covered a certain fraction of the course in one race, the runner is not allowed to carry over the gain to the next race; the runner's remaining time to cover the rest of the course remains Exp(1), as a result of resetting the race. Figure 4.1 illustrates the renewals and additions for the urn

$$\begin{pmatrix} 0 & 3 \\ 1 & 1 \end{pmatrix},$$

starting with one white and two blue balls, and at the first two epochs t_1, where a blue ball (bullet) is withdrawn, and t_2, when a white ball (circle) is withdrawn.

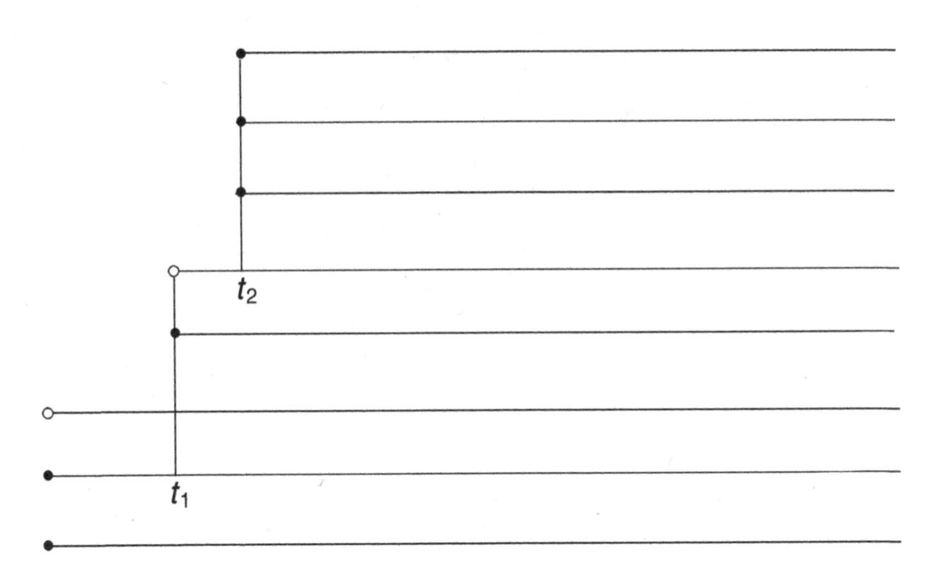

FIGURE 4.1
Ball additions at the epochs of a Pólya process.

Let Poi(λ) denote a Poisson random variable with parameter λ. With each individual process (runner) existing at time t we associate a random variable to represent the number of renewals it gives by time t. From basic properties of the Poisson process a runner entering the race by time $t' \le t$ gives Poi$(t - t')$ renewals by time t. We call the joint process $\mathbf{R}(t)$. That is,

$$\mathbf{R}(t) := (W(t), B(t))^T$$

is our official definition of a two-colored Pólya process. We point out that we define $W(t)$ (and similarly $B(t)$) to count the number of white (blue) balls at time t, including a change or jump in the number of white (blue) balls at time t.

We can formulate two simultaneous incremental equations for the process $(W(t), B(t))^T$. Consider the process at time $t + \Delta t$, where Δt is an infinitesimal increment of time. The number of white balls (given $\mathbf{R}(t)$) at time $t + \Delta t$ is what it was at time t, plus the number of white balls contributed by the various teams within the infinitesimal period $(t, t + \Delta t]$. Each member of the white team follows a Poisson process with parameter 1, and thus generates Poi(Δt) renewals in an interval of length Δt. Likewise, each member of the blue team generates Poi(Δt) renewals in an interval of length Δt. In turn, each newly born child in that interval may generate additional children by time $t + \Delta t$. Altogether, the number of children generated by all the new runners combined in the period $(t, t + \Delta t]$ is a remainder $r(\Delta t) = O_p((\Delta t)^2)$.[1]

---

[1] The notation O_p stands for a random variable that is O in probability, which means in this instance that, for some positive constant C, $\mathbf{P}(r(\Delta t) \le C(\Delta t)^2) \to 1$, as $\Delta t \to 0$. Some authors take this to be the strict sense of O_p.

This remainder is conditionally independent from the variables that generate it, with an average that is $O((\Delta t)^2)$. Each renewal by a white process increases the white team by a runners, and each renewal by a blue process increases the white team by c runners. A similar argument holds for the blue team. We have

$$\mathbf{E}\big[e^{uW(t+\Delta t)+vB(t+\Delta t)} \mid \mathbf{R}(t)\big] = \exp\left(\left[W(t) + a\sum_{i=1}^{W(t)} X_i + c\sum_{j=1}^{B(t)} Y_j\right] u + \left[B(t) + b\sum_{i=1}^{W(t)} X_i + d\sum_{j=1}^{B(t)} Y_j\right] v + O(r(\Delta t))\right). \tag{4.1}$$

For $X_1, \ldots, X_{W(t)}, Y_1, \ldots, Y_{B(t)}$, and $r(\Delta t)$ (conditionally) independent random variables, and the X_i's and Y_j's are $\mathrm{Poi}(\Delta t)$.

PROPOSITION 4.1

(Balaji and Mahmoud, 2006). The moment generating function $\phi(t, u, v) := \mathbf{E}[\exp(uW(t) + vB(t))]$ of the joint Pólya process satisfies

$$\frac{\partial \phi}{\partial t} + (1 - e^{au+bv})\frac{\partial \phi}{\partial u} + (1 - e^{cu+dv})\frac{\partial \phi}{\partial v} = 0.$$

PROOF Let $p_{km}(t) = \mathbf{P}(W(t) = k,\, B(t) = m)$. Taking expectations of Eq. (4.1) and computing it by conditioning, we see that

$$\phi(t + \Delta t, u, v) = \sum_{k,m} \mathbf{E}\left[e^{uk+vm} \exp\left(\left[a\sum_{i=1}^{k} X_i + c\sum_{j=1}^{m} Y_j\right] u + \left[b\sum_{i=1}^{k} X_i + d\sum_{j=1}^{m} Y_j\right] v + O(r(\Delta t))\right) \,\Big|\, W(t) = k,\, B(t) = m\right] p_{km}(t).$$

By the conditional independence of the X_i's, Y_j's and $r(\Delta t)$, and the identical distribution of X_i and Y_j, we have

$$\begin{aligned}\phi(t + \Delta t, u, v) = \sum_{k,m} e^{uk+vm} &\mathbf{E}[e^{(au+bv)X_1}] \times \cdots \times \mathbf{E}[e^{(au+bv)X_k}] \\ &\times \mathbf{E}[e^{(cu+dv)Y_1}] \times \cdots \times \mathbf{E}[e^{(cu+dv)Y_m}] \\ &\times \mathbf{E}[e^{O(r(\Delta t))}]\, p_{km}(t).\end{aligned}$$

The combined term $\mathbf{E}[e^{(au+bv)X_i}]$ is the moment generating function of X_i, evaluated at $au + bv$. With X_i being Poi(Δt), the combined term is $\exp(\Delta t\,(e^{au+bv} - 1))$. Similarly, $\mathbf{E}[e^{(cu+dv)Y_j}]$ is $\exp(\Delta t\,(e^{cu+dv} - 1))$. We can proceed with the identical distribution of the X_i's and Y_j's to get

$$\begin{aligned}\phi(t+\Delta t, u, v) &= \sum_{k,m} e^{uk+vm} \left(\mathbf{E}[e^{(au+bv)X_1}]\right)^k \\ &\quad \times (\mathbf{E}[e^{(cu+dv)Y_1}])^m \times \mathbf{E}[e^{O(r(\Delta t))}]\, p_{km}(t) \\ &= \sum_{k,m} e^{uk+vm} e^{k\Delta t(e^{au+bv}-1)} e^{m\Delta t(e^{cu+dv}-1)} \mathbf{E}[e^{O(r(\Delta t))}]\, p_{km}(t).\end{aligned}$$

We can expand all exponential functions locally around $u = v = 0$, using the usual Taylor series

$$e^z = 1 + z + \frac{z^2}{2!} + \cdots + \frac{z^k}{k!} + O(z^{k+1}).$$

Because our error terms are quadratic, it does not help the accuracy any further to go beyond a quadratic error term in the series, and we obtain

$$\begin{aligned}\phi(t+\Delta t, u, v) &= \sum_{k,m} e^{uk+vm}([1 + k\Delta t(e^{au+bv} - 1) + O((\Delta t)^2)] \\ &\quad \times [1 + m\Delta t(e^{cu+dv} - 1) + O((\Delta t)^2)] \\ &\quad \times \mathbf{E}[1 + O(r(\Delta t))]\, p_{km}(t) \\ &= \phi(t, u, v) + \Delta t(e^{au+bv} - 1) \sum_{k,m} k e^{ku+mv} p_{km}(t) \\ &\quad + \Delta t(e^{cu+dv} - 1) \sum_{k,m} m e^{ku+mv} p_{km}(t) + O((\Delta t)^2).\end{aligned}$$

The proposition follows upon reorganizing the expression above—move $\phi(t, u, v)$ to the left-hand side, divide by Δt throughout, and take the limit, as $\Delta t \to 0$. ∎

Moments follow mechanically from Proposition 4.1. Taking the jth derivative of both sides of the partial differential equation once with respect to one of the two dummy variables in the moment generating function, and then annulling all the dummy variables gives us a functional equation for the jth moment. Let us do this first for $j = 1$ to derive a functional equation for the first moment. By subjecting the equation to the operator $\frac{\partial}{\partial u}$, we obtain

$$\frac{\partial}{\partial u}\left(\frac{\partial \phi}{\partial t}\right) + (1 - e^{au+bv})\frac{\partial^2 \phi}{\partial u^2} - a e^{au+bv}\frac{\partial \phi}{\partial u} + (1 - e^{cu+dv})\frac{\partial^2 \phi}{\partial u \partial v} - c e^{cu+dv}\frac{\partial \phi}{\partial v} = 0.$$

Evaluation at $u = v = 0$ yields

$$\frac{\partial}{\partial t}\left(\frac{\partial \phi}{\partial u}\right)\Big|_{u=v=0} = \left[a\frac{\partial \phi}{\partial u} + c\frac{\partial \phi}{\partial v}\right]_{u=v=0}.$$

Noting that

$$\frac{\partial \phi}{\partial u} = \frac{\partial}{\partial u}\mathbf{E}[e^{uW(t)+vB(t)}] = \mathbf{E}\left[\frac{\partial}{\partial u}e^{uW(t)+vB(t)}\right] = \mathbf{E}[W(t)e^{uW(t)+vB(t)}],$$

evaluation at $u = v = 0$ gives

$$\frac{\partial \phi}{\partial u}\Big|_{u=v=0} = \mathbf{E}[W(t)].$$

Likewise,

$$\frac{\partial \phi}{\partial v}\Big|_{u=v=0} = \mathbf{E}[B(t)].$$

So, we have the functional equation

$$\frac{d}{dt}\,\mathbf{E}[W(t)] = a\,\mathbf{E}[W(t)] + c\,\mathbf{E}[B(t)]. \tag{4.2}$$

Symmetrically, by carrying out the operation on v, by first taking the partial derivative with respect to v, then evaluating at $u = v = 0$, we obtain the mirror image equation:

$$\frac{d}{dt}\,\mathbf{E}[B(t)] = b\,\mathbf{E}[W(t)] + d\,\mathbf{E}[B(t)]. \tag{4.3}$$

The couple of functional Eqs. (4.2) and (4.3) can be written in matrix form

$$\frac{d}{dt}\begin{pmatrix}\mathbf{E}[W(t)]\\ \mathbf{E}[B(t)]\end{pmatrix} = \begin{pmatrix}a & c\\ b & d\end{pmatrix}\begin{pmatrix}\mathbf{E}[W(t)]\\ \mathbf{E}[B(t)]\end{pmatrix} = \mathbf{A}^T\begin{pmatrix}\mathbf{E}[W(t)]\\ \mathbf{E}[B(t)]\end{pmatrix},$$

where $\mathbf{A}^T$ is the transpose of $\mathbf{A}$. This system of first-order ordinary differential equations has a standard solution.

THEOREM 4.1

Let $\mathbf{A}$ *be the schema of a two-color Pólya process of white and blue balls. At time* t*, the average number of white and blue balls in the process is*

$$\begin{pmatrix}\mathbf{E}[W(t)]\\ \mathbf{E}[B(t)]\end{pmatrix} = e^{\mathbf{A}^T t}\begin{pmatrix}W(0)\\ B(0)\end{pmatrix}.$$

One way of dealing with the matric exponential function in Theorem 4.1 is via series expansion. Another way is through the eigenvalues of $\mathbf{A}$. The eigenvalue approach goes through very smoothly in the case of two distinct eigenvalues. Minor complications (only of computational nature) arise in the case of repeated eigenvalues. We shall look at specific examples and give practice exercises.

4.2 Poissonization of Multicolor Deterministic Schemes

Suppose we have an urn that contains up to k colors with a deterministic schema

$$\mathbf{A} = \begin{pmatrix} a_{1,1} & a_{1,2} & \dots & a_{1,k} \\ a_{2,1} & a_{2,2} & \dots & a_{2,k} \\ \vdots & \vdots & \ddots & \vdots \\ a_{k,1} & a_{k,2} & \cdots & a_{k,k} \end{pmatrix}. \tag{4.4}$$

The equations we derived for two colors ($k = 2$) generalize easily to Pólya urns with deterministic schemata on a general number of colors.

Let the colors be numbered $1, \dots, k$, and let $X_j(t)$ be the number of balls of color j at time t, for $1 \le j \le k$. Here we need k dummy variables in the definition of the generating function:

$$\phi(t; u_1, \dots, u_k) = \mathbf{E}[e^{u_1 X_1(t) + u_2 X_2(t) + \cdots + u_k X_k(t)}].$$

The partial differential equations that corresponds to the two-color equation we found in Proposition 4.1 is

$$\frac{\partial \phi}{\partial t} + \sum_{j=1}^{k} (1 - e^{a_{j,1}u_1 + \cdots + a_{j,k}u_k}) \frac{\partial \phi}{\partial u_j} = 0, \tag{4.5}$$

and follows from a similar derivation.

Moreover, the computation for moments follows the same principles as for the two-color process. For example, for the mean we differentiate both sides of the partial differential equation once with respect to each dummy variable u_j, for $j = 1, \dots, k$, then set $u_1 = u_2 = \dots = u_k = 0$. Each derivative gives us one ordinary differential equation. When we solve this simultaneous system of k differential equations, we still get the form

$$\begin{pmatrix} \mathbf{E}[X_1(t)] \\ \mathbf{E}[X_2(t)] \\ \vdots \\ \mathbf{E}[X_k(t)] \end{pmatrix} = e^{\mathbf{A}^T t} \begin{pmatrix} X_1(0) \\ X_2(0) \\ \vdots \\ X_k(0) \end{pmatrix},$$

where $\mathbf{A}^T$ is the transpose of the $k \times k$ schema.

4.3 The Deterministic Monochromatic Pólya Process

We look here at one of the simplest forms of the Pólya process; one that involves only one color, say white. The schema in Eq. (4.4) is reduced to one entry, and we consider the deterministic case when that entry is a positive integer. Under this deterministic 1×1 schema we have a starting number $W(0)$ of white balls, each carrying its own independent Exp(1) exponential clock. At the time of a renewal in any Poisson process associated with a ball, $a := a_{11}$ balls are added to the urn, each new ball is equipped with its own independent Poisson process. In literature not necessarily concerned with urn schemes this process is called the *Yule process*.

Let $W(t)$ be the random number of balls in the urn at time t, with moment generating function

$$\phi(t, u) = \mathbf{E}[e^{uW(t)}].$$

For the monochromatic scheme the partial differential Eq. (4.5) becomes

$$\frac{\partial \phi}{\partial t} + (1 - e^{au})\frac{\partial \phi}{\partial u} = 0,$$

and is to be solved under the initial condition $\phi(0, u) = e^{W(0)u}$. It can be readily checked that the equation has the solution

$$\phi(t, u) = \left(\frac{e^{a(u-t)}}{e^{a(u-t)} - e^{au} + 1}\right)^{W(0)/a}. \tag{4.6}$$

Such a solution can be obtained by a variety of standard methods, such as the method of characteristics, which constitutes a chapter in many textbooks on partial differential equations; see Levine (1997), for example.

This distribution has interesting interpretations. Let us first take the case $a = 1$, and $W(0) = 1$. In this case

$$\phi(t, u) = \frac{e^{-t}e^{u}}{1 - (1 - e^{-t})e^{u}}.$$

Recall that the Geo(p) random variable has the moment generating function

$$\frac{pe^{u}}{1 - qe^{u}},$$

with $q = 1 - p$, and is defined only for $u < \ln \frac{1}{q}$. Matching the two forms, we immediately realize that

$$W(t) \overset{\mathcal{D}}{=} \text{Geo}(e^{-t}).$$

Furthermore, $\phi(t, u)$, as a moment generating function, imposes the restriction

$$u \le \ln\left(\frac{1}{1 - e^{-t}}\right), \tag{4.7}$$

and for every $t > 0$, there is a neighborhood of 0 on the u scale for which the moment generating function exists.

Subsequently, if $a = 1$, and $W(0) \geq 1$,

$$\phi(t, u) = \left(\frac{e^{-t} e^{u}}{1 - (1 - e^{-t}) e^{u}} \right)^{W(0)},$$

which is the moment generating function of the convolution of $W(0)$ independent geometric random variables, each with parameter e^{-t}. This makes perfect sense, because in this case $W(0)$ independent one-color Pólya processes run in parallel.

If $a \geq 1$, we can interpret the form in Eq. (4.6) as

$$\phi(t, u) = \sqrt[a]{\left(\frac{e^{-at} e^{au}}{1 - (1 - e^{-at}) e^{au}} \right)^{W(0)}}.$$

The function under the ath root sign is a convolution of $W(0)$ independent scaled geometric random variables, each distributed like $a\,\mathrm{Geo}(e^{-at})$. Geometric random variables are known to be infinitely divisible, that is to say, if G is $\mathrm{Geo}(p)$, we can always find a representation of G as a convolution of n independent identically distributed random variables—for each $n \geq 1$, there are n independent identically distributed random variables $Y_{1,n}, \ldots, Y_{n,n}$, such that

$$G \overset{\mathcal{D}}{=} Y_{1,n} + \cdots + Y_{n,n}.$$

Let $\psi_X(u)$ be the moment generating function of a generic random variable X. For each $n \geq 1$, $\psi_G(u)$ can be represented as

$$\psi_G(u) = \psi_{Y_{1,n}}(u) \ldots \psi_{Y_{n,n}}(u) = (\psi_{X_n}(u))^n,$$

where X_n is distributed like $Y_{1,n}$. In other words, ψ_G is the nth power of the moment generating function of some random variable, or equivalently, for each $n \geq 1$, the nth root of ψ_G defines ψ_{X_n}, a moment generating function of some random variable. It is an easy exercise to show that a finite linear combination of independent infinitely divisible random variables remains infinitely divisible. Whence, $\phi(t, u)$ for general $W(0) \geq 1$, and general $a \geq 1$, is the ath root of an infinitely divisible convolution of $W(0)$ independent $a\,\mathrm{Geo}(e^{-at})$ random variables.

We next find moments in the poissonized monochromatic deterministic process. Starting with $W(0)$ balls, the number of white balls in the Pólya process at time t has mean

$$\mathbf{E}[W(t)] = W(0)e^{at}, \tag{4.8}$$

and variance

$$\mathbf{Var}[W(t)] = a\,W(0)(e^{2at} - ae^{at}), \tag{4.9}$$

which follow immediately from the usual machinery of differentiation of the moment generating function in Eq. (4.6) (with respect to u) and evaluating at $u = 0$.

We next pursue the limit distribution (under appropriate scaling). As we develop asymptotics, when t becomes very large, we must constrict u in a very small range near 0, according to Eq. (4.7). Specifically, for x fixed, we take $u = x/e^{at}$, where the scale e^{at} was chosen because it is the leading asymptotic term in the standard deviation (as $t \to \infty$). It follows that

$$\mathbf{E}\left[\exp\left(\frac{W(t)}{e^{at}}\, x\right)\right] = \phi\left(t, \frac{x}{e^{at}}\right) \to \frac{1}{(1-ax)^{W_0/a}}.$$

The right-hand side is the moment generating function of Gamma$(\frac{W_0}{a}, a)$, and so

$$\frac{W(t)}{e^{at}} \xrightarrow{\mathcal{D}} \text{Gamma}\left(\frac{W_0}{a}, a\right).$$

4.4 The Pólya-Eggenberger Process

Recall that the Pólya-Eggenberger urn operates under the schema

$$\mathbf{A} = \begin{pmatrix} s & 0 \\ 0 & s \end{pmatrix},$$

where s balls of the same color are added upon withdrawing a ball. We shall look at a poissonized version of this scheme, which we call the Pólya-Eggenberger process. Without much effort, we can extend whatever can be done to the Pólya-Eggenberger process to the a Pólya process with the schema

$$\mathbf{A} = \begin{pmatrix} a & 0 \\ 0 & d \end{pmatrix}, \tag{4.10}$$

with a and d being possibly different. The independence of all the Poisson processes existing at any moment in time makes it possible to develop a result, for such a model. If $a \neq d$, the discrete model is unwieldy, the dependence and influence exerted on one color by the opposite color makes the analysis convoluted, but does give a comparable result to the ratio of gamma random variables that we shall see shortly in the poissonized process.

Let $(W(t), B(t))^T$ be the joint vector of the number of white and blue balls in the Pólya process with the diagonal urn scheme in Eq. (4.10). This Pólya process is relatively easy because any renewals from the white processes add only to the white processes, and the same applies to the blue processes. And so, the numbers of white and blue processes are independent. The number of white balls in the urn depends only on $W(0)$ and a, and independently, the number of blue balls depends only on $B(0)$ and d. The joint distribution can be comprehended as a superposition of two parallel monochromatic schemes: the white processes with scheme

$$\mathbf{A}_W = \begin{pmatrix} a & 0 \\ 0 & 0 \end{pmatrix},$$

with $(W(0), 0)^T$ initial conditions, and the blue processes with scheme

$$\mathbf{A}_B = \begin{pmatrix} 0 & 0 \\ 0 & d \end{pmatrix},$$

with $(0, B(0))^T$ initial conditions.

Each of these two schemes is a monochromatic Pólya process, for which a result has been developed in Section 4.3. Viz, if the process starts with $(W_0, B_0)^T$, then as $t \to \infty$,

$$\begin{pmatrix} \dfrac{W(t)}{e^{at}} \\ \dfrac{B(t)}{e^{dt}} \end{pmatrix} \xrightarrow{\mathcal{D}} \begin{pmatrix} \text{Gamma}(\frac{W(0)}{a}, a) \\ \text{Gamma}(\frac{B(0)}{d}, d) \end{pmatrix},$$

and the two limiting gamma random variables are independent.

4.5 The Ehrenfest Process

Let us embed in continuous time the urn scheme with ball replacement matrix

$$\mathbf{A} = \begin{pmatrix} -1 & 1 \\ 1 & -1 \end{pmatrix},$$

which is the Ehrenfest model for the mixing of gases. We call such a poissonized process the *Ehrenfest process*.

The poissonized Ehrenfest process has a nice *invariant property*. The total number of say τ balls in the process is constant at all times. Whenever we pick a ball of a color, we instantaneously replace it in the urn but painted with the antithetical color. Even if there is more than one renewal at the same time (an event that occurs with probability 0) each renewed ball is replaced with one of the opposite color. Consequently, at all times

$$W(t) + B(t) = \tau.$$

We can study the number of white processes via its moment generating function

$$\mathbf{E}[e^{W(t)u}] := \psi(t, u).$$

Recalling the joint moment generating function $\phi(t, u, v)$ of white and blue balls, we have set $\psi(t, u) = \phi(t, u, 0)$. Let us take $v = 0$ in Proposition 4.1 and write a partial differential equation for ψ. We obtain

$$\frac{\partial \psi}{\partial t} + (1 - e^{-u})\frac{\partial \psi}{\partial u} + (1 - e^{u})\frac{\partial \psi}{\partial v}\bigg|_{v=0} = 0.$$

Note that

$$\frac{\partial \phi}{\partial v}\bigg|_{v=0} = \mathbf{E}[B(t)e^{W(t)u + B(t)v}]\bigg|_{v=0} = \mathbf{E}[B(t)e^{W(t)u}].$$

Subsequently,

$$\frac{\partial \psi}{\partial t} + (1 - e^{-u})\frac{\partial \psi}{\partial u} + (1 - e^{u})\,\mathbf{E}[(\tau - W(t))e^{W(t)u}] = 0.$$

We can reorganize this equation in the form

$$\frac{\partial \psi}{\partial t} + (e^{u} - e^{-u})\frac{\partial \psi}{\partial u} + \tau(1 - e^{u})\psi = 0. \tag{4.11}$$

This partial differential equation can be solved exactly. For instance, if $\tau = 2$, we have two balls in the urn at any time. Suppose we start with one white and one blue ball. The initial condition under which the equation is to be solved is $\psi(0, u) = e^{u}$, and the exact solution is

$$\psi(t, u) = \mathbf{E}[e^{W(t)u}] = \frac{1}{4}(1 - e^{-4t}) + \frac{1}{2}(1 + e^{-4t})e^{u} + \frac{1}{4}(1 - e^{-2t})e^{2u}.$$

As $t \to \infty$, the right-hand side tends to

$$\frac{1}{4} + \frac{1}{2}e^{u} + \frac{1}{4}e^{2u},$$

which is the moment generating function of the binomial random variable $\text{Bin}(2, \frac{1}{2})$. We note in passing that while the Ehrenfest discrete process described in Figure 1.1 did not have a limit because of dependence on parity, when poissonized we have a simple limit. The intuitive reason is that, while the poissonized counterpart still has parity considerations, the even and odd parities are well mixed. How many renewals are induced by a white ball initially in the system by time t? If there is an even number of renewals from the initial white ball, that ball is white, and if the number is odd, that ball is blue. That is, the number of white balls associated with an initial white ball is a Bernoulli random variable with parameter dependent on t. This Bernoulli random variable can be 0 or 1, and the two cases are mixed in a manner that gives a smooth averaging over the two distributions.

Even though we can find exact solutions, the expressions quickly get rather lengthy as τ increases. Exercise 4.2 gives a sense for how the case $\tau = 3$ compares with the case $\tau = 2$. For $\tau = 3$, the limit remains a binomial random variable $\text{Bin}(3, \frac{1}{2})$ with the same rate of success, but on three trials. For general $\tau \ge 1$, The exact individual probabilities $\mathbf{P}(W(t) = k)$ are more complicated for each higher τ, and generally include a constant and functions decaying at the rates, $e^{-2t}, e^{-4t}, \ldots, e^{-2\tau t}$ but in all cases we shall prove a binomial limit:

$$W(t) \xrightarrow{\mathcal{D}} \text{Bin}\left(\tau, \frac{1}{2}\right), \qquad \text{as t} \to \infty.$$

It can be justified by the basic elements of the theory of semi-Markov chains that $\psi(t, u)$ approaches a limit, say $\eta(u)$, where the distribution does not change with time any more. In this steady state, the derivative with respect to time vanishes. We also know the form of such distribution. Regardless of how we start, $W(t) \in \{0, 1, 2, \dots, \tau\}$, and the moment generating function at any moment t must be of the form

$$\psi(t, u) = P_0(t) + P_1(t)e^u + P_2(t)e^{2u} + \cdots + P_\tau(t)e^{\tau u},$$

with $P_0(t) + \cdots + P_\tau(t) = 1$. At the steady state ($t = \infty$), the probabilities $P_j(t)$ are fixed numbers. The moment generating function is differentiable, and approaches a differential (with respect to u) limit, that is $\frac{\partial}{\partial u}\psi(t, u)$ approaches $\frac{\partial}{\partial u}\eta(u)$. We find the limiting moment generating function η by taking $\frac{\partial}{\partial t}$ to be 0 and the limit of $\frac{\partial}{\partial u}$ to be the total derivative of the limit η; so η satisfies the limiting equation

$$(e^u - e^{-u})\frac{d\eta}{du} + \tau(1 - e^u)\eta = 0.$$

The solution of this limiting equation is

$$\eta(u) = \frac{1}{2^\tau}(1 + e^u)^\tau,$$

which is the moment generating function of the binomial random variable $\text{Bin}(\tau, \frac{1}{2})$.

As $t \to \infty$, we have convergence in distribution:

$$W(t) \xrightarrow{\mathcal{D}} \text{Bin}\left(\tau, \frac{1}{2}\right).$$

For a generalization and detailed discussion of several aspects, see Balaji, Mahmoud, and Watanabe (2006).

4.6 Poissonization of Schemes with Random Entries

Suppose we have a tenable urn scheme of white and blue balls, growing in real time. Let the generator be

$$\begin{pmatrix} U & X \\ Y & Z \end{pmatrix},$$

and the entries are integer-valued random variables. Each ball has associated with it a Poisson process with parameter 1. We have processes representing white balls (we call them white processes), and we have processes representing blue balls (we call them blue processes). All processes are independent. When a renewal occurs, the Pólya process emulates the behavior of the discrete process with the same extended schema. A certain number of balls is added, and that number depends on the color of the ball picked (the color of the process that induced the renewal). It is assumed that ball additions take place instantaneously at the renewal points. If a white process causes the renewal we add an independent copy of U white balls and an independent copy of X blue balls to the urn, and if a blue process causes the renewal we add an independent copy of Y white balls and an independent copy of Z blue balls to the urn.

Let $W(t)$ and $B(t)$ respectively be the number of white and blue balls at time t (including any jumps at the moment). We can formulate two simultaneous incremental equations for the process $\mathbf{R}(t) = (W(t), B(t))^T$. Consider the process at time $t + \Delta t$. The number of white balls (given $\mathbf{R}(t)$) at time $t + \Delta t$ is what it was at time t, plus the number of white balls contributed by the various colors within the infinitesimal period $(t, t + \Delta t]$. Each white process is a Poisson process with parameter 1, and thus generates $w_i \overset{\mathcal{D}}{=} \text{Poi}(\Delta t)$ renewals in an interval of length Δt, $i = 1, \dots, W(t)$. Corresponding to each renewal from the ith process is an independent realization $U_k^{(i)} \overset{\mathcal{D}}{=} U$ white additions ($k = 1, \dots, w_i$) and $X_k^{(i)} \overset{\mathcal{D}}{=} X$ independent blue additions, $k=1, \dots, w_i$. Likewise, each blue process generates $b_j \overset{\mathcal{D}}{=} \text{Poi}(\Delta t)$ renewals in an interval of length Δt, $j=1, \dots, B(t)$, with $Y_k^{(j)} \overset{\mathcal{D}}{=} Y$ independent white additions ($k = 1, \dots, b_j$) and $Z_k^{(j)} \overset{\mathcal{D}}{=} Z$ independent blue additions, $k = 1, \dots, b_j$. In turn, each newly born child in that interval may generate additional children by time $t + \Delta t$. Altogether the number of children generated by any new process added in the period $(t, t + \Delta t]$ is $r(\Delta t) = O((\Delta t)^2)$ in probability, with an average that is $O((\Delta t)^2)$. We have

$$\begin{aligned}
&\mathbf{E}\left[e^{uW(t+\Delta t)+vB(t+\Delta t)} \,\middle|\, \mathbf{R}(t); w_1, \dots, w_{W(t)}; b_1, \dots, b_{B(t)}\right] \\
&= \mathbf{E}\left[\exp\left(u\left[W(t) + \sum_{i=1}^{W(t)} \sum_{j=1}^{w_i} U_j^{(i)} + \sum_{i=1}^{B(t)} \sum_{j=1}^{b_i} Y_j^{(i)}\right]\right.\right. \\
&\qquad \left.\left. + v\left[B(t) + \sum_{i=1}^{W(t)} \sum_{j=1}^{w_i} X_j^{(i)} + \sum_{i=1}^{B(t)} \sum_{j=1}^{b_i} Z_j^{(i)}\right] + O(r(\Delta t))\right)\right]. \qquad (4.12)
\end{aligned}$$

We represent the joint behavior in terms of the joint moment generating functions of the random variables on the rows of the generator. In what follows $\Psi_{Z_1,Z_2}(z_1, z_2)$ stands for the joint moment generating function of Z_1 and Z_2.

LEMMA 4.1
The moment generating function $\phi(t, u, v) := \mathbf{E}[\exp(uW(t) + vB(t))]$ of the joint process satisfies

$$\frac{\partial \phi}{\partial t} + (1 - \psi_{U,X}(u, v))\frac{\partial \phi}{\partial u} + (1 - \psi_{Y,Z}(u, v))\frac{\partial \phi}{\partial v} = 0.$$

PROOF Let

$$\begin{aligned} p_{k,m,\xi_1,\ldots,\xi_k,\eta_1,\ldots,\eta_m}(t) := {} & P(W(t)=k,\, B(t)=m;\, w_1=\xi_1,\ldots,w_k=\xi_k;\\ & b_1=\eta_1,\ldots,b_m=\eta_m)\\ = {} & P(w_1=\xi_1,\ldots,w_k=\xi_k;\, b_1=\eta_1,\ldots,\\ & b_m=\eta_m \mid W(t)=k,\, B(t)=m)) \times q_{k,m}(t), \end{aligned}$$

where

$$q_{k,m}(t) = P(W(t)=k,\, B(t)=m).$$

And so,

$$\begin{aligned} p_{k,m,\xi_1,\ldots,\xi_k,\eta_1,\ldots,\eta_m}(t) = {} & \frac{(\Delta t)^{\xi_1}e^{-\Delta t}}{\xi_1!} \times \cdots \times \frac{(\Delta t)^{\xi_k}e^{-\Delta t}}{\xi_k!}\\ & \times \frac{(\Delta t)^{\eta_1}e^{-\Delta t}}{\eta_1!} \times \cdots \times \frac{(\Delta t)^{\eta_m}e^{-\Delta t}}{\eta_m!} \times q_{k,m}(t). \end{aligned}$$

Taking expectations of Eq. (4.12) we arrive at

$$\begin{aligned} \mathbf{E}[e^{uW(t+\Delta t)+vB(t+\Delta t)}] = {} & \sum_{\substack{k,m\\ \xi_1,\ldots,\xi_k\\ \eta_1,\ldots,\eta_m}} \mathbf{E}\left[\exp\left(u\left[k+\sum_{i=1}^{k}\sum_{j=1}^{\xi_i} U_j^{(i)} + \sum_{i=1}^{m}\sum_{j=1}^{\eta_i} Y_j^{(i)}\right]\right.\right.\\ & + v\left[m+\sum_{i=1}^{k}\sum_{j=1}^{\xi_i} X_j^{(i)} + \sum_{i=1}^{m}\sum_{j=1}^{\eta_i} Z_j^{(i)}\right]\\ & \left.\left. + \; O(r(\Delta t))\right)\right] \times p_{k,m,\xi_1,\ldots,\xi_k,\eta_1,\ldots\eta_m}(t), \end{aligned}$$

which we can express as

$$\begin{aligned} \phi(t+\Delta t,u,v) = {} & \sum_{\substack{k,m\\ \xi_1,\ldots,\xi_k\\ \eta_1,\ldots,\eta_m}} e^{ku+mv}\mathbf{E}\left[\exp\left(u\left[\sum_{i=1}^{k}\sum_{j=1}^{\xi_i} U_j^{(i)} + \sum_{i=1}^{m}\sum_{j=1}^{\eta_i} Y_j^{(i)}\right]\right.\right.\\ & + v\left[\sum_{i=1}^{k}\sum_{j=1}^{\xi_i} X_j^{(i)} + \sum_{i=1}^{m}\sum_{j=1}^{\eta_i} Z_j^{(i)}\right]\left(\frac{(\Delta t)^{\xi_1+\cdots+\xi_k}e^{-k\Delta t}}{\xi_1!\ldots\xi_k!}\right)\\ & \left.\left.\times\left(\frac{(\Delta t)^{\eta_1+\cdots+\eta_m}e^{-m\Delta t}}{\eta_1!\ldots\eta_m!}\right)\right)\right] \mathbf{E}[e^{O(r(\Delta t))}]\, q_{k,m}(t). \end{aligned}$$

Because each independent Poisson process tends to generate 0 renewals with probability approaching 1, and 1 event with probability proportional to Δt, the probability of more than 2 renewals in the infinitesimal interval is $O((\Delta t)^2)$, and the probability of new children generating renewals is also $O((\Delta t)^2)$. We also have $\mathbf{E}[e^{O(r(\Delta t))}] = 1 + O((\Delta t)^2)$. We can separate the sums into components:

$$\phi(t, u, v) = \sum_{\substack{k,m \\ \xi_1=\cdots=\xi_k=0 \\ \eta_1=\cdots=\eta_m=0}} + \sum_{\substack{k,m \\ \xi_1+\cdots+\xi_k=1 \\ \eta_1=\cdots=\eta_m=0}} + \sum_{\substack{k,m \\ \xi_1=\cdots=\xi_k=0 \\ \eta_1+\cdots+\eta_m=1}} + O((\Delta t)^2).$$

Over the first of these components with all zeros, the sum is

$$\begin{aligned}
&\sum_{\substack{k,m \\ \xi_1=\cdots=\xi_k=0 \\ \eta_1=\cdots=\eta_m=0}} e^{ku+mv} e^{-(m+k)\Delta t} \mathbf{E}[e^{O(r(\Delta t))}] q_{k,m}(t) \\
&\quad = \sum_{k,m} e^{ku+mv}(1 - (k+m)\Delta t + O((\Delta t)^2)) q_{k,m}(t), \\
&\quad = \phi(t, u, v) - \frac{\partial \phi(t, u, v)}{\partial u} \Delta t - \frac{\partial \phi(t, u, v)}{\partial v} \Delta t + O((\Delta t)^2).
\end{aligned}$$

We next handle the sum

$$\sum_{\substack{k,m \\ \xi_1+\cdots+\xi_k=1 \\ \eta_1=\cdots=\eta_m=0}}.$$

For the outer sum on k and m, there are k inner sums where only one ξ_i is one, and all the η_j's are zero. The expectation within is the same by identical distribution:

$$\begin{aligned}
\sum_{\substack{k,m \\ \xi_1+\cdots+\xi_k=1 \\ \eta_1=\cdots=\eta_m=0}} &= \sum_{k,m} e^{ku+mv} \mathbf{E}\left[\exp\left(uU_1^{(1)} + vX_1^{(1)}\right)\right] \\
&\quad \times \mathbf{E}[1 + O(\Delta t)^2)]((\Delta t) e^{-k\Delta t}) \times (e^{-m\Delta t})]\, q_{k,m}(t) \\
&= \Delta t\, \mathbf{E}[e^{uU+vX}] \left(\sum_{k,m} k e^{ku+mv}(1 - (k+m)\Delta t\, q_{k,m}(t)) + O((\Delta t)^2)\right), \\
&= \psi_{U,X}(u, v) \frac{\partial \phi(t, u, v)}{\partial u} \Delta t + O((\Delta t)^2).
\end{aligned}$$

A similar argument holds for the third sum to give

$$\sum_{\substack{k,m \\ \xi_1=\cdots=\xi_k=0 \\ \eta_1+\cdots+\eta_m=1}} = \psi_{Y,Z}(u, v) \frac{\partial \phi(t, u, v)}{\partial v} \Delta t + O((\Delta t)^2).$$

The lemma follows upon reorganizing the expression above: move $\phi(t, u, v)$ to the left-hand side, divide by Δt throughout, and take the limit, as $\Delta t \to 0$. ■

Lemma 4.1 extends easily to multicolor schemes, with generator $[U_{i,j}]_{1\le i,j\le k}$, where the entries are random. Let $X_i(t)$ be the number of balls of color i at time t (including any jumps at the moment). If we let the moment generating function of the joint process be $\phi(t; u_1, \dots, u_k) := \mathbf{E}[\exp(u_1 X_1(t) + \dots + u_k X_k(t)]$, the differential equation takes the form

$$\frac{\partial \phi}{\partial t} + \sum_{i=1}^{k} (1 - \psi_{U_{i,1},\dots,U_{i,k}}(u_1, \dots, u_k)) \frac{\partial \phi}{\partial u_i} = 0,$$

where $\psi_{U_{i,1},\dots,U_{i,k}}(u_1, \dots, u_k)$ is the joint moment generating function of the random variables on the ith row. These differential equations provide moments and can be solved in some simple instance.

Exercises

4.1 (Balaji, Mahmoud, and Watanabe, 2006). Let us add weights to the action of withdrawing a ball in the Ehrenfest process of white and blue balls, and consider the generalization

$$\mathbf{A} = \begin{pmatrix} -\alpha & \beta \\ \alpha & -\beta \end{pmatrix},$$

with positive integers α and β. Let $W(t)$ and $B(t)$ be the number of white and blue balls at time t.

(a) Under what starting conditions does this process remain tenable at all times?

(b) Show that

$$\frac{W(t)}{\alpha} + \frac{B(t)}{\beta}$$

is invariant. Argue that the invariant is an integer.

(c) What is the average number of white balls at time t?

4.2 Suppose a poissonized Ehrenfest process starts with two white balls and one blue. Show that the probability distribution of $W(t)$, the number of white balls at time t, is

$$\mathbf{P}(W(t) = 0) = \frac{1}{8}(1 - e^{-2t} - e^{-4t} + e^{-6t}),$$

$$\mathbf{P}(W(t) = 1) = \frac{1}{8}(3 - e^{-2t} + e^{-4t} - 3e^{-6t}),$$

$$\mathbf{P}(W(t) = 2) = \frac{1}{8}(3 + e^{-2t} + e^{-4t} + 3e^{-6t}),$$

$$\mathbf{P}(W(t) = 3) = \frac{1}{8}(1 + e^{-2t} - e^{-4t} - e^{-6t}).$$

What are the mean and variance exactly and asymptotically?

4.3 Let

$$\mathbf{A} = \begin{pmatrix} 2 & 3 \\ 1 & 4 \end{pmatrix}$$

be the schema of a poissonized instance of the Bagchi-Pal urn of white and blue balls. What is the exact average of the number of white balls in the urn at time t?

4.4 An urn of white, blue, and red balls evolves under the schema

$$\mathbf{A} = \begin{pmatrix} 1 & 0 & 0 \\ 1 & 2 & 0 \\ 1 & 1 & 2 \end{pmatrix},$$

where the rows and columns correspond to the ball colors in the order given. First show that

$$e^{\mathbf{A}^T t} = te^{2t} \begin{pmatrix} 0 & 0 & 1 \\ 0 & 0 & 1 \\ 0 & 0 & 0 \end{pmatrix} + e^{2t} \begin{pmatrix} 0 & 1 & 0 \\ 0 & 1 & 0 \\ 0 & 0 & 1 \end{pmatrix} + e^{t} \begin{pmatrix} 1 & -1 & 0 \\ 0 & 0 & 0 \\ 0 & 0 & 0 \end{pmatrix},$$

then derive the average number of balls of each color.

4.5 Let **A** be a schema with random entries for a tenable two-color Pólya process of white and blue balls. Let $W(t)$ and $B(t)$ respectively be the number of white and blue balls in the process at time t. Show that, at time t, the average number of white and blue balls in the process is

$$\begin{pmatrix} \mathbf{E}[W(t)] \\ \mathbf{E}[B(t)] \end{pmatrix} = e^{\mathbf{E}[\mathbf{A}^T]t} \begin{pmatrix} W(0) \\ B(0) \end{pmatrix}.$$

4.6 A Pólya process grows in real time under the schema

$$\begin{pmatrix} X & 1 - X \\ 1 - X & X \end{pmatrix},$$

where X is a Ber(p) random variable. Let $W(t)$ and $B(t)$ respectively be the number of white and blue balls in the process at time t. What are the average values of $W(t)$ and $B(t)$?

5

The Depoissonization Heuristic

We have seen some similarities between discrete urn schemes and the continuous-time Pólya processes. The forms of the results are somewhat similar. To name one, let us recall the binomial random variable $\mathrm{Bin}(\tau, \frac{1}{2})$ as a stationary distribution for the number of white balls in the discrete Ehrenfest scheme (when starting from randomized initial conditions) and as a limit distribution for the number of white balls in the Ehrenfest process of white and blue balls. Recall also that the limit may not exist in the discrete scheme, if we start with a fixed number of balls of each color.

Another similarity in form is the aforementioned gamma random variables in the Pólya-Eggenberger-like scheme

$$\begin{pmatrix} a & 0 \\ 0 & d \end{pmatrix},$$

when $a \neq d$. This result is easy to establish in the continuous case, but it remained elusive for a long time in the discrete scheme. Attention was drawn to these urns as early as Savkevich (1940), but a distributional result remained defiant for the discrete scheme until Janson (2004), who closed an important chapter on research in urn models.

To put the similarities in perspective, let us consider the standard Pólya-Eggenberger urn of white and blue balls with the schema $\begin{pmatrix} 1 & 0 \\ 0 & 1 \end{pmatrix}$. Let W_n and B_n be respectively the number of white and blue balls after n draws from the urn. We have seen that, as $n \to \infty$,

$$\mathbf{E}[W_n] \sim \frac{W_0}{W_0 + B_0} n,$$

$$\mathbf{Var}[W_n] \sim \frac{W_0 B_0 n^2}{(W_0 + B_0)^2 (W_0 + B_0 + 1)};$$

see Corollary 3.1. The result for the Pólya process with the same scheme is

$$\mathbf{E}\big[W(t)\big] = W(0)e^t.$$

$$\mathbf{Var}\big[W(t)\big] \sim W(0)e^{2t},$$

as $t \to \infty$ (cf. Eqs. (4.8) and (4.9)). The two systems have the same rates of growth for the mean and variance, under the interpretation that t is $\ln n$.

Notwithstanding the similarities between the Pólya process and the discrete urn process, there are also differences. For example, in the Pólya-Eggenberger urn with the schema $\begin{pmatrix} 1 & 0 \\ 0 & 1 \end{pmatrix}$, by time n in the discrete scheme, $W_n \le W_0 + n$, thus a limit (as $n \to \infty$) for W_n/n must have support on $[0, 1]$; in this case the limit is $\beta(W_0, B_0)$. By contrast, in a Pólya process with the same scheme we have $W(t)/e^t$ converging to a Gamma$(W(0), 1)$ random variable (as $t \to \infty$), with support on $[0, \infty)$. This happens because by time t, the number of renewals is not restricted and can be indefinitely large, having generated a large number of ball additions.

The Pólya process can furnish heuristics to understand the discrete-time urn scheme. Embedding a discrete urn in a continuous-time Poisson process, or poissonization, is an idea that goes back to Athreya and Karlin (1968). One can view the embedding operation as a mathematical transform. We discuss one heuristic route for inverting back into the discrete domain, or depoissonization.

We illustrate the heuristic on tenable white-blue processes with the *deterministic* schema

$$\mathbf{A} = \begin{pmatrix} a & b \\ c & d \end{pmatrix},$$

but generalization to $k \ge 3$ colors follows easily the same principles.

Let t_n be the (random) moment of time at which the nth renewal in the Pólya process occurs. Let W_n and B_n be respectively the number of white and blue balls of the discrete Pólya urn scheme with the same ball replacement matrix after n draws. Let us go back to the race analogy (Section 4.1) to establish a connection to the Pólya process at the renewal point t_n. By the independent identical distribution of running times in the Pólya process, any of the runners is equally likely to win the next race, that is,

$$\mathbf{P}(\text{white team wins the } (n+1)\text{st race} \mid W_n, B_n) = \frac{W_n}{W_n + B_n},$$

and subsequently, a white balls and b blue balls are added to the urn, and

$$\mathbf{P}(\text{blue team wins the } (n+1)\text{st race} \mid W_n, B_n) = \frac{B_n}{W_n + B_n},$$

and subsequently, c white balls and d blue balls are added to the urn, and constituting a growth rule at the renewal times (epochs) in the number of runners in the Pólya process identical to that of the growth under random sampling in discrete time from a Pólya urn with schema $\mathbf{A}$. In other words, the two processes evolve in exactly the same way at the points

$n = 0, 1, 2, \ldots$ in the discrete scheme, and t_n in the Pólya process:

$$\mathbf{R}_n := \begin{pmatrix} W_n \\ B_n \end{pmatrix} \overset{\mathcal{D}}{=} \begin{pmatrix} W(t_n) \\ B(t_n) \end{pmatrix} =: \mathbf{R}(t_n),$$

if the two processes start with identical initial conditions

$$\begin{pmatrix} W_0 \\ B_0 \end{pmatrix} = \begin{pmatrix} W(0) \\ B(0) \end{pmatrix}.$$

However, $\mathbf{R}_n$ is only a discretized form of a continuous renewal process with rewards, the renewals of which are the starting whistle of the races, and the rewards of which at every renewal are determined by $\mathbf{A}$. We can think of $\mathbf{R}_n$, for $n = 1, 2, \ldots$, as a series of snap shots in time of the continuous process at the moments when a renewal takes place in it.

A depoissonization heuristic may be based on the observation that t_n is highly concentrated around its mean value. This can be proved rigorously in several special cases, and is observed experimentally in all the cases. We illustrate this by an example.

Example 5.1

Consider a Pólya-Eggenberger process with the schema $\begin{pmatrix} 1 & 0 \\ 0 & 1 \end{pmatrix}$, and starting with one ball of each color. Let t_n be the time of the nth renewal in the process. At time t let us number the balls $1, 2, \ldots, \tau(t)$, where $\tau(t)$ is the total number of balls by time t. Moreover, let $X_1, \ldots, X_{\tau(t)}$ be the associated independent Exp(1) running times at t. Within the parallel of races, there are two runners competing to win the first race, the race is won at the time $t_1 = \min(X_1, X_2)$, and a new runner joins the next race. The minimum of two exponential random variables with parameter 1 each is also exponential but with parameter $\frac{1}{2}$. More generally, the minimum of n independent Exp(1) random variables is $\text{Exp}(\frac{1}{n})$. The first renewal in the process occurs at the time $t_1 \overset{\mathcal{D}}{=} \text{Exp}(\frac{1}{2})$. The second race is immediately begun after the first ends, and takes an additional $\text{Exp}(\frac{1}{3})$ time, till its finish time. The second renewal takes place after $\text{Exp}(\frac{1}{2}) + \text{Exp}(\frac{1}{3})$ units of time, and the two exponentials in it are independent. Further renewals follow the pattern, and generally for $n \geq 1$,

$$t_n \overset{\mathcal{D}}{=} \text{Exp}\left(\frac{1}{2}\right) + \text{Exp}\left(\frac{1}{3}\right) + \cdots + \text{Exp}\left(\frac{1}{n+1}\right),$$

and all the exponential random variables in the expression are independent. So,

$$\begin{aligned} \mathbf{E}[t_n] &= \mathbf{E}\left[\text{Exp}\left(\frac{1}{2}\right)\right] + \mathbf{E}\left[\text{Exp}\left(\frac{1}{3}\right)\right] + \cdots + \mathbf{E}\left[\text{Exp}\left(\frac{1}{n+1}\right)\right] \\ &= \frac{1}{2} + \frac{1}{3} + \cdots + \frac{1}{n+1} \\ &= H_{n+1} - 1, \end{aligned}$$

where

$$H_k = 1 + \frac{1}{2} + \cdots + \frac{1}{k}$$

is the kth harmonic number. By the independence of the exponential running times we also have

$$\begin{aligned}\mathbf{Var}[t_n] &= \mathbf{Var}\left[\mathrm{Exp}\left(\frac{1}{2}\right)\right] + \mathbf{Var}\left[\mathrm{Exp}\left(\frac{1}{3}\right)\right] + \cdots + \mathbf{Var}\left[\mathrm{Exp}\left(\frac{1}{n+1}\right)\right] \\ &= \frac{1}{2^2} + \frac{1}{3^2} + \cdots + \frac{1}{(n+1)^2} \\ &= H_{n+1}^{(2)} - 1,\end{aligned}$$

where

$$H_k^{(2)} = 1 + \frac{1}{2^2} + \cdots + \frac{1}{k^2}$$

is the kth harmonic number of the second degree. We have at our disposal well-known asymptotics for the harmonic numbers:

$$H_k \sim \ln k, \qquad \text{as } k \to \infty,$$

$$H_k^{(2)} \sim \frac{\pi^2}{6}, \qquad \text{as } k \to \infty;$$

the sequence $H_k^{(2)}$ increases to its limit. Thus, the mean of t_n is asymptotic to $\ln n$ and the variance remains bounded as n grows to be very large.

For a law at large n, we apply Chebyshev's inequality

$$\mathbf{P}(|t_n - \mathbf{E}[t_n]) \ge \varepsilon) \le \frac{\mathbf{Var}[t_n]}{\varepsilon^2},$$

for any $\varepsilon > 0$. Let us replace ε with $\varepsilon\mathbf{E}[t_n]$, to get

$$\mathbf{P}\left(\left|\frac{t_n}{\mathbf{E}[t_n]} - 1\right| \ge \varepsilon\right) \le \frac{\pi^2/6 - 1}{\varepsilon^2(H_{n+1} - 1)^2} = O\left(\frac{1}{\ln^2 n}\right) \to 0.$$

So,

$$\frac{t_n}{\mathbf{E}[t_n]} \xrightarrow{P} 1,$$

In view of the convergence $\mathbf{E}[t_n]/\ln n \to 1$, the statement of Slutsky's theorem (see Billingsley, 1995, for example) allows us to multiply the two relations and get

$$\frac{t_n}{\ln n} \xrightarrow{P} 1. \tag{5.1}$$

We see that t_n is sharply concentrated about its logarithmic average. ◇

5.1 Depoissonization of Dichromatic Pólya Processes

Toward depoissonization of tenable two-color Pólya-processes we first establish connections between the number of draws from each of the two colors and the number of balls of each color. In the Pólya process set $\tilde{W}(t)$ to be the number of white ball drawings (number of times the white team has won a race by time t), set $\tilde{B}(t)$ to be the number of blue ball drawings (number of times the blue team has won a race by time t), and let $\tilde{\mathbf{R}}(t) = (\tilde{W}(t), \tilde{B}(t))^T$. As each white ball drawing accrues a white balls and each blue ball drawing accrues c white balls, we have the relation

$$W(t) = W(0) + a\,\tilde{W}(t) + c\,\tilde{B}(t).$$

Similarly,

$$B(t) = B(0) + b\tilde{W}(t) + d\,\tilde{B}(t).$$

In matric form these equations are

$$\mathbf{R}(t) = \mathbf{B}\tilde{\mathbf{R}}(t) + \mathbf{R}(0),$$

where $\mathbf{B} = \mathbf{A}^T$, and on average

$$\mathbf{E}[\mathbf{R}(t)] = \mathbf{B}\,\mathbf{E}[\tilde{\mathbf{R}}(t)] + \mathbf{R}(0).$$

Thus, at the random time t_n,

$$\mathbf{B}\,\mathbf{E}[\tilde{\mathbf{R}}(t_n) \mid t_n] + \mathbf{R}(0) = \mathbf{E}[\mathbf{R}(t_n) \mid t_n],$$

the expectation of which is

$$\mathbf{B}\,\mathbf{E}[\tilde{\mathbf{R}}(t_n)] + \mathbf{R}(0) = \mathbf{E}[\mathbf{R}(t_n)]. \tag{5.2}$$

Note how the influence of $\mathbf{A}$ appears through its transpose $\mathbf{B}$.

Generalization to k colors follows the same lines, and in fact the same equation holds, but now the matrix $\mathbf{B}$ is the transpose of the $k \times k$ schema $\mathbf{A}$, and all the vectors are $k \times 1$.

To get a sense for what we are about to do in the dichromatic Pólya process, let us consider first a monochromatic Pólya (Yule) process with schema $\mathbf{A} = [a]$. Suppose the single color in the monochromatic process is white, and $W(t)$ is the number of balls at time t, and $\tilde{W}(t)$ is the number of draws by time t. Trivially in this case all the draws are of white balls and each adds a white balls. That is, $\tilde{W}(t_n) = n$, and $W(t_n) = W(0) + an$. It was demonstrated that in the monochromatic process $\mathbf{E}[W(t)] = W(0)e^{at}$. This form is for fixed t, and a starting condition $W(0)$ that coincides with the number of balls in the discrete scheme. We wish to use this exponential form as an approximation at the random time t_n. We can establish, as we did in Eq. (5.1) for $a = 1$ that t_n is sharply concentrated around its mean value $\ln n$. This follows from the representation

$$t_n \overset{\mathcal{D}}{=} \mathrm{Exp}\left(\frac{1}{\tau_0}\right) + \mathrm{Exp}\left(\frac{1}{\tau_0 + a}\right) + \cdots + \mathrm{Exp}\left(\frac{1}{\tau_0 + (n-1)a}\right),$$

with mean

$$\mathbf{E}[t_n] = \frac{1}{\tau_0} + \frac{1}{\tau_0 + a} + \cdots + \frac{1}{\tau_0 + (n-1)a} \sim \frac{1}{a} \ln n,$$

and variance

$$\mathbf{Var}[t_n] = \frac{1}{\tau_0^2} + \frac{1}{(\tau_0 + a)^2} + \cdots + \frac{1}{(\tau_0 + (n-1)a)^2} = O(1).$$

Hence, $t_n / \frac{\ln n}{a} \xrightarrow{P} 1$, and we would expect an approximation to $\mathbf{E}[W(t_n)]$ via an exponential function to hold for t_n with small errors. Technically speaking, we are seeking an average measure of time $\bar{t}_n$ so that

$$\mathbf{E}[W(t_n)] = an + W_0 = W_0 e^{a \bar{t}_n}. \tag{5.3}$$

Therfore, if we define $\bar{t}_n$ to be

$$\frac{1}{a} \ln \left(\frac{an + W_0}{W_0}\right),$$

the expression $W_0 e^{a \bar{t}_n}$ is an *exact* depoissonization for the average number of balls in the discrete monochromatic scheme as obtained from the poissonized mean expression in the monochromatic Pólya process. When we go for more colors such technique will provide only an approximation as the exactitude of the monochromatic case cannot be achieved.

5.2 Depoissonization of Invertible Processes

We now go back to two dimensions. We pick up the discussion at Eq. (5.2). One can think of the monochromatic depoissonization in Eq. (5.3) as a corollary of the generalized mean-value theorem. Indeed it is the case for the one-color urn that there is a unique number $\bar{t}_n$ for which the relation Eq. (5.3) holds *exactly*. However, generally for more colors (where the linear algebra is in higher dimensions), because the Eq. (5.2) is a vectorial relation, each component of the vector might need a different average measure, but they are all of the order $\ln n$.

Certain algebraic operations will be performed to achieve the desired depoissonization. These operations involve the inversion of the transposed schema. The heuristic in the sequel will only apply if $\mathbf{B}$ is invertible, and we shall call the class of Pólya processes possessing invertible $\mathbf{B} = \mathbf{A}^T$ *invertible Pólya processes*. For an invertible schema the relation Eq. (5.2) is the same as

$$\mathbf{E}[\tilde{\mathbf{R}}(t_n)] = \mathbf{B}^{-1}(\mathbf{E}[\mathbf{R}(t_n)] - \mathbf{R}(0)),$$

and $\mathbf{B}^{-1}$ exists.

For an invertible Pólya process we are seeking an average measure $\bar{t}_n$, so that

$$\mathbf{E}[\mathbf{R}(t_n)] \approx \mathbf{E}[\mathbf{R}(\bar{t}_n)] = e^{\mathbf{B}\bar{t}_n}\mathbf{R}(0).$$

On any stochastic path whatever, we have n races by time t_n; the components of $\tilde{\mathbf{R}}(t_n)$ must add up to n. Let $\mathbf{J}$ be the 1×2 row vector of ones. Premultiplying a 2×1 vector by $\mathbf{J} = (1 \quad 1)$ adds up the two components of the vector. So,

$$\begin{aligned} n &= \mathbf{E}[\mathbf{J}\,\tilde{\mathbf{R}}(t_n)] \\ &\approx \mathbf{J}\mathbf{B}^{-1}e^{\mathbf{B}\bar{t}_n}\mathbf{R}(0) + o(n). \end{aligned}$$

The exponential matric form $e^{\mathbf{B}\bar{t}_n}$ can be computed via eigenvalues and matrices called *idempotents*. We take up a case of two real distinct eigenvalues, and leave an instance of repeated eigenvalues to a forthcoming example. With distinct eigenvalues $\lambda_1 > \lambda_2$ and for any real number x, we have the expansion

$$e^{\mathbf{B}x} = e^{\lambda_1 x}\boldsymbol{\mathcal{E}}_1 + e^{\lambda_2 x}\boldsymbol{\mathcal{E}}_2, \tag{5.4}$$

where the two matrices $\boldsymbol{\mathcal{E}}_1$ and $\boldsymbol{\mathcal{E}}_2$ are the idempotents of $\mathbf{B}$.

For large n,

$$\begin{aligned} n &\approx \mathbf{J}\mathbf{B}^{-1}(e^{\lambda_1\bar{t}_n}\boldsymbol{\mathcal{E}}_1 + e^{\lambda_2\bar{t}_n}\boldsymbol{\mathcal{E}}_2)\mathbf{R}(0) + o(n). \\ &\approx \mathbf{J}\mathbf{B}^{-1}\boldsymbol{\mathcal{E}}_1\mathbf{R}(0)e^{\lambda_1\bar{t}_n}. \end{aligned}$$

It follows that

$$e^{\mathbf{B}\bar{t}_n} \approx e^{\lambda_1\bar{t}_n}\boldsymbol{\mathcal{E}}_1 = \frac{n}{\mathbf{J}\mathbf{B}^{-1}\boldsymbol{\mathcal{E}}_1\mathbf{R}(0)}\boldsymbol{\mathcal{E}}_1.$$

Finally, we arrive at the approximation

$$\mathbf{E}\big[\mathbf{R}_n\big] \approx e^{\mathbf{B}\bar{t}_n}\mathbf{R}(0) = \frac{1}{\mathbf{J}\mathbf{B}^{-1}\boldsymbol{\mathcal{E}}_1\mathbf{R}(0)}\,\boldsymbol{\mathcal{E}}_1\mathbf{R}(0)\,n. \tag{5.5}$$

5.3 Depoissonization Examples

We illustrate the depoissonization heuristic with four examples.

Example 5.2
Consider a Bernard Friedman's process of white and blue balls, with the schema

$$\begin{pmatrix} 0 & 1 \\ 1 & 0 \end{pmatrix}.$$

For this symmetric schema $\mathbf{B} = \mathbf{A}^T = \mathbf{A}$,

$$e^{\mathbf{B}t} = \frac{1}{2}\begin{pmatrix} 1 & 1 \\ 1 & 1 \end{pmatrix} e^t + \frac{1}{2}\begin{pmatrix} 1 & -1 \\ -1 & 1 \end{pmatrix} e^{-t},$$

and

$$\mathcal{E}_1 = \frac{1}{2}\begin{pmatrix} 1 & 1 \\ 1 & 1 \end{pmatrix}.$$

Hence,

$$\mathcal{E}_1\mathbf{R}(0) = \frac{1}{2}\begin{pmatrix} 1 & 1 \\ 1 & 1 \end{pmatrix}\begin{pmatrix} W(0) \\ B(0) \end{pmatrix} = \frac{1}{2}(W(0) + B(0))\begin{pmatrix} 1 \\ 1 \end{pmatrix};$$

$$\mathbf{J}\mathbf{B}^{-1}\mathcal{E}_1\mathbf{R}(0) = W(0) + B(0).$$

Plugging these calculations in Eq. (5.5), we obtain

$$\begin{pmatrix} \mathbf{E}[W_n] \\ \mathbf{E}[B_n] \end{pmatrix} \approx \frac{1}{2}\begin{pmatrix} 1 \\ 1 \end{pmatrix} n,$$

in accord with Proposition 3.1. ◇

Example 5.3
Consider a Bagchi-Pal process of white and blue balls, with the schema

$$\begin{pmatrix} 2 & 2 \\ 1 & 3 \end{pmatrix}.$$

For this schema

$$\mathcal{E}_1 = \frac{1}{3}\begin{pmatrix} 1 & 1 \\ 2 & 2 \end{pmatrix}.$$

By calculations similar to those in the previous example we find

$$\begin{pmatrix} \mathbf{E}[W_n] \\ \mathbf{E}[B_n] \end{pmatrix} \approx \frac{4}{3}\begin{pmatrix} 1 \\ 2 \end{pmatrix} n,$$

in accord with Proposition 3.1. ◇

Example 5.4
Consider a Bagchi-Pal process of white and blue balls, with the schema

$$\begin{pmatrix} 2 & 0 \\ 1 & 1 \end{pmatrix}.$$

Here

$$\mathcal{E}_1 \mathbf{R}(0) = \begin{pmatrix} W(0) + B(0) \\ 0 \end{pmatrix},$$

The heuristic approximation gives

$$\begin{pmatrix} \mathbf{E}[W_n] \\ \mathbf{E}[B_n] \end{pmatrix} \approx \begin{pmatrix} 2 \\ 0 \end{pmatrix} n;$$

the 0 needs some interpretation. In all the asymptotics applied in the depoissonization heuristic we neglected terms that are of order lower than n. The sublinear term can be any function that is $o(n)$. A correct interpretation is that

$$\mathbf{E}[W_n] \sim 2n$$
$$\mathbf{E}[B_n] = o(n),$$

as $n \to \infty$. In the next example we shall see how to determine the asymptotic order of sublinear terms. ◇

Example 5.5
The urn of white and blue balls with the schema

$$\begin{pmatrix} 1 & 0 \\ 1 & 1 \end{pmatrix}$$

was considered in Kotz, Mahmoud, and Robert (2000). There is something different and more challenging about this urn scheme. The eigenvalues of the replacement matrix are repeated (both are 1), but the scheme remains within the invertible class.

In this scheme we always add one white ball after each draw, and $W_n = W(t_n) = n + W(0)$. However, there is variability in the number of blue balls—we add a blue ball only if a blue ball is withdrawn; occasionally a draw of a white ball does not change the number of blue balls. The urn scheme is trivial if it contains only white balls at the start. In this case after n draws there are $n + W(0)$ white balls and no blue balls ever appear. Let us assume $B(0) > 0$. We proceed with the representation

$$e^{\mathbf{B}t} = \begin{pmatrix} e^t & te^t \\ 0 & e^t \end{pmatrix} = \begin{pmatrix} 0 & 1 \\ 0 & 0 \end{pmatrix} te^t + \begin{pmatrix} 1 & 0 \\ 0 & 1 \end{pmatrix} e^t.$$

Note the extra t in the expansion, which is a departure from Eq. (5.4). Let

$$\mathcal{E}_1 := \begin{pmatrix} 0 & 1 \\ 0 & 0 \end{pmatrix}, \qquad \mathcal{E}_2 := \begin{pmatrix} 1 & 0 \\ 0 & 1 \end{pmatrix}.$$

When we proceed with the inversion, the two components of the vector of the number of draws must still add up to n, and so

$$\begin{aligned} n &= \mathbf{E}\big[\mathbf{J}\,\tilde{\mathbf{R}}(t_n)\big] \\ &\approx \mathbf{J}\mathbf{B}^{-1}e^{\mathbf{B}\bar{t}_n}\mathbf{R}(0) + o(n) \\ &\approx \mathbf{J}\mathbf{B}^{-1}(\bar{t}_n e^{\bar{t}_n}\mathcal{E}_1 + e^{\bar{t}_n}\mathcal{E}_2)\mathbf{R}(0) + o(n) \\ &\approx \mathbf{J}\mathbf{B}^{-1}\mathcal{E}_1\mathbf{R}(0)\bar{t}_n e^{\bar{t}_n}. \end{aligned}$$

Thus,

$$\bar{t}_n e^{\bar{t}_n} \approx \frac{n}{\mathbf{J}\mathbf{B}^{-1}\mathcal{E}_1\mathbf{R}(0)} = \frac{n}{B(0)},$$

and subsequently

$$\bar{t}_n + \ln \bar{t}_n \approx \ln n.$$

The increasing function $\bar{t}_n$ must be asymptotic to $\ln n$, and we also have

$$e^{\bar{t}_n} \approx \frac{n}{\mathbf{J}\mathbf{B}^{-1}\mathcal{E}_1\mathbf{R}(0)\bar{t}_n} \sim \frac{n}{B(0)\ln n}.$$

It follows that

$$e^{\mathbf{B}\bar{t}_n} \approx \begin{pmatrix} 0 & 1 \\ 0 & 0 \end{pmatrix}\frac{n}{B(0)} + \begin{pmatrix} 1 & 0 \\ 0 & 1 \end{pmatrix}\frac{n}{B(0)\ln n}.$$

Finally, we arrive at the approximation

$$\mathbf{E}[\mathbf{R}_n] \approx e^{\mathbf{B}\bar{t}_n}\mathbf{R}(0) \approx \begin{pmatrix} n \\ \dfrac{n}{\ln n} \end{pmatrix}.$$

Exercises

5.1 Apply the depoissonization heuristic to derive the average number of balls in a Pólya-Eggenberger dichromatic urn scheme growing in discrete time after n draws, where s balls of the same color are added after each draw.

5.2 A hyper Pólya-Eggenberger urn scheme of white, blue, and red balls evolves in discrete time under the schema

$$\mathbf{A} = \begin{pmatrix} 1 & 0 & 0 \\ 0 & 1 & 0 \\ 0 & 0 & 1 \end{pmatrix},$$

where the rows and columns correspond to the ball colors in the order given. Find the approximate asymptotic average number of balls of each color after n draws. Extend the result to k colors.

5.3 An urn scheme of white and blue balls evolves in discrete time under the Pólya-Eggenberger-like schema

$$\mathbf{A} = \begin{pmatrix} a & 0 \\ 0 & d \end{pmatrix},$$

where $a > d \geq 1$, and the rows and columns correspond to the ball colors in the order given. Show that after n draws:

(a) If $W_0 \geq 0$, the average number of white balls is given by

$$\mathbf{E}[W_n] \approx an - O(n^{d/a}).$$

(b) If $B_0 \geq 0$, the average number of blue balls is given by

$$\mathbf{E}[B_n] \approx O(n^{d/a}).$$

5.4 An urn scheme of white, blue, and red balls evolves in discrete time under the schema

$$\mathbf{A} = \begin{pmatrix} 1 & 0 & 0 \\ 1 & 2 & 0 \\ 1 & 1 & 2 \end{pmatrix},$$

where the rows and columns correspond to the ball colors in the order given. Apply the depoissonization heuristic to estimate the leading asymptotic term in the average number of balls of each color after n draws. (Hint: Exercise 4.4 gives the result for the poissonized version.)

6

Urn Schemes with Random Replacement

In this chapter we extend several of the ideas and methods found in the previous chapters. A broader theory is presented, and it turns out to be the theory needed for a myriad of applications that are not restricted to deterministic schemata. We dealt with two-color deterministic ball addition schemes with normal behavior, such as the Bagchi-Pal urn. We expand the scope by considering dichromatic generators with random entries that are within the normal class. We then present results for a class of normal schemes with $k \geq 2$ colors.

A basic tool is martingale theory, which can deal with dependent random variables with a certain dependency structure like that in urns. We devote a subsection for an overview of the definition and the results needed.

6.1 Martingales

The derivations for laws of large numbers and central limit theorems in urn schemes will rely on martingale theory. We sketch here some preliminaries, such as the definition and standard results that will be used later.

A sequence of random variables X_n is a martingale if

(i) For each $n \geq 1$, X_n is absolutely integrable, that is, $\mathbf{E}[|X_n|] < \infty$;
(ii) $\mathbf{E}[X_n \mid X_{n-1}] = X_{n-1}$.

In (ii) the given is all the information that can be gleaned from knowing the random variables X_{n-1}, or the so-called sigma field generated by the $(n-1)$st variable in the sequence, denoted by $\mathcal{F}_{n-1}$. A variant of the definition uses an arbitrary sequence of increasing sigma fields $\mathcal{G}_1 \subseteq \mathcal{G}_2 \subseteq \mathcal{G}_3 \ldots$.

Example 6.1
An example of martingales arises in fair gambling. Let X_n be the gamblers's fortune after betting n times on a fair game. At each bet the gambler puts down all his money as a bet, and this fortune is either doubled by winning the game, or the money is gone by losing it. The probabilities of a win or a

loss are both $\frac{1}{2}$ (because the game is fair), and

$$\mathbf{E}[X_n \mid X_{n-1}] = 2X_{n-1} \times \frac{1}{2} + 0 \times \frac{1}{2} = X_{n-1}.$$

Additionally, $\mathbf{E}[|X_n|] = \mathbf{E}[X_n] = 0$, and X_n is a martingale. ◇

The following is a simplified version of a technical proposition (proved in Hall and Heyde (1980), p. 36). It provides useful arguments.

PROPOSITION 6.1
Let X_n be a martingale (with respect to the increasing sigma fields $\mathcal{F}_{n-1}$), that is stochastically dominated by an integrable random variable. Then[1]

$$\frac{1}{n}\sum_{i=1}^{n}(X_i - \mathbf{E}[X_i \mid \mathcal{F}_{i-1}]) \xrightarrow{P} 0.$$

Sufficient conditions for the martingale central limit theorem are the *conditional Lindeberg condition* and the *conditional variance condition* on the martingale differences $\nabla X_n = X_n - X_{n-1}$; (see Hall and Hyde (1980), see Theorem 3.2 and Corollary 3.1, p. 58 in that source).

The conditional Lindeberg condition requires that, for all $\varepsilon > 0$,

$$\sum_{k=1}^{n} \mathbf{E}[(\nabla X_k)^2 \mathbf{1}_{\{|\nabla X_k| > \varepsilon\}} \mid \mathcal{F}_{k-1}] \xrightarrow{P} 0.$$

A Z-conditional variance condition requires that

$$\sum_{k=1}^{n} \mathbf{E}[(\nabla X_k)^2 \mid \mathcal{F}_{k-1}] \xrightarrow{P} Z.$$

The following standard theorem is demonstrated in Hall and Heyde (1980); see Theorem 3.2 and Corollary 3.1, p. 58 in that source.

THEOREM 6.1
Let X_n be a martingale satisfying the conditional Lindeberg condition and a Z-conditional variance condition. Then $X_n/\sqrt{n}$ converges in distribution to a random variable with characteristic function $\mathbf{E}[\exp(-Zt^2/2)]$.

---

[1] Under certain conditions, such as the independence of X_n, or the stochastic bound is square integrable, the convergence is strengthened to be almost sure.

6.2 Extended Urn Schemes

In Theorem 3.5 we presented a subclass of Bagchi-Pal urn schemes with normal behavior. In a Bagchi-Pal scheme with schema

$$\begin{pmatrix} a & b \\ c & d \end{pmatrix},$$

a sufficient condition for normality is $\lambda_2 = (a - c) < \frac{1}{2}(a + b) = \lambda_1$. We shall extend this class to a larger class of multicolor urn schemes. We can make the presentation in terms of the urn parameters, $a, b, c,$ and d, however, we shall switch to the notation of eigenvalues λ_1 and λ_2, as such presentation is more compact and more easily generalized.

Consider a multicolor urn with the generator

$$\mathbf{A} = \begin{pmatrix} U_{1,1} & U_{1,2} & \dots & U_{1,k} \\ U_{2,1} & U_{2,2} & \dots & U_{2,k} \\ \vdots & \vdots & \ddots & \vdots \\ U_{k,1} & U_{k,2} & \dots & U_{k,k} \end{pmatrix},$$

where the entries of the generator are integer-valued random variables (some entries are possibly of support containing negative integers). We arrange the k eigenvalues according to their decreasing real parts:

$$\Re\,\lambda_1 \ge \Re\,\lambda_2 \ge \ldots \ge \Re\,\lambda_k.$$

We call λ_1 the *principal eigenvalue*, and we call the corresponding eigenvector the *principal eigenvector*. The urn scheme will be called *extended*, if it satisfies the following conditions:

(i) The scheme is tenable.
(ii) All the entries in the generator have finite variances.
(iii) The average generator $\mathbf{E}[\mathbf{A}]$ has constant row sum.
(iv) The average generator has one real positive principal eigenvalue (which must then be the same as the constant row sum).
(v) The components of the principal eigenvector are all strictly positive.
(vi) $\Re\,\lambda_2 < \frac{1}{2}\Re\,\lambda_1$.

The last condition is a technical one that suffices to make the extended class normal.

6.3 Deterministic Dichromatic Extended Urn Schemes

Consider a tenable urn scheme of white and blue balls, where the schema is

$$\mathbf{A} = \begin{pmatrix} a & b \\ c & d \end{pmatrix};$$

as usual white precedes blue in the indexing of rows and columns. For transparency of proof, let us first develop results for the balanced case of constant row sum. We already saw a proof of normality based on the method of moments in the case of the Bagchi-Pal urn scheme. We present here an alternative proof, due to Smythe (1996). The advantage of this latter proof is the ease of adaptability to deterministic schemes with $k \ge 2$ colors, and to schemes with random generators, with minimal structural change in the proof.

Let the constant row sum be K. We are considering schemata with

$$a + b = c + d = K.$$

Note that the schema has the eigenvalues $\lambda_1 = K$ and $\lambda_2 = a - c$. The case $\lambda_2 = 0$ is trivial, as in this case $a = c$, and the schema is

$$\begin{pmatrix} a & K - a \\ a & K - a \end{pmatrix},$$

with no randomness, as we always add a white balls and $K - a$ blue balls after each drawing. We shall assume $\lambda_2 \ne 0$, that is, $a \ne c$.

6.3.1 Laws of Large Numbers

We discuss in this subsection weak laws of large numbers associated with deterministic dichromatic urn schemes. The results go back to Athreya and Karlin (1968), who used poissonization–depoissonization. The proof we present is due to Smythe (1996), who extended the class of urns to which the theory applies.

Let W_n and B_n be respectively the number of white balls and blue balls present in the urn after n draws, and let $\tau_n = W_n + B_n$ be the total number of balls after n draws. In the simple case of balanced schemes, we add the same number of balls (namely K) after each drawing. That is, $\tau_n = Kn + \tau_0$, and trivially we have the convergence

$$\frac{\tau_n}{n} \xrightarrow{a.s.} \lambda_1, \tag{6.1}$$

which, as we shall see later, generalizes to balanced dichromatic cases with random entries, where the average generator has constant row sum.

The last relation captures the role of λ_1, the principal eigenvalue of the schema. The complete picture takes into account the role of the second eigenvalue λ_2, which controls the rates of convergence. We shall call the left (row) eigenvectors[2] corresponding to λ_1 and λ_2, respectively, $\mathbf{v} := (v_1, v_2)$ and $\mathbf{u} := (u_1, u_2)$.

Let

$$\mathbf{Z}_n = \begin{pmatrix} W_n \\ B_n \end{pmatrix},$$

and

$$X_n = \mathbf{u}\mathbf{Z}_n.$$

Note that for large n, $|X_n|$ is linearly bounded, as can be seen from

$$\begin{aligned}
|X_n| &= |u_1 W_n + u_2 B_n| \\
&\le |u_1| W_n + |u_2| B_n \\
&\le \max(|u_1|, |u_2|)(W_n + B_n) \\
&= \max(|u_1|, |u_2|)\tau_n \\
&= \max(|u_1|, |u_2|)(\lambda_1 n + \tau_0) \\
&\le \max(|u_1|, |u_2|)(\lambda_1 n + \lambda_1 n), \qquad \text{for large } n \\
&= 2\max(|u_1|, |u_2|)\lambda_1 n \\
&=: C_1 n. \qquad (6.2)
\end{aligned}$$

Let $\mathcal{F}_j$ be the sigma field generated by W_j. So,

$$\begin{aligned}
\mathbf{E}[\triangledown X_n \mid \mathcal{F}_{n-1}] &= \mathbf{E}[X_n - X_{n-1} \mid \mathcal{F}_{n-1}] \\
&= \mathbf{E}[X_n \mid \mathcal{F}_{n-1}] - X_{n-1} \\
&= \mathbf{E}[\mathbf{u}\mathbf{Z}_n \mid \mathcal{F}_{n-1}] - X_{n-1} \\
&= \mathbf{E}[u_1 W_n + u_2 B_n \mid \mathcal{F}_{n-1}] - X_{n-1} \\
&= u_1 \left(W_{n-1} + a\,\frac{W_{n-1}}{\tau_{n-1}} + c\,\frac{B_{n-1}}{\tau_{n-1}} \right) \\
&\qquad + u_2 \left(B_{n-1} + b\,\frac{W_{n-1}}{\tau_{n-1}} + d\,\frac{B_{n-1}}{\tau_{n-1}} \right) - X_{n-1} \\
&= (u_1 W_{n-1} + u_2 B_{n-1}) + \frac{1}{\tau_{n-1}} \mathbf{u}\mathbf{A}^T \mathbf{Z}_{n-1} - X_{n-1} \\
&= X_{n-1} + \frac{\lambda_2}{\tau_{n-1}} \mathbf{u}\mathbf{Z}_{n-1} - X_{n-1} \\
&= \frac{\lambda_2}{\tau_{n-1}} X_{n-1}.
\end{aligned}$$

[2] All eigenvectors are normalized—the scale is chosen such that the components add up to 1.

Therefore,

$$\mathbf{E}\left[\nabla X_n - \frac{\lambda_2}{\tau_{n-1}} X_{n-1} \,|\, \mathcal{F}_{n-1}\right] = 0,$$

and

$$M_n := \nabla X_n - \frac{\lambda_2}{\tau_{n-1}} X_{n-1}.$$

are martingale differences. Note that $\mathbf{E}[M_n] = 0$.

We shall seek constants $\beta_{j,n}$ such that

$$V_n := \sum_{j=1}^{n} \beta_{j,n} M_j$$

are good approximations of X_n. That is,

$$V_n = \sum_{j=1}^{n} \beta_{j,n} M_j = X_n + \varepsilon_n,$$

where ε_n are small error terms that do not affect the asymptotics. Then, an asymptotic results for V_n will also be one for X_n. Note that $\mathbf{E}[V_n] = 0$.

This approximation is possible if we match the coefficients:

$$\begin{aligned}
X_n + \varepsilon_n &= V_n \\
&= \beta_{n,n}\left(X_n - X_{n-1} - \frac{\lambda_2}{\tau_{n-1}} X_{n-1}\right) + \beta_{n-1,n}\left(X_{n-1} - X_{n-2} - \frac{\lambda_2}{\tau_{n-2}} X_{n-2}\right) \\
&\quad + \beta_{n-2,n}\left(X_{n-2} - X_{n-3} - \frac{\lambda_2}{\tau_{n-3}} X_{n-3}\right) \\
&\quad \vdots \\
&\quad + \beta_{1,n}\left(X_1 - X_0 - \frac{\lambda_2}{\tau_0} X_0\right).
\end{aligned}$$

We take $\beta_{n,n} = 1$, and annul the coefficients of $X_{n-1}, X_{n-2}, \dots, X_1$, recursively giving the desired values $\beta_{j,n}$. Namely, we set

$$-\beta_{n,n} X_{n-1} - \beta_{n,n} \frac{\lambda_2}{\tau_{n-1}} X_{n-1} + \beta_{n-1,n} X_{n-1} = 0,$$

which gives

$$\beta_{n-1,n} = 1 + \frac{\lambda_2}{\tau_{n-1}},$$

then set

$$-\beta_{n-1,n} X_{n-2} - \beta_{n-1,n} \frac{\lambda_2}{\tau_{n-2}} X_{n-2} + \beta_{n-2,n} X_{n-2} = 0,$$

which gives

$$\beta_{n-2,n} = \Big(1 + \frac{\lambda_2}{\tau_{n-2}}\Big)\beta_{n-1,n} = \Big(1 + \frac{\lambda_2}{\tau_{n-2}}\Big)\Big(1 + \frac{\lambda_2}{\tau_{n-1}}\Big),$$

and so on in backward inductive steps to get in general

$$\beta_{j,n} = \prod_{k=j}^{n-1}\Big(1 + \frac{\lambda_2}{\tau_k}\Big).$$

After all the backward induction is completed, we see that there is a leftover error

$$\varepsilon_n = -X_0\left(1 + \frac{\lambda_2}{\tau_0}\right)\beta_{1,n} = -X_0\left(1 + \frac{\lambda_2}{\tau_0}\right)\prod_{k=1}^{n-1}\left(1 + \frac{\lambda_2}{\tau_k}\right).$$

The rate of growth of the coefficients $\beta_{j,n}$ determines the convergence laws for the proportions of each color. We shall take a look at these numbers. We have

$$\begin{aligned}
\beta_{j,n} &= \prod_{k=j}^{n-1}\frac{k\lambda_1 + \tau_0 + \lambda_2}{k\lambda_1 + \tau_0} \\
&= \prod_{k=j}^{n-1}\frac{k + \dfrac{\tau_0 + \lambda_2}{\lambda_1}}{k + \dfrac{\tau_0}{\lambda_1}} \\
&= \frac{\Gamma\left(n + \dfrac{\tau_0 + \lambda_2}{\lambda_1}\right)}{\Gamma\left(n + \dfrac{\tau_0}{\lambda_1}\right)} \times \frac{\Gamma\left(j + \dfrac{\tau_0}{\lambda_1}\right)}{\Gamma\left(j + \dfrac{\tau_0 + \lambda_2}{\lambda_1}\right)} \\
&= \left(\frac{n}{j}\right)^{\lambda_2/\lambda_1} + O(n^{\lambda_2/\lambda_1 - 1}),
\end{aligned}$$

where we applied Stirling's approximation for the ratio of Gamma functions. In particular,

$$\begin{aligned}
\varepsilon_n &= -X_0\left(1 + \frac{\lambda_2}{\tau_0}\right)\beta_{1,n} \\
&= -X_0\left(1 + \frac{\lambda_2}{\tau_0}\right)\frac{\Gamma\left(n + \dfrac{\tau_0 + \lambda_2}{\lambda_1}\right)}{\Gamma\left(n + \dfrac{\tau_0}{\lambda_1}\right)} \times \frac{\Gamma\left(1 + \dfrac{\tau_0}{\lambda_1}\right)}{\Gamma\left(1 + \dfrac{\tau_0 + \lambda_2}{\lambda_1}\right)} \\
&= O(n^{\lambda_2/\lambda_1}).
\end{aligned} \tag{6.3}$$

In preparation for weak laws of large numbers for the proportions of each color, we first show via a lemma that both $V_n/n \xrightarrow{P} 0$ and $X_n/n \xrightarrow{P} 0$, under the technical condition $\lambda_2 < \frac{1}{2}\lambda_1$.

LEMMA 6.1
If $\lambda_2 < \frac{1}{2}\lambda_1$, then

$$\frac{X_n}{n} \xrightarrow{P} 0.$$

PROOF Compute the variance

$$\begin{aligned}
\mathbf{Var}[V_n] &= \mathbf{E}[V_n^2] \\
&= \left(\sum_{j=1}^{n} \beta_{j,n} M_j\right)^2 \\
&= \mathbf{E}\left[\sum_{j=1}^{n} \beta_{j,n}^2 M_j^2\right] + 2\mathbf{E}\left[\sum_{1\le r<s\le n} \beta_{r,n}\beta_{s,n} M_r M_s\right].
\end{aligned}$$

The double sum has zero expectation because,

$$\mathbf{E}[\beta_{r,n}\beta_{s,n} M_r M_s \mid \mathcal{F}_r] = \beta_{r,n}\beta_{s,n} M_r \mathbf{E}[M_s \mid \mathcal{F}_r] = 0.$$

Hence, using the linear bound on X_j (cf. Eq. (6.2)) we see that

$$\begin{aligned}
\mathbf{Var}[V_n] &= \sum_{j=1}^{n} \beta_{j,n}^2 \mathbf{E}[M_j^2] \\
&= \sum_{j=1}^{n} \beta_{j,n}^2 \mathbf{E}\left[\left((X_j - X_{j-1}) - \frac{\lambda_2 X_{j-1}}{\tau_{j-1}}\right)^2\right] \\
&\le \sum_{j=1}^{n} \beta_{j,n}^2 \mathbf{E}\left[\left(|X_j - X_{j-1}| + \frac{\lambda_2 |X_{j-1}|}{\tau_{j-1}}\right)^2\right] \\
&\le \sum_{j=1}^{n} \beta_{j,n}^2 \mathbf{E}\left[\left(|\mathbf{u}(\mathbf{Z}_n - \mathbf{Z}_{n-1})| + \frac{\lambda_2 C_1 (j-1)}{\lambda_1 (j-1) + \tau_0}\right)^2\right],
\end{aligned}$$

and each component in the difference $|\mathbf{Z}_n - \mathbf{Z}_{n-1}| = \begin{pmatrix} |\nabla W_n| \\ |\nabla B_n| \end{pmatrix}$ is at most $\max(|a|, |b|, |c|, |d|)$. Whence, for some positive constants C_2, and C_3,

$$\begin{aligned}
\mathbf{Var}[V_n] &\le C_2 \sum_{j=1}^{n} \beta_{j,n}^2 \\
&= C_2 \sum_{j=1}^{n} \left(\left(\frac{n}{j}\right)^{\lambda_2/\lambda_1} + O(n^{\lambda_2/\lambda_1 - 1})\right)^2
\end{aligned}$$

$$
\begin{aligned}
&= C_2 \left[\sum_{j=1}^{n} \left(\left(\frac{n}{j} \right)^{2\lambda_2/\lambda_1} + O\left(\frac{n^{2\lambda_2/\lambda_1 - 1}}{j^{\lambda_2/\lambda_1}} \right) + O(n^{2\lambda_2/\lambda_1 - 2}) \right) \right] \\
&\le C_3 n,
\end{aligned}
$$

according to the technical condition $2\lambda_2 < \lambda_1$.

By Chebyshev's inequality:

$$
\mathbf{P}\left(\left| \frac{V_n}{n} \right| > \varepsilon \right) \le \frac{C_3 n}{\varepsilon^2 n^2} \to 0,
$$

and $V_n/n \xrightarrow{P} 0$.

An immediate corollary is that

$$
\frac{X_n}{n} = \frac{V_n - \varepsilon_n}{n} \xrightarrow{P} 0,
$$

as we have shown in Eq. (6.3) that $\varepsilon_n = O(n^{\lambda_2/\lambda_1}) = o(n)$. ■

THEOREM 6.2

(Smythe, 1996). Let W_n and B_n be respectively the number of white and blue balls in an extended dichromatic urn with a schema with principal eigenvalue λ_1, and corresponding left eigenvector (v_1, v_2). Then,

$$
\begin{aligned}
\frac{W_n}{n} &\xrightarrow{P} \lambda_1 v_1, \\
\frac{B_n}{n} &\xrightarrow{P} \lambda_1 v_2.
\end{aligned}
$$

PROOF For two certain constants α_1 and α_2 a general row vector $\mathbf{y} = (y_1 \; y_2)$ can be written as

$$
\mathbf{y} = \alpha_1 (1 \;\; 1) + \alpha_2 \mathbf{u}.
$$

Then

$$
\mathbf{y}\mathbf{v} = \alpha_1 (1 \;\; 1)\mathbf{v} + \alpha_2 \mathbf{u}\mathbf{v} = \alpha_1 (v_1 + v_2) + 0 = \alpha_1;
$$

the eigenvectors $\mathbf{v}$ and $\mathbf{u}$ are orthogonal, as the two eigenvalues are distinct (the vectors $\mathbf{v}$ and $\mathbf{u}$ are also normalized). So,

$$
\begin{aligned}
\frac{1}{n}\mathbf{y}\mathbf{Z}_n &= \frac{1}{n}\alpha_1 (1 \;\; 1)\mathbf{Z}_n + \frac{\alpha_2}{n}\mathbf{u}\mathbf{Z}_n \\
&= \frac{1}{n}\mathbf{y}\mathbf{v}(1 \;\; 1)\begin{pmatrix} W_n \\ B_n \end{pmatrix} + \alpha_2 \frac{X_n}{n} \\
&= \frac{1}{n}\mathbf{y}\mathbf{v}(W_n + B_n) + \alpha_2 \frac{X_n}{n}.
\end{aligned}
$$

We have $(W_n + B_n)/n = \tau_n/n = (\lambda_1 n + \tau_0)/n \to \lambda_1$, and we have shown in Lemma 6.1 that $X_n/n \xrightarrow{P} 0$, hence

$$\frac{1}{n}\mathbf{y}\mathbf{Z}_n \xrightarrow{P} \lambda_1 \mathbf{y}\mathbf{v}.$$

The vector $\mathbf{y}$ is arbitrary, let us take it first equal to $(1 \ 0)$ to get the convergence law for W_n/n, then take it equal to $(0 \ 1)$ to get the convergence law for B_n/n, as stated. ■

6.3.2 Asymptotic Normality

Let us recall the notation of the preceding subsection. We are still dealing with an extended deterministic dichromatic scheme, with schema that has two real eigenvalues $\lambda_1 > 2\lambda_2$. We denoted by W_n and B_n respectively, the number of white and blue balls after n draws.

We found weak laws of large numbers for W_n/n, which is a convergence in probability relation. We shall next identify the rate of that convergence, and find therein a central limit theorem for extended schemes with fixed schemata. The machinery is as the one used for the linear combination $X_n = u_1 W_n + u_2 B_n$, with $\mathbf{u} = (u_1 \ u_2)$ being the nonprincipal left eigenvector of the schema.

Let

$$W_n^* = W_n - v_1 \tau_n,$$

that is, they are the numbers W_n asymptotically centered (this centering is not exact, and may leave small errors, relative to the scale factors that will be chosen). We shall come up with a representation

$$V_n^* = \sum_{j=1}^{n} \beta_{j,n}^* M_j^* = W_n^* - \varepsilon_n^*,$$

where M_j^* are martingales, ε_n^* is a small error, and choose the coefficients $\beta_{j,n}^*$ so that V_n^* is a good approximation for W_n^*. The advantage here is that W_n^* is a sum of well-behaved random variables, which invites the machinery for sums of random variables, particularly the martingale central limit theorem.

We have seen the details on X_n, and we shall allow some liberty of being less detailed here. Note that W_n is linearly bounded (by the linear total), and for some constant C_4, and every $n \ge 1$,

$$W_n \le \tau_n \le C_4 n.$$

The asymptotically centered numbers W_n^* are "second-order terms" of small magnitude, as contrasted to the "first-order" raw values of W_n. While W_n needs linear scaling to converge to a limit (in probability), W_n^* require a smaller scale to converge (in distribution). Indeed, if we scale W_n^* by n, we get

$$\frac{W_n^*}{n} = \frac{W_n}{n} - v_1 \frac{\tau_n}{n},$$

and both terms on the right converge to $\lambda_1 v_1$, (for the convergence of the first term see Theorem 6.2 and for the second see Eq. (6.1)), that is,

$$\frac{W_n^*}{n} \xrightarrow{P} 0. \tag{6.4}$$

That appropriate scale for W_n^* to converge to a normal distribution will turn out to be $\sqrt{n}$, as is usual in central limit theorems.

Let $\mathbf{1}_n^W$ be the indicator of the event of picking a white ball in the nth sample. Conditionally, we can write the incremental changes

$$\begin{aligned}
\mathbf{E}[W_n \mid \mathcal{F}_{n-1}] &= W_{n-1} + a\,\mathbf{P}(\mathbf{1}_n^W = 1 \mid \mathcal{F}_{n-1}) + c\,\mathbf{P}(\mathbf{1}_n^W = 0 \mid \mathcal{F}_{n-1}) \\
&= W_{n-1} + a\,\frac{W_{n-1}}{\tau_{n-1}} + c\,\frac{B_{n-1}}{\tau_{n-1}} \\
&= W_{n-1} + a\,\frac{W_{n-1}}{\tau_{n-1}} + c\,\frac{\tau_{n-1} - W_{n-1}}{\tau_{n-1}}.
\end{aligned}$$

We have the relation

$$\begin{aligned}
\mathbf{E}[W_n^* + v_1\tau_n \mid \mathcal{F}_{n-1}] = W_{n-1}^* + v_1\tau_{n-1} + a\,&\frac{W_{n-1}^* + v_1\tau_{n-1}}{\tau_{n-1}} \\
+ c\,&\frac{\tau_{n-1} - W_{n-1}^* - v_1\tau_{n-1}}{\tau_{n-1}},
\end{aligned}$$

which we reorganize as

$$\begin{aligned}
\mathbf{E}\big[W_n^* - W_{n-1}^* \mid \mathcal{F}_{n-1}\big] &= v_1(\tau_{n-1} - \tau_n) + (a - c)\,\frac{W_{n-1}^*}{\tau_{n-1}} + (a - c)v_1 + c \\
&= -\lambda_1 v_1 + (a - c)\,\frac{W_{n-1}^*}{\tau_{n-1}} + (a - c)v_1 + c \\
&= \lambda_2\,\frac{W_{n-1}^*}{\tau_{n-1}};
\end{aligned}$$

recall that $\lambda_1 = K$, $\lambda_2 = a - c$, $\mathbf{v} = (v_1 \;\; v_2)$, and $v_1 = c/(K + c - a)$, and $v_2 = (K - a)/(K + c - a)$. Then,

$$M_n^* := \triangledown W_n^* - \frac{\lambda_2}{\tau_{n-1}} W_{n-1}^*$$

are martingale differences. These differences are uniformly bounded in view of

$$\begin{aligned}
|M_n^*| &\le |\triangledown W_n^*| + \frac{|\lambda_2|}{\tau_{n-1}} |W_{n-1}^*| \\
&= |\triangledown W_n - v_1 \triangledown \tau_n| + \frac{|\lambda_2|}{\tau_{n-1}} |W_{n-1} - v_1\tau_{n-1}| \\
&\le |\triangledown W_n| + v_1 \triangledown \tau_n + \frac{|\lambda_2|}{\tau_{n-1}} (W_{n-1} + v_1\tau_{n-1})
\end{aligned}$$

$$\leq \max\left(|a|, |c|\right) + \lambda_1 v_1 + |\lambda_2|(1 + v_1)$$
$$=: C_5. \tag{6.5}$$

We can also compute the conditional second moments of these differences (which is helpful to identify the variance):

$$\begin{aligned}
\mathbf{E}[(M_n^*)^2 \,|\, \mathcal{F}_{n-1}] &= \mathbf{E}\left[(\nabla W_n^*)^2 \,|\, \mathcal{F}_{n-1}\right] + \frac{\lambda_2^2}{\tau_{n-1}^2}\, \mathbf{E}\left[(W_{n-1}^*)^2 \,|\, \mathcal{F}_{n-1}\right] \\
&\quad - 2\frac{\lambda_2 W_{n-1}^*}{\tau_{n-1}}\, \mathbf{E}\left[\nabla W_n^* \,|\, \mathcal{F}_{n-1}\right] \\
&= \mathbf{E}[(\nabla(W_n - v_1\tau_n))^2 \,|\, \mathcal{F}_{n-1}] + \frac{\lambda_2^2 (W_{n-1}^*)^2}{\tau_{n-1}^2} \\
&\quad - 2\frac{\lambda_2 W_{n-1}^*}{\tau_{n-1}} \mathbf{E}[(\nabla W_n - v_1 \nabla \tau_n) \,|\, \mathcal{F}_{n-1}] \\
&= \mathbf{E}[(\nabla W_n - v_1 \nabla \tau_n)^2 \,|\, \mathcal{F}_{n-1}] + \frac{\lambda_2^2 (W_{n-1}^*)^2}{\tau_{n-1}^2} \\
&\quad - 2\frac{\lambda_2 W_{n-1}^*}{\tau_{n-1}} \mathbf{E}[(\nabla W_n - v_1 \nabla \tau_n) \,|\, \mathcal{F}_{n-1}] \\
&= \mathbf{E}\left[(\nabla W_n)^2 - 2\lambda_1 v_1 \nabla W_n + \lambda_1^2 v_1^2) \,|\, \mathcal{F}_{n-1}\right] + \frac{\lambda_2^2 (W_{n-1}^*)^2}{\tau_{n-1}^2} \\
&\quad - 2\frac{\lambda_2 W_{n-1}^*}{\tau_{n-1}} \mathbf{E}[(\nabla W_n - v_1 \nabla \tau_n) \,|\, \mathcal{F}_{n-1}] \\
&= \left(a^2 \frac{W_{n-1}}{\tau_{n-1}} + c^2 \frac{B_{n-1}}{\tau_{n-1}}\right) \\
&\quad - \left(2\lambda_1 v_1 + 2\frac{\lambda_2 W_{n-1}^*}{\tau_{n-1}}\right)\left(a\frac{W_{n-1}}{\tau_{n-1}} + c\frac{B_{n-1}}{\tau_{n-1}}\right) \\
&\quad + \lambda_1^2 v_1^2 + \frac{\lambda_2^2 (W_{n-1}^*)^2}{\tau_{n-1}^2} + 2\frac{\lambda_1 \lambda_2 v_1 W_{n-1}^*}{\tau_{n-1}} \\
&\xrightarrow{P} a^2 v_1 + c^2 v_2 - (2\lambda_1 v_1 + 0)(a v_1 + c v_2) + \lambda_1^2 v_1^2 + \lambda_2^2 \times 0 \\
&\quad + 2\lambda_1 \lambda_2 v_1 \times 0 \\
&= a v_1 (a - 2\lambda_1 v_1) + c v_2 (c - 2\lambda_1 v_1) + \lambda_1^2 v_1^2 \\
&=: C_6;
\end{aligned}$$

we used Eqs. (6.1) and (6.4) to replace W_{n-1}^*/τ_{n-1} with a zero in the limit.

In the sum $V_n^* = \sum_{j=1}^n \beta_{j,n}^* M_j^*$, we take the constants $\beta_{j,n}^*$ recursively so that

$$W_n^* = V_n^* + \varepsilon_n^*,$$

for an asymptotically negligible error ε_n^*. As was done in deriving the weak law, we take $\beta_{n,n}^* = 1$, and annul the coefficients of $W_{n-1}^*, W_{n-2}^*, \ldots, W_1^*$, recursively giving the desired values $\beta^*_{j,n}$. They coincide with $\beta_{j,n}$ (which have already been determined), because the recurrences are basically the same. Hence,

$$\beta^*_{j,n} = \prod_{k=j}^{n-1}\left(1 + \frac{\lambda_2}{\tau_k}\right).$$

with rate of growth $(\frac{n}{j})^{\lambda_2/\lambda_1}$ for these coefficients, and in particular,

$$\varepsilon^*_n = -W^*_0\left(1 + \frac{\lambda_2}{\tau_0}\right)\beta^*_{1,n} = O(n^{\lambda_2/\lambda_1}).$$

We check the two conditions sufficient for Theorem 6.1 to hold. By the uniform bound $M^*_j \le C_5$, see Eq. (6.5) we have

$$\begin{aligned}\frac{1}{n}\sum_{j=1}^{n}\mathbf{E}\big[(\beta^*_{j,n}M^*_j)^2 \mid \mathcal{F}_{j-1}\big] &= \frac{1}{n}\sum_{j=1}^{n}(\beta^*_{j,n})^2\mathbf{E}\big[(M^*_j)^2 \mid \mathcal{F}_{j-1}\big] \sim_P \frac{1}{n}\sum_{j=1}^{n}\Big(\frac{n}{j}\Big)^{2\lambda_2/\lambda_1}C_6\\ &\xrightarrow{P} \frac{av_1(a-2\lambda_1v_1)+cv_2(c-2\lambda_1v_1)+\lambda_1^2v_1^2}{1-2\lambda_2/\lambda_1}\\ &=: \sigma^2.\end{aligned}$$

Consequently,[3]

$$\frac{1}{n}\sum_{j=1}^{n}\mathbf{E}\big[(\beta^*_{j,n}M^*_j)^2 \mid \mathcal{F}_{n-1}\big] \xrightarrow{P} \sigma^2,$$

and V^*_n satisfies a σ^2–conditional variance condition.

We have also found the rates of growth of the coefficients $\beta^*_{j,n}$, namely

$$\beta^*_{j,n} = \Big(\frac{n}{j}\Big)^{\lambda_2/\lambda_1} + O(n^{\lambda_2/\lambda_1 - 1}).$$

We proved that M^*_j are uniformly bounded (see Eq. (6.5)), and consequently,

$$\max_{1\le j\le n}\frac{\mathbf{E}[(\beta^*_{j,n}M^*_j)^2 \mid \mathcal{F}_{j-1}]}{n} \le \max_{1\le j\le n}\frac{C_5^2\,\mathbf{E}\big[(\beta^*_{j,n})^2 \mid \mathcal{F}_{j-1}\big]}{n}.$$

[3] We replaced the terms in the sum with their in-probability asymptotic equivalent. In general, the sum of random variables converging in probability does not converge to the sum of their limits, but here they do, because in fact this convergence occurs in the stronger L_1 sense, as can be shown from the rate of growth of the mean and variance.

The subject of maximization is $O((\frac{n}{j})^{2\lambda_2/\lambda_1})/n$, which converges to 0 almost surely, under the imposed technical condition that $\lambda_2/\lambda_1 < \frac{1}{2}$, and it follows that

$$\max_{1\le j\le n} \frac{\mathbf{E}[(\beta^*_{j,n} M^*_j)^2 \,|\, \mathcal{F}_{j-1}]}{n} \xrightarrow{P} 0,$$

which is a condition equivalent to the conditional Lindeberg's condition (see Theorem 3.2 and Corollary 3.1, p. 58 in Hall and Hyde (1980)).

According to the martingale central limit theorem, the sum $V^*_n/\sqrt{n} = \sum_{j=1}^{n} \beta^*_{j,n} M^*_j/\sqrt{n}$ converges in distribution to a random variable with characteristic function $e^{-\sigma^2 t^2/2}$, that is, to a normal $\mathcal{N}(0, \sigma^2)$ random variable. Thus,

$$\frac{W^*_n - \varepsilon^*_n}{\sqrt{n}} = \frac{V^*_n}{\sqrt{n}} \xrightarrow{\mathcal{D}} \mathcal{N}(0, \sigma^2),$$

and, on the other hand, $-\frac{\varepsilon^*_n}{\sqrt{n}} = O(n^{\lambda_2/\lambda_1})/\sqrt{n} \xrightarrow{a.s.} 0$, because in extended urns $\lambda_2/\lambda_1 < \frac{1}{2}$. Adding up the central limit result to the latter almost sure convergence (by Slutsky's theorem), we get

$$\frac{W_n - v_1 \tau_n}{\sqrt{n}} \xrightarrow{\mathcal{D}} \mathcal{N}(0, \sigma^2).$$

But then again, $\tau_n/n \to \lambda_1$, and by another adjustment via Slutsky's theorem we finally get the central limit result

$$\frac{W_n - \lambda_1 v_1 n}{\sqrt{n}} \xrightarrow{\mathcal{D}} \mathcal{N}(0, \sigma^2),$$

$$\sigma^2 = \frac{a v_1(a - 2\lambda_1 v_1) + c v_2(c - 2\lambda_1 v_1) + \lambda_1^2 v_1^2}{1 - 2\lambda_2/\lambda_1},$$

which coincides with the Bagchi-Pal variance in Theorem 3.5, but is organized quite differently.

6.4 Dichromatic Extended Urn Schemes with Random Entries

Consider an extended urn scheme of white and blue balls, where the generator is

$$\begin{pmatrix} U & X \\ Y & Z \end{pmatrix},$$

and the entries are integer-valued random variables; as usual white precedes blue in the indexing of rows and columns. In this dichromatic model, the four numbers U, X, Y, and Z are random—when we pick a white ball, we add to the urn an independent copy of U white balls, and an independent copy of X blue balls, and when we pick a blue ball, we add to the urn an independent copy of Y white balls, and an independent copy of Z blue balls. By independent we mean, independent of everything else, such as the status of the urn, and the past history of additions of all kinds.

In this extended urn scheme, if the average row sum is, say, K, we are considering schemata with

$$\mathbf{E}[U] + \mathbf{E}[X] = \mathbf{E}[Y] + \mathbf{E}[Z] = K.$$

Note that the average generator has the eigenvalues $\lambda_1 = K$, and $\lambda_2 = \mathbf{E}[U] - \mathbf{E}[Y]$.

These urns (and their $k \times k$) extensions where considered in Athreya and Karlin (1968), and Smythe (1996), and more recently in Janson (2004). Each step in these research endeavors took the subject into a more general framework.

The arguments are mostly the same as the case of balanced schemata with fixed entries. The main difference is that τ_n is replaced with an almost sure asymptotic equivalent, which suffices to go through with all the above proofs—we saw that we only need asymptotic estimates. We justify this technicality first.

THEOREM 6.3

(Smythe, 1996). Suppose we have a dichromatic extended urn with average generator with principal eigenvalue λ_1. Let τ_n be the total number of balls in the urn after n draws. Then,

$$\frac{\tau_n}{n} \xrightarrow{a.s.} \lambda_1.$$

PROOF Let W_n and B_n be the number of white and blue balls in the urn after n draws, and let $\mathcal{F}_n$ be the sigma field generated by W_n and B_n. Let $\mathbf{1}_n^W$ be the indicator of the event of picking a white ball in the nth sample. Conditionally, we can write the incremental changes

$$\begin{aligned}\mathbf{E}[\tau_n \mid \mathcal{F}_{n-1}] &= \tau_{n-1} + \mathbf{E}[U + X \mid \mathcal{F}_{n-1}]\,\mathbf{P}\big(\mathbf{1}_n^W = 1 \mid \mathcal{F}_{n-1}\big) \\ &\qquad + \mathbf{E}[Y + Z \mid \mathcal{F}_{n-1}]\,\mathbf{P}\big(\mathbf{1}_n^W = 0 \mid \mathcal{F}_{n-1}\big) \\ &= \tau_{n-1} + (\mathbf{E}[U] + \mathbf{E}[X])\,\frac{W_{n-1}}{\tau_{n-1}} + (\mathbf{E}[Y] + \mathbf{E}[Z])\,\frac{B_{n-1}}{\tau_{n-1}};\end{aligned}$$

a relation like $\mathbf{E}[U \mid \mathcal{F}_{n-1}] = \mathbf{E}[U]$ follows by independence. The conditional change in the total is

$$\mathbf{E}[\nabla \tau_n \mid \mathcal{F}_{n-1}] = \lambda_1 \frac{W_{n-1}}{\tau_{n-1}} + \lambda_1 \frac{B_{n-1}}{\tau_{n-1}} = \lambda_1.$$

The differences of martiangles are themselves martiangles. In our case $\nabla \tau_n$ are stochastically baunded by square integrable random variables (at each sampling step we add a copy $U + X$ or $Y + Z$, and these variables have finite variances.) According to proposition 6.1,

$$\frac{1}{n}\sum_{i=1}^{n}(\nabla\tau_i) - \mathbf{E}[\nabla\tau_i \mid \mathcal{F}_{n-1}]) \xrightarrow{a.s.} 0,$$

so we can say

$$\frac{\tau_n - \tau_0}{n} - \lambda_1 \xrightarrow{a.s.} 0,$$

or

$$\frac{\tau_n}{n} \xrightarrow{a.s.} \lambda_1. \quad \blacksquare$$

6.5 Extended Multicolor Urn Schemes

Extensions to k-color schemes follow the same lines as in the two-color case. If the scheme has a $k \times k$ average generator, its eigenvalues may be complex. By following the same principles as in the dichromatic case one gets the following law of large numbers.

THEOREM 6.4
(Smythe, 1996). Suppose a k–color extended Pólya urn scheme has an average generator with the principal eigenvalue $\lambda_1 > 0$, and the corresponding (normalized) principal left eigenvector is $\mathbf{v} = (v_1 \ v_2 \ \dots \ v_k)$. Let $X_{j,n}$ be the number of balls of color j after n draws, for $j = 1, \dots, k$. For each component, $j = 1, \dots, k$,

$$\frac{X_{j,n}}{n} \xrightarrow{P} \lambda_1 v_j.$$

For central limit theorems, we can also follow the same basic principles as in the two-color case. In that latter case we derived a central limit theorem for the centered and scaled number $W_n^* = (W_n - \lambda_1 v_1)/\sqrt{n}$ of white balls, which of course gives a central limit theorem for the companion centered and scaled number $B_n^* = (B_n - \lambda_1 v_2)/\sqrt{n}$ of blue balls. In fact, we could very slightly adapt the proof to find a central limit for any linear combination of the form $\alpha_1 W_n^* + \alpha_2 B_n^*$, for real coefficients α_1 and α_2, not both zero. The coefficients $\beta_{j,n}^*$ that we found in the dichromatic case are now functions of α's. For work in k-dimensions, it suffices to consider an arbitrary linear combination of only $k-1$ colors, say the first $k-1$. The strategy is to take the linear combination

$$\alpha_1 X_{1,n} + \alpha_2 X_{2,n} + \dots + \alpha_{k-1} X_{k-1,n}$$

for any real coefficients $\alpha_1, \alpha_2, \dots, \alpha_{k-1}$ (not all zero), and show by the same techniques that the linear combination is asymptotically normally distributed.

The variance–covariance matrix is of course a function of the α's. The Cramér-Wold device then tells us that each marginal distribution is normally distributed, too. The main theorem is the following.

THEOREM 6.5

(Smythe, 1996). Suppose a k–color extended Pólya urn scheme has an average generator with the principal eigenvalue $\lambda_1 > 0$, and the corresponding (normalized) principal left eigenvector is $\mathbf{v} = (v_1 \ v_2 \ \ldots \ v_k)$. Let $X_{j,n}$ be the number of balls of color j after n draws, for $j = 1, \ldots, k$. The vector $\mathbf{X}_n = (X_{1,n}, \ldots, X_{n,k})^T$ has a limiting multivariate normal distribution:

$$\frac{\mathbf{X}_n - \lambda_1 n \mathbf{v}}{\sqrt{n}} \xrightarrow{\mathcal{D}} \mathcal{N}_k(\mathbf{0}_k, \Sigma_k),$$

where $\mathcal{N}_k$ is the multivariate k-dimensional jointly normal random vector with mean $\mathbf{0}_k$ (vector of k zeros) and an effectively computable variance–covariance matrix Σ_k.

The variances and covariances are in general rather complicated.

Several theorems in this chapter have strong (almost sure) versions that apply at least to some large subclass of extended urn schemes. These are proved in sources like Athreya and Karlin (1968) and Janson (2004). These sources consider classes of urn schemes that are a little different form extended urn schemes. For example, Athreya and Karlin (1968) relaxes the condition that the urn is balanced (with average generator with constant row sum), but requires that the diagonal elements be all at least -1. There is a large intersection, however, among these classes of urns, and some interesting examples fall in all the classes. For example, the urns we shall see in the application on random recursive trees (in Section 8.2) are extended, and are also within the Athreya-Karlin class and Janson's broader schemes.

Exercises

6.1 An extended Pólya urn scheme of white and blue balls progresses in time by sampling a ball at random and adding white and blue balls according to a given fixed schema. Let $\tilde{W}_n$ and $\tilde{B}_n$ be the number of times a white or respectively a blue ball has been sampled. Derive weak laws for $\tilde{W}_n/n$ and $\tilde{B}_n/n$.

6.2 A Pólya urn scheme on white and blue balls grows in discrete time under the schema

$$\begin{pmatrix} X & Y \\ X & Y \end{pmatrix},$$

where $X \overset{\mathcal{D}}{=} Y \overset{\mathcal{D}}{=} \mathrm{Ber}(p)$, and $\mathrm{Ber}(p)$ is the Bernoulli random variable with rate of success p.

(a) Determine the exact distribution of the number of white balls after n draws.

(b) Find a central limit representation for the number of white balls, as $n \to \infty$.

6.3 The contents of a Bernard-Friedman-like urn grow in discrete time under the schema

$$\begin{pmatrix} X & 1-X \\ 1-X & X \end{pmatrix},$$

where X is a Ber$(\frac{1}{3})$ random variable. Find a central limit representation for the number of white balls, as $n \to \infty$.

7

Analytic Urns

We consider in this chapter a different approach to urns. So far we have looked at purely probabilistic methods. The subject can be approached from a combinatorial angle with analytic tools such as formal generating functions. The methods are innovative and have only been considered very recently. The main method is enumerative, in the sense of finding the number of all possible (equally likely) histories and the proportion therein that satisfies a certain criterion, such as the number of white balls being a prespecified number.

The title *analytic urns* refers to a class of urns amenable to these analytic methods. We shall focus on 2×2 balanced schemes on two colors (white and blue as usual). A few extensions are possible including the triangular case on more colors. This new line is pioneered by Flajolet, Dumas, and Puyhaubert (2006).

7.1 Enumerating Histories in Balanced Schemes

Suppose our urn follows the scheme

$$\mathbf{A} = \begin{pmatrix} 2 & 1 \\ 0 & 3 \end{pmatrix}$$

of white and blue balls (white preceding blue, as usual). Let W_n be the number of white balls in the urn after n draws, and B_n be the number of blue balls in it. This is an example of a balanced urn scheme, where the row sum is constant. We shall enumerate "histories" or sample paths. To instantiate the enumeration, suppose this scheme starts out with $W_0 = 2$ white balls, and $B_0 = 1$ blue balls. One can describe the state of the urn after n draws with a string of W's and B's specifying, respectively, the number of balls of each color. For example, the starting state is WWB. One possible history of evolution may correspond to the drawing of a white ball, in which case we put it back in the urn together with two new white balls and a new blue ball. There are initially two white balls, and drawing either one of them gives one such history. The second drawing may be that of the new blue ball, in which case we replace it in the urn together with three new blue balls, and so on. Indicating the ball drawn at a stage with $*$ above it, we have one possible history described by

$$W\overset{*}{W}B \Longrightarrow WWWWB\overset{*}{B} \Longrightarrow WWWWBBBBB.$$

Note that we are treating the particular ball chosen at a stage to be part of the history at a later stage. For example, we consider

$$\overset{*}{W}WB \Longrightarrow WWWWB\overset{*}{B} \Longrightarrow WWWWBBBBB$$

to be a different history, even though the urn is in the exact same state after two draws, but having arrived at it via a different stochastic path. According to this view, all histories are equally likely.

Let τ_n be the total number of balls after n draws. Also let $h_n = h_{n,W_0,B_0}$ be the number of possible histories after n draws from an urn. We define the (exponential) generating function of histories:

$$\mathcal{H}(z) = \mathcal{H}_{W_0,B_0}(z) = \sum_{n=0}^{\infty} h_n \frac{z^n}{n!}.$$

Both h_n and the function $\mathcal{H}(z)$ depend on W_0 and B_0, but we suppress that dependency for brevity.

PROPOSITION 7.1

(Flajolet, Dumas, and Puyhaubert, 2006). For a tenable balanced urn with the scheme

$$\mathbf{A} = \begin{pmatrix} a & b \\ c & d \end{pmatrix},$$

with total row sum $a + b = c + d = K > 0$, we have

$$h_n = \tau_0(\tau_0 + K)(\tau_0 + 2K)\dots(\tau_0 + (n-1)K),$$

and

$$\mathcal{H}(z) = \frac{1}{(1 - Kz)^{\tau_0/K}}.$$

PROOF There are τ_0 choices for the first ball, with $h_1 = \tau_0$, on top of which there are $\tau_1 = \tau_0 + K$ choices for the ball in the second draw, giving after two draws a total number of histories equal to $\tau_0\tau_1 = \tau_0(\tau_0 + K)$, and so on. The total number of histories can be written as

$$\begin{aligned} h_n &= \tau_0\tau_1 \dots \tau_{n-1} \\ &= \tau_0(\tau_0 + K)\dots(\tau_0 + (n-1)K) \end{aligned}$$

$$= \frac{\tau_0}{K}\left(\frac{\tau_0}{K}+1\right)\dots\left(\frac{\tau_0}{K}+n-1\right) K^n$$

$$= \frac{\left(\frac{\tau_0}{K}+n-1\right)_n}{n!}\, n!\, K^n,$$

or

$$\frac{h_n}{n!} = \binom{\frac{\tau_0}{K}+n-1}{n} K^n.$$

But these numbers are the coefficients of z^n in the expansion $(1 - Kz)^{-\tau_0/K}$. ■

The asymptotic number of histories is a corollary of Proposition 7.1—one can express the rising factorial involved in terms of Gamma functions, then apply Stirling's approximation:

$$\frac{1}{n!}\left\langle\frac{\tau_0}{K}\right\rangle_n = \frac{1}{n!} \times \frac{\tau_0}{K}\left(\frac{\tau_0}{K}+1\right)\dots\left(\frac{\tau_0}{K}+n-1\right)$$

$$= \frac{\Gamma\left(n+\frac{\tau_0}{K}\right)}{\Gamma(n+1)\,\Gamma\left(\frac{\tau_0}{K}\right)}$$

$$= \frac{n^{\tau_0/K-1}}{\Gamma\left(\frac{\tau_0}{K}\right)}\left(1+O\left(\frac{1}{n}\right)\right).$$

Thus, the asymptotic number of histories is

$$h_n \sim \frac{n!\, K^n n^{\tau_0/K-1}}{\Gamma\left(\frac{\tau_0}{K}\right)}.$$

The way we view histories (with the particular ball chosen being part of it) makes all histories equally likely, and they will appear as denominators in probability calculations of certain events, where the numerators are the number of histories favorable to these events. Therefore, there is a need for a count of histories that are specific to an event. Let $h_n(w, b) = h_{n,W_0,B_0}(w, b)$ be the number of histories, which after n draws correspond to an urn with w white balls and b blue balls, and let

$$\mathcal{H}(x, y, z) = \mathcal{H}_{W_0,B_0}(x, y, z) = \sum_{n=0}^{\infty} \sum_{0\le w,b\le\infty} h_n(w, b) x^w y^b \frac{z^n}{n!}$$

be its exponential generating function. Both $h_n(w, b)$ and the generating function $\mathcal{H}(x, y, z)$ depend on W_0 and B_0, but we suppress that dependency for brevity.

It will be quite useful to get functional forms for the trivariate generating function $\mathcal{H}(x, y, z)$, such as solvable differential equations. It will then be only a matter of extracting coefficients from the solution to find the desired probabilities. We shall use the extractor notation $[z_1^{n_1} z_2^{n_2} \dots z_k^{n_k}]g(z_1, \dots, z_k)$ to extract the coefficient of $z_1^{n_1} z_2^{n_2} \dots z_k^{n_k}$ in a multivariate generating function $g(z_1, \dots, z_k)$.

All histories are equally likely; the probability of $W_n = w$ and $B_n = b = Kn + \tau_0 - w$ is just the proportion of histories that correspond to urns with w white balls and b blue balls. The number of such histories is $h_n(w, b)$, and the total number of histories is h_n. The total count has the exponential generating function

$$\mathcal{H}(z) = \mathcal{H}(1, 1, z),$$

and thus

$$P(W_n = w, B_n = b) = \frac{h_n(w, b)}{h_n} = \frac{h_n(w, b)/n!}{h_n/n!} = \frac{[x^w y^b z^n]\mathcal{H}(x, y, z)}{[z^n]\mathcal{H}(1, 1, z)}.$$

We could also work directly with the marginal distribution of the number of balls of one color through interpretations of the history generating function. Let $h_n(w)$ be the number of histories giving w white balls after n draws. So,

$$h_n(w) = \sum_{b=0}^{\infty} h_n(w, b) = h_n(w, Kn + \tau_0 - w),$$

and

$$\mathcal{H}(x, 1, z) = \sum_{n=0}^{\infty} \sum_{0 \le w, b \le \infty} h_n(w, b) x^w \frac{z^n}{n!} = \sum_{n=0}^{\infty} \sum_{w=0}^{\infty} h_n(w) x^w \frac{z^n}{n!}.$$

The univariate distribution is

$$\mathbf{P}(W_n = w) = \frac{h_n(w)}{h_n} = \frac{h_n(w)/n!}{h_n/n!} = \frac{[x^w z^n]\mathcal{H}(x, 1, z)}{[z^n]\mathcal{H}(1, 1, z)}. \tag{7.1}$$

The interplay between some differential operators will be of great help in determining the required generating function. Let us consider the following classical combinatorial interpretation of the operator $\frac{\partial}{\partial x}$. Applied to x^j, we can view this operator as j individual operations acting at j different positions

$$\frac{\partial}{\partial x} x^j = j x^{j-1} = \not{x} x x \dots x + x \not{x} x \dots x + x x \not{x} \dots x + \dots + x x x \dots \not{x}.$$

This mimics the one-step histories in sampling without replacement from an urn with j equally likely balls in it. To make such an operation completely analogous to the one-step progress of a dichromatic Pólya urn, with a replacement scheme

$$\mathbf{A} = \begin{pmatrix} a & b \\ c & d \end{pmatrix},$$

we need a differential operator on two variables, say x and y, and we need to add appropriate powers (balls) of x and y to emulate the replacement scheme $\mathbf{A}$. The operator

$$\mathcal{D} = x^{a+1} y^b \frac{\partial}{\partial x} + x^c y^{d+1} \frac{\partial}{\partial y}$$

is most suited for the task. The role of the individual differential operators is to select a ball from among however many there are. If the ball chosen is white, the derivative with respect to x is the one that kicks in—we put the white ball back in the urn plus a additional white balls (hence the power $a+1$ for x), and b blue balls (hence the additional power b for y). The role of the derivative with respect to y and the associated powers is symmetrical. Thus, $\mathcal{D}^n$, or n applications of the operator $\mathcal{D}$ to the monomial $x^{W_0} y^{B_0}$, captures the histories of the urn evolution over n draws under the scheme $\mathbf{A}$, when it starts with W_0 white balls and B_0 blue balls. The histories generating function is

$$\mathcal{H}(x, y, z) = \sum_{n=0}^{\infty} \mathcal{D}^n (x^{W_0} y^{B_0}) \frac{z^n}{n!}. \tag{7.2}$$

The main theorem (the so-called *basic isomorphism theorem*), goes via series expansion and a connection between the above operator and a different kind of operator, also differential in a parametric representation via an auxiliary variable. When admissible, the Taylor series expansion (around $t = 0$) of a function $x^j(t+z)$ is

$$x^j(t+z) = x^j(t) + \frac{\partial}{\partial t} x^j(t) z + \frac{\partial^2}{\partial t^2} x^j(t) \frac{z^2}{2} + \cdots.$$

The product of expandable functions $x^j(t+z)\, y^k(t+z)$ is

$$\begin{aligned} x^j(t+z)\, y^k(t+z) &= \left(x^j(t) + \frac{\partial}{\partial t} x^j(t) z + \frac{\partial^2}{\partial t^2} x^j(t) \frac{z^2}{2} + \cdots \right) \\ &\quad \times \left(y^k(t) + \frac{\partial}{\partial t} y^k(t) z + \frac{\partial^2}{\partial t^2} y^k(t) \frac{z^2}{2} + \cdots \right) \\ &= \sum_{n=0}^{\infty} \sum_{r=0}^{n} \binom{n}{r} \left(\frac{\partial^r}{\partial t^r} x^j(t) \right) \left(\frac{\partial^{n-r}}{\partial t^{n-r}} y^k(t) \right) \frac{z^n}{n!} \\ &= \sum_{n=0}^{\infty} \frac{\partial^n}{\partial t^n} (x^j(t)\, y^k(t)) \frac{z^n}{n!}. \end{aligned}$$

We shall call the operator $\frac{\partial}{\partial t}$ by the name $\tilde{\mathcal{D}}$, and as usual denote its nth iterate by $\tilde{\mathcal{D}}^n$.

THEOREM 7.1
(Flajolet, Dumas, and Puyhaubert, 2006). Suppose the balanced urn

$$\mathbf{A} = \begin{pmatrix} a & b \\ c & d \end{pmatrix},$$

of white and blue balls with total row sum $a + b = c + d = K > 0$ starts out nonempty with W_0 white balls and B_0 blue balls. Then the generating function of the urn histories is

$$\mathcal{H}(x(0), y(0), z) = x^{W_0}(z)\, y^{B_0}(z),$$

where the pair $x(z)$ and $y(z)$ is the solution to the differential system of equations

$$\begin{aligned} x'(t) &= x^{a+1}(t)\, y^b(t), \\ y'(t) &= x^c(t)\, y^{d+1}(t); \end{aligned}$$

this histories representation is valid whenever $x(0)\, y(0) \neq 0$.[1]

PROOF The bulk of the work has been done in the interpretations of the differential operators before the theorem. We have[2]

$$x^{W_0}(t+z)\, y^{B_0}(t+z) = \sum_{n=0}^{\infty} \tilde{\mathcal{D}}^n(x^{W_0}(t)\, y^{B_0}(t))\, \frac{z^n}{n!}.$$

As established in Eq. (7.2), this is also

$$\mathcal{H}(x(t), y(t), z) = \sum_{n=0}^{\infty} \mathcal{D}^n(x^{W_0}(t)\, y^{B_0}(t))\, \frac{z^n}{n!}.$$

These two expansions coincide, with the consequence that $\mathcal{H}(x(t), y(t), z) = x^{W_0}(t+z)\, y^{B_0}(t+z)$, if we let

$$\mathcal{D} = \tilde{\mathcal{D}},$$

or

$$\left(x^{a+1}(t)\, y^b(t) \frac{\partial}{\partial x(t)} + x^c(t)\, y^{d+1}(t) \frac{\partial}{\partial y(t)}\right) (x^j(t)\, y^k(t)) = \frac{\partial}{\partial t}(x^j(t)\, y^k(t)),$$

---

[1] This is only a technical condition to guarantee the analyticity in the case of diminishing urns.

[2] A classic theorem of Cauchy and Kovaleskaya guarantees the existence of these expansions near the origin.

which is the same as letting

$$j x^{a+j}(t)\, y^{b+k}(t) + k x^{c+j}(t)\, y^{d+k}(t) = j x^{j-1}(t)\, y^{k}(t) \frac{\partial}{\partial t} x(t) + k x^{j}(t)\, y^{k-1}(t) \frac{\partial}{\partial t} y(t).$$

This is possible if

$$\frac{\partial}{\partial t} x(t) = x^{a+1}(t)\, y^{b}(t),$$
$$\frac{\partial}{\partial t} y(t) = x^{c}(t)\, y^{d+1}(t).$$

Applying the equality at $t = 0$, we get

$$\mathcal{H}(x(0), y(0), z) = x^{W_0}(t+z)\, y^{B_0}(t+z)\big|_{t=0} = x^{W_0}(z)\, y^{B_0}(z). \qquad \blacksquare$$

7.2 Analytic Urn Examples

Several of the classical cases fit well within the analytic framework. In several cases the differential system described in Theorem 7.1 is simple enough to admit an exact solution. Hence an exact probability distribution is obtained, and subsequently limit distributions follow.

7.2.1 Analytic Pólya-Eggenberger Urn

Let us return to the classical example

$$\mathbf{A} = \begin{pmatrix} 1 & 0 \\ 0 & 1 \end{pmatrix},$$

and treat it with the analytic method. The associated differential system is

$$x'(t) = x^2(t),$$
$$y'(t) = y^2(t).$$

Each individual equation is a Riccati differential equation. The solution is

$$x(t) = \frac{x(0)}{1 - x(0)t},$$
$$y(t) = \frac{y(0)}{1 - y(0)t}.$$

According to Theorem 7.1,

$$\mathcal{H}(x(0), y(0), z) = \left(\frac{x(0)}{1 - x(0)z}\right)^{W_0} \left(\frac{y(0)}{1 - y(0)z}\right)^{B_0},$$

for any pair $x(0)$ and $y(0)$ of complex numbers with nonzero product. For convenience we rewrite the histories generating function as

$$\mathcal{H}(u, v, z) = \Big(\frac{u}{1 - uz}\Big)^{W_0} \Big(\frac{v}{1 - vz}\Big)^{B_0}, \tag{7.3}$$

for any pair u and v of complex numbers, with $uv \neq 0$ Then,

$$\mathcal{H}(1, 1, z) = \frac{1}{(1 - z)^{W_0+B_0}} = \sum_{n=0}^{\infty} (-1)^n \binom{-\tau_0}{n} z^n,$$

and

$$\begin{aligned}
[z^n]\, \mathcal{H}(1, 1, z) &= (-1)^n \binom{-\tau_0}{n} \\
&= (-1)^n \frac{(-\tau_0)(-\tau_0 - 1) \ldots (-\tau_0 - n + 1)}{n!} \\
&= \frac{(\tau_0 + n - 1)_n}{n!} \\
&= \binom{n + \tau_0 - 1}{\tau_0 - 1}.
\end{aligned}$$

A similar extraction of coefficients from Eq. (7.3), recalling the restriction $w + b = W_0 + B_0 + n$, yields

$$\begin{aligned}
[u^w v^b z^n]\, \mathcal{H}(u, v, z) &= [u^w v^b z^n] \left(\left(u^{W_0} \sum_{j=0}^{\infty} \binom{j + W_0 - 1}{W_0 - 1} (uz)^j \right) \right. \\
&\quad \left. \times \left(v^{B_0} \sum_{k=0}^{\infty} \binom{k + B_0 - 1}{B_0 - 1} (vz)^k \right) \right) \\
&= \binom{w - 1}{W_0 - 1} \binom{b - 1}{B_0 - 1}.
\end{aligned}$$

Hence,

$$\begin{aligned}
\mathbf{P}(W_n = w,\, B_n = b) &= \frac{h_n(w, b)}{h_n} \\
&= \frac{h_n(w, b)/n!}{h_n/n!}
\end{aligned}$$

$$= \frac{[u^w v^b z^n]\,\mathcal{H}(u, v, z)}{[z^n]\,\mathcal{H}(1, 1, z)}$$

$$= \frac{\binom{w-1}{W_0-1}\binom{b-1}{B_0-1}}{\binom{n+\tau_0-1}{\tau_0-1}}.$$

The two joint events in this probability are the same, and we can write it simply as

$$\mathbf{P}(W_n = w) = \frac{\binom{w-1}{W_0-1}\binom{n+\tau_0-w-1}{B_0-1}}{\binom{n+\tau_0-1}{\tau_0-1}},$$

for $w = W_0, W_0 + 1, \ldots, W_0 + n$. Of course, this result is the same as that in Theorem 3.1, but the mechanical way in which it is derived from analytic generating function is quite pleasant, and indicative that this analytic method is quite promising for many other urn schemes too. We shall see a few additional examples.

7.2.2 Analytic Bernard Friedman's Urn

Recall the classical example

$$\mathbf{A} = \begin{pmatrix} 0 & 1 \\ 1 & 0 \end{pmatrix},$$

and let us take up the scheme when it starts with one white ball (the case that will arise in the analysis of the number of leaves in a recursive tree in Subsection 8.2.1) via the analytic method. The associated differential system is

$$x'(t) = x(t)y(t),$$
$$y'(t) = x(t)y(t).$$

Here $x'(t) = y'(t)$, and $y(t)$ is the same as $x(t)$ up to translation by a constant determined by the initial conditions. That is $y(t) = x(t) - x(0) + y(0)$. We wish then to solve

$$x'(t) = x(t)\,(x(t) - x(0) + y(0)).$$

This is a Riccati differential equation, with solution

$$x(t) = \frac{x(0)(x(0) - y(0))}{x(0) - y(0)\exp((x(0) - y(0))t)},$$

Substituting back in $y(t)$ we find the symmetrical form

$$y(t) = \frac{y(0)\,(y(0) - x(0))}{y(0) - x(0)\exp((y(0) - x(0))t)}.$$

According to Theorem 7.1 (writing $u = x(0)$ and $v = y(0)$),

$$\begin{aligned}\mathcal{H}(u, v, z) &= \left(\frac{u(u - v)}{u - v\exp((u - v)z)}\right)^{W_0} \left(\frac{v(v - u)}{v - u\exp((v - u)z)}\right)^{B_0} \bigg|_{W_0=1,\, B_0=0} \\ &= \frac{u(u - v)}{u - v\exp((u - v)z)}.\end{aligned}$$

We compute $\mathcal{H}(1, 1, z)$ via appropriate limits: First let v approach u, then set $u = 1$, one application of l'Hôpital's rule gives

$$\begin{aligned}\mathcal{H}(1, 1, z) &= \lim_{u\to 1}\lim_{v\to u} \frac{u(u - v)}{u - v\exp((u - v)z)} \\ &= \lim_{u\to 1} \frac{-u}{0 + vz\exp((u - v)z) - \exp((u - v)z)}\bigg|_{v=u} \\ &= \frac{1}{1 - z} \\ &= 1 + z + z^2 + \cdots,\end{aligned}$$

for $|z| < 1$, and

$$[z^n]\,\mathcal{H}(1, 1, z) = 1.$$

Recall that this coefficient is $h_n/n!$. That is, $h_n = n!$, which we could have computed directly from Proposition 7.1. To simplify matters, we shall set $v = 1$ and use the univariate approach, as in Eq. (7.1):

$$\mathcal{H}(u, 1, z) = \frac{u(u - 1)}{u - \exp((u - 1)z)}.$$

This is the exponential generating function of the Eulerian numbers $\langle {n \atop k} \rangle$.

We shall say here a few words about these combinatorial counts before engaging in extracting them as coefficients in the histories generating function. An Eulerian number of the first kind counts the number of ascents or increasing runs in a permutation of $\{1, \ldots, n\}$. In a permutation $(\pi_1\ \pi_2 \ldots \pi_n)$ the number of positions where $\pi_{j+1} < \pi_j$ defines the number of points where an increasing streak breaks, and the number of *ascents* in the permutation is one plus the number of these breaking points. The Eulerian number of the first kind $\langle {n \atop k} \rangle$ is the number of permutations of $\{1, \ldots, n\}$ with exactly k ascents. The permutations of $\{1, 2, 3\}$, and the corresponding number of ascents are shown in Table 7.1. For instance, in the permutation (1 3 2) one finds two increasing blocks of numbers 1 3 and 2, with one breaking points between them.

TABLE 7.1

Permutations of $\{1, 2, 3\}$ and their Number of Ascents

Permutation	Number of Ascents
1 2 3	1
1 3 2	2
2 1 3	2
2 3 1	2
3 1 2	2
3 2 1	3

This permutation has two ascents in it. There is one permutation with one ascent, four with two ascents, and one with three ascents. Hence

$$\left\langle {3 \atop 1} \right\rangle = 1, \qquad \left\langle {3 \atop 2} \right\rangle = 4, \qquad \left\langle {3 \atop 3} \right\rangle = 1.$$

The Eulerian numbers of the first kind satisfy the recurrence

$$\left\langle {n \atop k} \right\rangle = k \left\langle {n-1 \atop k} \right\rangle + (n-k+1) \left\langle {n-1 \atop k-1} \right\rangle. \tag{7.4}$$

This can be seen in a count obtained by placing n at the end of an ascent of a permutation of $\{1, \dots, n-1\}$ with k ascents (in k ways), in which case the number of ascents remains the same but one of them is extended in length by one, or by breaking up an ascent of a permutation of $\{1, \dots, n-1\}$ with $k-1$ ascents at a nonending point to place n, or placing n at the beginning (and there are $n-k+1$ such insertion positions). Graham, Knuth, and Patashnik (1994) gives a complete account of these combinatorial counts including properties, recurrence, and generating functions. They are also discussed in Knuth (1998) from an angle geared toward searching and sorting.[3]

Let us now go back to the extraction of the coefficients from the histories generating function to find

$$[u^k z^n]\,\mathcal{H}(u, 1, z) = \frac{1}{n!} \left\langle {n \atop k} \right\rangle.$$

Thus, by Eq. (7.1) we have

$$\mathbf{P}(W_n = k) = \frac{[u^k z^n]\,\mathcal{H}(u, 1, z)}{[z^n]\,\mathcal{H}(1, 1, z)} = \frac{1}{n!} \left\langle {n \atop k} \right\rangle.$$

[3] The reader should be aware that Knuth (1998) used $\left\langle {n \atop k+1} \right\rangle$ for what Graham, Knuth, and Patashnik (1994) calls $\left\langle {n \atop k} \right\rangle$.

Calculations for other starting conditions are more involved, but follow the same general principles.

7.2.3 Analytic Ehrenfest Urn

We go back to the classic urn

$$\mathbf{A} = \begin{pmatrix} -1 & 1 \\ 1 & -1 \end{pmatrix},$$

from an analytic view. This urn scheme arises as a model for the exchange of gases between two chambers. The associated differential system is

$$x'(t) = y(t),$$
$$y'(t) = x(t).$$

We can eliminate $y(t)$ between the two equations via a second derivative: $x''(t) = y'(t) = x(t)$. The equation

$$x''(t) = x(t)$$

is a classical linear homogeneous equation with a linear combination of e^t and e^{-t} as solution. Choosing the constants to match the initial conditions we have the solution

$$x(t) = \frac{1}{2}(x(0) + y(0))e^t + \frac{1}{2}(x(0) - y(0))e^{-t},$$
$$y(t) = \frac{1}{2}(x(0) + y(0))e^t - \frac{1}{2}(x(0) - y(0))e^{-t}.$$

Let us solve a case in the flavor of the Blom, Holst, and Sandell (1993) problem which we discussed in Section 3.5, with $W_0 > 0$ and $B_0 = 0$. As a model for the mixing of gases, this choice corresponds to a system initially in a state where all the particles are in one chamber. Let us compute $\mathbf{P}(W_n = W_0)$, that is, the probability of returning to the initial state after n draws.

For simplicity, we replace the initial conditions $x(0)$ and $y(0)$, respectively, with u and v. According to Theorem 7.1,

$$\mathcal{H}(u, v, z) = x^{W_0}(z)\, y^0(x) = \left(\frac{1}{2}(u + v)e^z + \frac{1}{2}(u - v)e^{-z}\right)^{W_0}.$$

So,

$$\mathcal{H}(1, 1, z) = e^{W_0 z} = \sum_{n=0}^{\infty} W_0^n \frac{z^n}{n!},$$

and

$$[z^n]\, \mathcal{H}(1, 1, z) = \frac{W_0^n}{n!}.$$

That is, $h_n/n! = W_0^n/n!$. By the nth draw, there are $h_n = W_0^n$ histories. Of these histories there are $h_n(W_0, 0)$ histories in favor of the desired event after n draws. It is impossible to have W_0 white balls, and $b \ge 1$ blue balls in the urn, as the total number of balls remains constant in the Ehrenfest urn scheme. There are no histories taking the urn into the state $W_n = w = W_0$, $B_0 = b \ge 1$. In other words $h_n(W_0, b) = 0$, for all $b \ge 1$. We can take advantage of the meaning of the histories generating function when the first parameter is set to 1 and the second to 0:

$$\begin{aligned}
\mathcal{H}(1, 0, z) &= \sum_{n=0}^{\infty} \sum_{0 \le w, b \le \infty} h_n(w, b) u^w v^b \left. \frac{z^n}{n!} \right|_{u=1, v=0} \\
&= \sum_{n=0}^{\infty} \sum_{w=0}^{\infty} h_n(w, 0) u^w \left. \frac{z^n}{n!} \right|_{u=1} \\
&= \sum_{n=0}^{\infty} h_n(W_0, 0) \frac{z^n}{n!}.
\end{aligned}$$

Consequently, we have

$$\begin{aligned}
\mathbf{P}(W_n = W_0) &= \frac{h_n(W_0, 0)}{h_n} \\
&= \frac{h_n(W_0, 0)/n!}{h_n/n!} \\
&= \frac{[z^n]\, \mathcal{H}(1, 0, z)}{h_n/n!} \\
&= \frac{n!}{W_0^n} [z^n] \left(\frac{e^z + e^{-z}}{2} \right)^{W_0} \\
&= \frac{n!}{2^{W_0} W_0^n} [z^n] \sum_{j=0}^{W_0} \binom{W_0}{j} (e^z)^j (e^{-z})^{W_0 - j} \\
&= \frac{n!}{2^{W_0} W_0^n} [z^n] \sum_{j=0}^{W_0} \binom{W_0}{j} e^{(2j - W_0) z} \\
&= \frac{n!}{2^{W_0} W_0^n} [z^n] \sum_{j=0}^{W_0} \sum_{k=0}^{\infty} \binom{W_0}{j} (2j - W_0)^k \frac{z^k}{k!} \\
&= \frac{1}{2^{W_0} W_0^n} \sum_{j=0}^{W_0} \binom{W_0}{j} (2j - W_0)^n.
\end{aligned}$$

7.2.4 Analytic Coupon Collector's Urn

In Section 2.3 we discussed a class of urn problems associated with coupon collection. We viewed the problem there as m different balls in an urn; the balls are sampled at random (with replacement) at each stage, and the sampling is continued until all m colors are observed. The chief concern is *How long does it take to observe all the colors (coupons) in the samples?*

Another Pólya urn scheme, that monitors which balls have been picked, admits an easy formulation as an analytic urn. We can view the problem as an urn of balls of two colors: One color is indicative of that the ball has not been drawn before, say the white color. The other color (say blue) is indicative of that the ball has been observed before in a previous sampling. The urn starts with m white balls. Whenever a ball of the white color is drawn, we paint it with blue and put it back in the urn, and whenever a blue ball is withdrawn, we just put it back as is. This is a Pólya urn with the scheme

$$\begin{pmatrix} -1 & 1 \\ 0 & 0 \end{pmatrix},$$

and initial conditions $W_0 = m$, and $B_0 = 0$. This scheme is amenable to the analytic method. The associated differential system is

$$x'(t) = y(t),$$
$$y'(t) = y(t).$$

The differential equation for $y(t)$ has the solution $y(t) = y(0)e^t$, and the solution for $x(t)$ follows: It is $x(t) = x(0) - y(0) + y(0)e^t$.

According to Theorem 7.1,

$$\mathcal{H}(x(0), y(0), z) = (x(0) - y(0) + y(0)e^z)^{W_0}(y(0)e^z)^{B_0},$$

which we can rewrite as

$$\mathcal{H}(u, v, z) = (u - v + ve^z)^{W_0}(ve^z)^{B_0} = (u - v + ve^z)^m.$$

Using the univariate approach in Eq. (7.1), we get

$$\begin{aligned}\mathbf{P}(B_n = b) = \mathbf{P}(W_n = m - b) &= \frac{h_n(m-b)}{h_n} \\ &= \frac{[u^{m-b}z^n]\,\mathcal{H}(u, 1, z)}{[z^n]\,\mathcal{H}(1, 1, z)}.\end{aligned}$$

For the denominator we have

$$\mathcal{H}(1, 1, z) = e^{mz} = \sum_{n=0}^{\infty} m^n \frac{z^n}{n!},$$

giving

$$[z^n]\,\mathcal{H}(1, 1, z) = \frac{m^n}{n!}.$$

For the numerator, we have

$$\begin{aligned}[u^{m-b}z^n]\,\mathcal{H}(u, 1, z) &= [u^{m-b}z^n]\,(u + e^z - 1)^m \\ &= [u^{m-b}z^n] \sum_{k=0}^{m} \binom{m}{k} u^{m-k}(e^z - 1)^k \\ &= \binom{m}{b} [z^n](e^z - 1)^b.\end{aligned}$$

The function $(e^z - 1)^b$ generates the sequence involving the Stirling numbers of the second kind. We shall present a sketch of these numbers first.

The number $\left\{ {r \atop j} \right\}$ is the number of ways to partition a set of r elements into j nonempty sets, or in terms of urns, it is the number of ways to distribute r distinguishable balls among j indistinguishable urns so that no urn is empty.[4] For example, the number $\left\{ {4 \atop 3} \right\}$ is 6 as we can come up with six distinct ways of partitioning a set like $\{1, 2, 3, 4\}$ into three nonempty subsets; these partitions are:

$$\begin{array}{lll} \{1\}, & \{2\}, & \{3, 4\}, \\ \{1\}, & \{3\}, & \{2, 4\}, \\ \{1\}, & \{4\}, & \{2, 3\}, \\ \{2\}, & \{3\}, & \{1, 4\}, \\ \{2\}, & \{4\}, & \{1, 3\}, \\ \{3\}, & \{4\}, & \{1, 2\}. \end{array}$$

The Stirling numbers of the second kind satisfy the recurrence

$$\left\{ {r \atop j} \right\} = j \left\{ {r-1 \atop j} \right\} + \left\{ {r-1 \atop j-1} \right\}. \tag{7.5}$$

This can be seen in a count of the number of partitions of $\{1, \ldots, r\}$ obtained by delaying the placement of a particular number, say r, in a set: We can come up with a partition of $r - 1$ distinguishable balls into $j - 1$ indistinguishable urns in $\left\{ {r-1 \atop j-1} \right\}$ ways, then place r by itself in a new urn, or we can come up with a partition of $r - 1$ distinguishable balls into j indistinguishable

[4] The large curly brackets notation was introduced by Karamata (see Karamata, 1935) and promoted by Knuth (see Knuth 1992) to parallel the binomial coefficient symbolism.

urns in $\left\{ {r-1 \atop j} \right\}$ ways, then choose one of them in j ways to receive r. Graham, Knuth, and Patashnik (1994) gives a complete account of these combinatorial counts including properties, recurrence, and generating functions. They are also discussed in Knuth (1998) in connection with searching and sorting.

The function $(e^z - 1)^b$ appears in $\mathcal{H}(u, 1, z)$. Let us show that it generates the sequence $b!\left\{ {n \atop b} \right\}/n!$, where $\left\{ {n \atop b} \right\}$ are the Stirling numbers of the second kind. We shall demonstrate by induction on b that the generating function

$$f_b(z) = b! \sum_{n=1}^{\infty} \left\{ {n \atop b} \right\} \frac{z^n}{n!}$$

is $(e^z - 1)^b$. At the basis $b = 1$, we have

$$f_1(z) = \sum_{n=1}^{\infty} \left\{ {n \atop 1} \right\} \frac{z^n}{n!},$$

and the coefficients $\left\{ {n \atop 1} \right\}$ are 1, because there is only one way to partition a set into one part (throw all the balls in one urn). That is,

$$f_1(z) = \sum_{n=1}^{\infty} \frac{z^n}{n!} = e^z - 1.$$

Assume, as an induction hypothesis, that $f_{b-1}(z) = (e^z - 1)^{b-1}$. Starting with a version of the recurrence in Eq. (7.5) in which n replaces r and b replaces j, we multiply all sides by $b!\, z^n/n!$, and sum over $n \ge 1$, to get

$$b! \sum_{n=1}^{\infty} \left\{ {n \atop b} \right\} \frac{z^n}{n!} = b! \sum_{n=1}^{\infty} b \left\{ {n-1 \atop b} \right\} \frac{z^n}{n!} + b! \sum_{n=1}^{\infty} \left\{ {n-1 \atop b-1} \right\} \frac{z^n}{n!}.$$

Take the derivative with respect to z, to obtain

$$\frac{d}{dz}\left(b! \sum_{n=1}^{\infty} \left\{ {n \atop b} \right\} \frac{z^n}{n!} \right)$$
$$= b\left(b! \sum_{n=1}^{\infty} \left\{ {n-1 \atop b} \right\} \frac{z^{n-1}}{(n-1)!} \right) + (b-1)!\, b \sum_{n=1}^{\infty} \left\{ {n-1 \atop b-1} \right\} \frac{z^{n-1}}{(n-1)!},$$

which is the functional equation

$$f_b'(z) = b f_b(z) + b f_{b-1}(z) = b f_b(z) + b(e^z - 1)^{b-1},$$

according to the induction hypothesis. The solution to this differential equation that is consistent with the boundary condition is $f_b(z) = (e^z - 1)^b$.

Going back to the extraction of coefficients from $f_b(z)$, we find the distribution of the number of blue balls to be

$$\mathbf{P}(B_n = b) = \frac{[u^{m-b} z^n]\, \mathcal{H}(u, 1, z)}{[z^n]\, \mathcal{H}(1, 1, z)} = \binom{m}{b} \frac{[z^n]\,(e^z - 1)^b}{m^n/n!} = \frac{b!}{m^n} \binom{m}{b} \left\{ {n \atop b} \right\}.$$

Exercises

7.1 What is the total number of histories for a balanced two-color urn scheme where the total row sum of the schema is $K = 0$? What is the corresponding generating function?

7.2 What is the total number of histories for a balanced diminishing two-color urn scheme where the total row sum of the schema is $K < 0$? What is the corresponding generating function?

7.3 (Flajolet, Dumas, and Puyhaubert, 2006). Suppose we sample white and blue balls (one at a time) without replacement from an urn containing only balls of these two colors. This is the Pólya urn scheme with the replacement matrix $\begin{pmatrix} -1 & 0 \\ 0 & -1 \end{pmatrix}$. Let W_n and B_n be, respectively, the number of white and blue balls in the urn after n draws. Using the analytic method, show that

$$\mathbf{P}(W_n = w,\, B_n = b) = \frac{\binom{W_0}{w}\binom{B_0}{n-w}}{\binom{W_0 + B_0}{n}}.$$

7.4 (Flajolet, Dumas, and Puyhaubert, 2006). Consider the hyper-Pólya urn with the replacement matrix

$$\begin{pmatrix} 1 & 0 & 0 \\ 0 & 1 & 0 \\ 0 & 0 & 1 \end{pmatrix}.$$

Derive the associated system of differential equations and solve it to find the exact distribution of the number of balls of each color.

8

Applications of Pólya Urns in Informatics

Trees are graphs without loops or cycles. They abound in computer science and engineering applications as models: Certain forms are used as data structures, other forms are useful as a backbone for analyzing algorithms, and other forms model the growth of certain structures, such as the Internet or social networks. In many cases they grow in a random manner, by adding new vertices in an uncertain way. A variety of probability models exist to reflect aspects of the growth or the data stored.

When a tree grows by having one of its vertices attract a new vertex (leaf), the degree of the attracting vertex increases by one, and the new leaf is a vertex of degree one. It is evident that adding nodes to a tree affects a certain change in the profile of degree sequences in the tree. If we paint the nodes of a certain degree, say k, with a particular color (say color k), the vertices can then be viewed as colored balls in a Pólya urn. We review in this chapter some applications of Pólya urns in random trees.

8.1 Search Trees

The binary search tree is a fundamental construct of computer science that is used for efficient storage of data and the modeling of many algorithms. For definitions and combinatorial properties see Mahmoud (1992); for applications in sorting see Knuth (1998) or Mahmoud (2000).

Under the constraints of modern technology, external computer memory is much cheaper than internal memory. That is why internal memory is usually small. Speedwise, data in internal memory are accessed much faster than data residing outside. In applications involving large volumes of data special data structures are preferred to store the bulk of data outside the computer on (slow) secondary storage. Fragments therein are brought into internal memory upon request for fast local searching. These data structures involve generalizations of the binary search tree into data structures the nodes of which are *blocks* or *buckets*. Two such generalizations are presented. The analysis of both is amenable to urn models. Surprisingly, they both exhibit an interesting phase transition in their distribution.

To handle data in high dimensions, the binary tree is generalized in yet another direction. Quad trees and k–d trees are suitable storage media for geometry algorithms, such as nearest neighbor queries. The analysis of these trees, too, is amenable to Pólya urn models. In the following subsections we shall outline the algorithmic aspects and uses of these trees in informatics and related areas.

The results presented below had all been obtained mostly by other disparate methods, but it has recently become clear that they can all be obtained via urn models.

8.1.1 Binary Search Trees

A *binary tree* is a structure of nodes each having no children, one left child, one right child, or two children (one left and one right). (This structure is not a tree in the graph-theoretic sense that does not make a distinction between different orientations in the plane or drawings of the same tree.) In practice, the tree carries labels (data) and is endowed with a *search property*. According to the search property, the label of any node is larger than the labels in its left subtree and no greater than any label in its right subtree, and this property permeates the structure recursively.

Several models of randomness are commonly used on binary trees. The uniform model in which all trees are equally likely is useful in formal language studies, compilers, computer algebra, etc. Kemp (1984) is a good source for this subject. The *random permutation model* comports more faithfully to sorting and data structures. In the random permutation probability model, we assume that the tree is built from permutations of $\{1, \ldots, n\}$, where a uniform probability model is imposed on the permutations instead of the trees. When all $n!$ permutations are equally likely or *random*, binary search trees are not equally likely. It is well known in enumerative combinatorics that the number of distinct binary trees is $(n+1)^{-1}\binom{2n}{n} < n!$ (see Knuth (1998)). By the pigeon-hole principle, several permutations must "collide" and correspond to the same tree, but there are permutations that each correspond to precisely one tree. The random permutation model does not give rise to a uniform distribution on binary trees. Generally the model is biased toward shorter trees, which is a desirable properly for fast searching applications (see Mahmoud (1992)). The random permutation model covers a wide variety of real-world applications, such as sampling data from *any* continuous distribution, where the data ranks almost surely form a random permutation.

The term *random binary search tree* will refer to a binary tree built from a random permutation. A tree grows from a permutation $(\pi_1 \ldots \pi_n)$ of $\{1, \ldots, n\}$ as follows. The first element π_1 is inserted in an empty tree, a root of a new nonempty tree is allocated for it. Each subsequent element π_j is taken to the left or right subtree according as whether it is smaller or larger than the root. In the relevant subtree, the element is inserted recursively. The search continues until an empty subtree is found where the element is inserted into a new node, just like π_1 was initially inserted in the root. Space is allocated for the

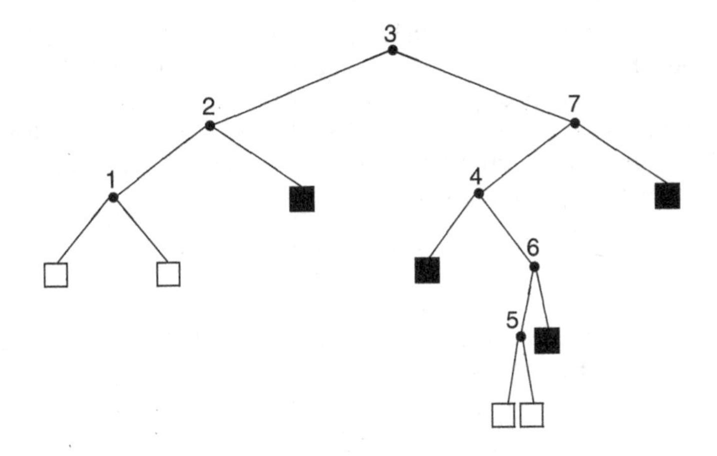

FIGURE 8.1
An extended binary tree on 7 internal nodes.

new node and the node is linked as a left or right child of the last node on the search path, depending on whether the new key is smaller or larger than the content of that last node.

A binary search tree is often *extended* by supplying each node with a sufficient number of distinguished nodes called *external* to uniformly make the outdegree of each original tree node (now called *internal*) equal to 2. Figure 8.1 shows an extended binary search tree grown from the permutation $3, 7, 4, 2, 6, 1, 5$. In the figure internal nodes are depicted as bullets and external nodes are depicted as squares. For a future white-blue coloring scheme, the external nodes are shown to contain blank (white) or are darkened (blue).

It is well known that the random permutation model on binary search trees gives an evolutionary stochastic process in which the external nodes are equally likely (see Knuth (1998)). That is, if a tree is grown in steps by turning a randomly chosen external node into an internal node, the resulting binary tree shapes follow the distribution of binary trees grown from random permutations.

The number of internal nodes with k external children, $k = 0, 1, 2$, provides certain space efficiency measures on binary search trees. For example, leaves in the original tree (nodes with two external children in the extension) are wasteful nodes that are allocated pointers which are not used (in a real implementation they carry nill value). This efficiency measure can be found from the following.

THEOREM 8.1

(Devroye, 1991). Let L_n be the number of leaves in a binary search tree of size n. Then,

$$\frac{L_n - \frac{1}{3}n}{\sqrt{n}} \xrightarrow{\mathcal{D}} \mathcal{N}\left(0, \frac{2}{45}\right).$$

PROOF Color each external node with an external sibling with white, and color all the other external nodes with blue. Figure 8.1 illustrates an instance of this coloring. When insertion hits a white external node, the node is turned into an internal one, with two white external children; its sibling turns blue. If insertion hits a blue external node, it turns into an internal node with two white external children. The schema

$$\begin{pmatrix} 0 & 1 \\ 2 & -1 \end{pmatrix}$$

underlies this growth rule. If W_n is the number of white balls in the urn after n draws, it follows from Theorem 3.5 that

$$\frac{W_n - \frac{2}{3}n}{\sqrt{n}} \xrightarrow{\mathcal{D}} \mathcal{N}\left(0, \frac{8}{45}\right).$$

But then $W_n = 2L_n$. ■

Theorem 6.5 gives a bit more; it specifies a multivariate joint distribution for the profile of nodes of various outdegrees in the tree, of which the distribution of the leaves is only one marginal distribution.

8.1.2 Fringe-Balanced Trees

Binary trees are quite efficient (see Mahmoud, 1992) for numerous applications). To improve the speed of retrieval even more, height reduction heuristics known as *balancing* are employed. One such heuristic is the local balancing of the fringe (Poblete and Munro, 1985)). Fringe-balancing is a local "rotation" that compresses subtrees near the leaves into shorter shrubs. The operation is performed when the insertion of a new node falls under an internal node that happens to be the only child of its parent in the binary tree. When this occurs a compression operation called *rotation* promotes the median of these three nodes to become the parent and repositions the other two nodes under it in a manner preserving the search property. Figure 8.2 illustrates some cases of fringe balancing. The cases not shown in Figure 8.2 are only mirror images of those depicted.

The depth of a node in the tree is its distance (in terms of edges) from the root, and the total path length is the sum of all the depths. The introduction of rotations reduces the total path length, but the cost of rotation becomes a factor in the construction of the tree.

To model rotations, we color the external nodes according to a scheme in which the number of times the external nodes of a particular color are chosen for insertion corresponds to rotations. Color 1 generally represents balance. If insertion falls at an external node of color 1, a tolerable degree of imbalance appears in the tree and is coded by colors 2 and 3: Color 2 is for the external node, the sibling of which is internal; the two children of this internal node are colored with color 3 (Figure 8.2(a)). If insertion falls at an external node of

(a) Imbalance appears when insertion hits an external node of color 1.

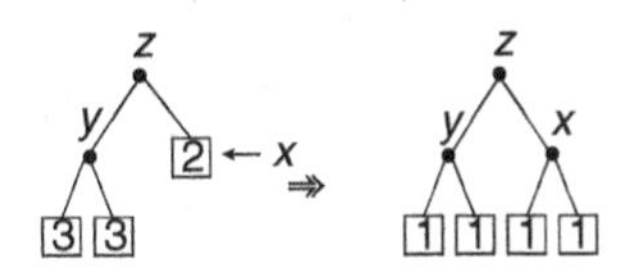

(b) Insertion at an external node of color 2 fills out a subtree.

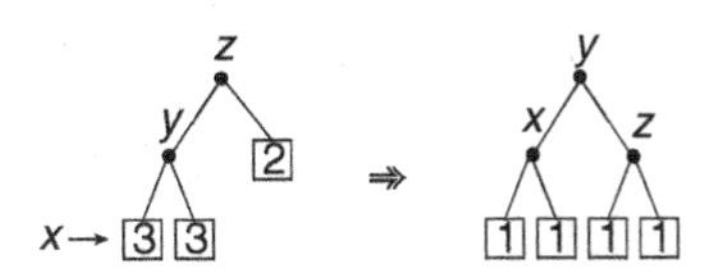

(c) A rotation is executed when insertion hits an external node of color 3.

FIGURE 8.2
Insertions in a fringe-balanced binary search tree.

color 2, the insertion fills out a subtree forming a perfectly balanced shrub of height 2. The four balanced external nodes of this perfectly balanced shrub are colored with color 1 (Figure 8.2(b)). If instead insertion falls at an external node of color 3, more imbalance is imposed, and balance is restored by a rotation and again a perfectly balanced shrub of height 2 appears; the four external nodes of this shrub are colored with color 1 (Figure 8.2(c)).

After n insertions in an empty tree (one initial external node), the number of times insertion falls at an external node of color 3 is the number of rotations, to be denoted by R_n. The underlying urn has the schema

$$\mathbf{A} = \begin{pmatrix} -2 & 1 & 2 \\ 4 & -1 & -2 \\ 4 & -1 & -2 \end{pmatrix}.$$

THEOREM 8.2
(Mahmoud, 1998, Panholzer and Prodinger, 1998). Let R_n be the number of rotations after n insertions in an initially empty fringe-balanced binary search tree. Then,

$$\frac{R_n - \frac{2}{7}n}{\sqrt{n}} \xrightarrow{\mathcal{D}} \mathcal{N}\left(0, \frac{66}{637}\right).$$

PROOF *(sketch)* The extended schema underlying rotations has eigenvalues $\lambda_1 = 1$, $\lambda_2 = 0$, and $\lambda_3 = -6$, with $\Re\,\lambda_2 < \frac{1}{2}\lambda_1$. All we need to complete the proof is the principal eigenvector. For $\mathbf{A}^T$ the principal eigenvalue is 1 with corresponding eigenvector

$$\mathbf{v} = \begin{pmatrix} v_1 \\ v_2 \\ v_3 \end{pmatrix} = \frac{1}{7}\begin{pmatrix} 4 \\ 1 \\ 2 \end{pmatrix}.$$

So,

$$\frac{R_n}{n} \xrightarrow{P} v_3 = \frac{2}{7},$$

and

$$\frac{R_n - \frac{2}{7}n}{\sqrt{n}} \xrightarrow{\mathcal{D}} \mathcal{N}(0, \sigma^2)$$

for some $\sigma^2 > 0$ that can be found by a lengthy computation. ■

As mentioned before, there is no convenient method to find variances in Smythe-like schemes. So far, each instance of such an urn scheme has been handled in an ad-hoc manner in the original references. The variance in Theorem 8.2 was obtained by recurrence methods (in fact in exact form). (The same exact result is presented in Hermosilla and Olivos (1985)). Panholzer and Prodinger (1998) report on an analytic approach to Theorem 8.2.

8.1.3 *m*-Ary Search Trees

Search speed is reduced with increased branching, as data are distributed among more subtrees. The m-ary search tree has branching factor m, and each node holds up to $m-1$ keys. In a practical implementation the nodes are blocks of space, each having $m-1$ slots for data and m pointers. The tree grows from n keys according to a recursive algorithm: The first insertion falls in an empty tree; a root node is allocated for it. The second key joins, but the two keys are arranged in increasing order from left to right. Subsequent keys (up to the $(m-1)$st insertion) are likewise placed in the root node, and after each insertion the keys are sorted in increasing order. The $(m-1)$st insertion fills out the root node; a subsequent key goes to the ith subtree if it ranks i among the first m keys—a node is allocated for it and linked to the root by inserting a nexus to the new node at the ith pointer position in the root. Let $\mathcal{R}$ be the set of keys in the root. A subsequent key K ranking ith among the elements of the set $\{K\} \cup \mathcal{R}$ is sent to the ith subtree, and within the subtree the insertion procedure is applied recursively. Figure 8.3 shows a ternary tree grown from the permutation 6, 4, 1, 5, 3, 9, 2, 8, 7.

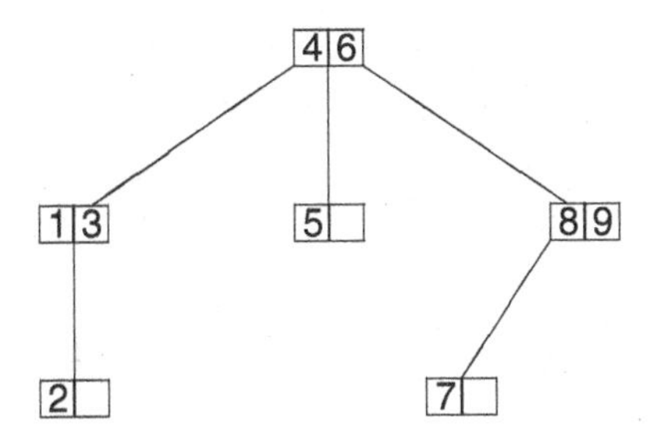

FIGURE 8.3
A ternary search tree ($m = 3$).

Unlike the binary case of deterministic size, for $m \geq 3$ the size of the m-ary tree is random. For instance Figure 8.4 shows a ternary tree grown from the permutation 7, 4, 1, 6, 5, 3, 9, 2, 8. This new permutation has the same keys as those in the old permutation that gave rise to the tree of Figure 8.3, but the corresponding new tree is of size smaller (by one node) than the tree of the old permutation.

The tree size is a measure of the space efficiency of the algorithm. The tree growth can be modeled by an urn consisting of balls of $m - 1$ different colors. Color i corresponds to a "gap" (an insertion position) between two keys of a leaf node with $i - 1$ keys, for $i = 1, \ldots, m - 1$ (and $i = 1$ corresponds to insertion in an empty tree). A leaf node containing $i - 1$ keys has i gaps ($i - 1$ real numbers cut up the real line into i intervals). A key falling in a leaf node with $i - 1 < m - 2$ keys will adjoin itself in the node, increasing its number of keys to i, and consequently the number of gaps to $i + 1$. The corresponding rule for the growth of the associated urn is to promote i balls of color i into $i + 1$ balls of color $i + 1$; this insertion affects colors i and $i + 1$ and no other.

The urn rule corresponding to filling out a node is a little different. The node already contains $m - 2$ keys ($m - 1$ gaps corresponding to $m - 1$ balls of color $m - 1$). The insertion falls in a gap and the node fills out defining m empty subtrees (at the next level in the tree). The m empty subtrees are

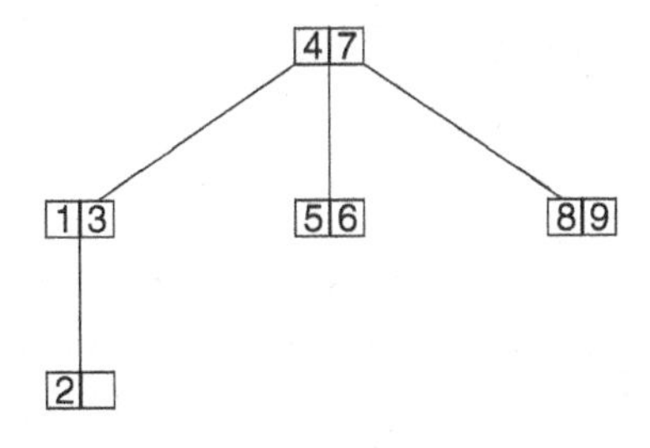

FIGURE 8.4
A ternary search tree ($m = 3$).

represented by m balls of color 1 in the urn. The schema is

$$\mathbf{A} = \begin{pmatrix} -1 & 2 & 0 & 0 & 0 & \dots & 0 & 0 \\ 0 & -2 & 3 & 0 & 0 & \dots & 0 & 0 \\ 0 & 0 & -3 & 4 & 0 & \dots & 0 & 0 \\ \vdots & \vdots & & & \ddots & & \vdots & \vdots \\ 0 & 0 & & & & \dots & -(m-2) & m-1 \\ m & 0 & & & & \dots & 0 & -(m-1) \end{pmatrix}.$$

The characteristic equation $|\mathbf{A}^T - \lambda\mathbf{I}| = 0$ can be obtained by expanding the first row—the top left entry in $\mathbf{A}^T - \lambda\mathbf{I}$ has a cofactor with zeros above its diagonal (and a determinant equal to the product of its diagonal elements), and the top right element has a cofactor with zeros below its diagonal (and a determinant equal to the product of its diagonal elements). The characteristic equation is

$$(-\lambda - 1)(-\lambda - 2)(-\lambda - 3)\dots(-\lambda - (m-1)) \\ + (-1)^m m \times 2 \times 3 \times \dots \times (m-1) = 0.$$

This simplifies to the characteristic equation

$$(\lambda + 1)(\lambda + 2)\dots(\lambda + m - 1) = m!.$$

Let us arrange the eigenvalues $\lambda_1, \lambda_2, \dots, \lambda_{m-1}$ by their decreasing real parts:

$$\Re\,\lambda_1 \ge \Re\lambda_2 \ge \dots \ge \Re\,\lambda_{m-1}.$$

The roots of the characteristic equation are simple and satisfy:

(i) The principal eigenvalue $\lambda_1 = 1$.
(ii) For odd m, the characteristic equation has one real negative root $\lambda_{m-1} = -m - 1$.
(iii) Other than the principal eigenvalue and a possible negative real root, all the other eigenvalues are complex with nonzero imaginary parts.
(iv) Complex roots occur in conjugate pairs.
(v) Each conjugate pair has a distinct real part.
(vi) $\Re\lambda_2 < \frac{1}{2}$, for $m \le 26$, and $\Re\lambda_2 > \frac{1}{2}$, for $m > 26$.

A proof of these statements can be found in Mahmoud and Smythe (1995).

The eigenvector $\mathbf{v} = (v_1, \dots, v_{m-1})^T$ corresponding to the principal eigenvalue of $\mathbf{A}^T$ has a particular structure. The components of $\mathbf{v}$ are in the proportion

$$v_1 : v_2 : \dots : v_{m-1} = \frac{2}{2} : \frac{2}{3} : \frac{2}{4} : \dots : \frac{2}{m}.$$

This can be seen by expanding the matrix equation $\mathbf{A}^T\mathbf{v} = \lambda_1\mathbf{v} = \mathbf{v}$ into a set of linear equations:

$$\begin{aligned} -v_1 + mv_{m-1} &= v_1, \\ 2v_1 - 2v_2 &= v_2, \\ &\vdots \\ (m-1)v_{m-2} - (m-1)v_{m-1} &= v_{m-2}, \end{aligned}$$

which unwind into the proportions mentioned above. Normalizing the scale of the length of $\mathbf{v}$ to 1, we get the additional constraint

$$v_1 + \frac{2}{3}v_1 + \frac{2}{4}v_1 + \cdots + \frac{2}{m}v_1 = 1,$$

and so

$$v_1 = \frac{1}{1 + 2\left(\frac{1}{3} + \frac{1}{4} + \cdots + \frac{1}{m}\right)} = \frac{1}{2(H_m - 1)},$$

where H_r is the rth harmonic number. Hence

$$v_i = \frac{1}{(i+1)(H_m - 1)}.$$

Let $X_n^{(i)}$ be the number of balls of color i after n draws from an initial urn with one ball of color 1. We shall consider in the sequel $3 \le m \le 26$, for which the corresponding urn scheme is an extended one. A consequence of Theorem 6.4 is

$$\frac{X_n^{(i)}}{n} \xrightarrow{P} \lambda_1 v_i = \frac{1}{(i+1)(H_m - 1)}.$$

Having computed the weak limits for the entire profile of various ball colors, we can work our way toward the size of the tree. We view the nodes with $m-1$ keys and at least one nonempty child as internal, and let I_n denote their number. Consider all the other actual nodes (nodes carrying keys) as actual leaves, let L_n be their number. Even though they are really dummy nodes that do not actually exist, we consider the empty subtrees (balls of color 1) as dummy leaves for the sake of analysis; there are $X_n^{(1)}$ of them. In a practical implementation an empty subtree is a nill pointer stored in a parent node one level above. Let L_n' be the number of virtual leaves (actual plus dummy leaves). The virtual leaves appear in the tree at all the right places to make the outdegree of any internal node m. We can consider the virtual leaves as the external nodes of an extended m–ary tree of size I_n. Therefore,

$$L_n' = (m-1)I_n + 1.$$

We are after the size S_n of the m–ary tree, that is, the actual number of nodes allocated. This does not include the virtual leaves:

$$\begin{aligned} S_n &= I_n + L_n \\ &= \frac{L'_n - 1}{m-1} + L_n \\ &= \frac{L'_n - 1}{m-1} + L'_n - X_n^{(1)} \\ &= \frac{m}{m-1} \sum_{i=1}^{m-1} \frac{X_n^{(i)}}{i} - X_n^{(1)} - \frac{1}{m-1}. \end{aligned}$$

Each ball color follows a weak law, and we see right away that the tree size also satisfies the weak convergence

$$\frac{S_n}{n} \xrightarrow{P} \frac{m}{m-1} \sum_{i=1}^{m-1} \frac{1}{i(i+1)(H_m - 1)} - \frac{1}{2(H_m - 1)} = \frac{1}{2(H_m - 1)}.$$

THEOREM 8.3
(Chern and Hwang, 2001). Let S_n be the size of an m–ary search tree grown from a random permutation of $\{1, \ldots, n\}$. If the branch factor $3 \le m \le 26$, then

$$\frac{S_n - n/(2(H_m - 1))}{\sqrt{n}} \xrightarrow{\mathcal{D}} \mathcal{N}\left(0, \sigma_m^2\right),$$

for some effectively computable constant σ_m^2.

The limiting variance σ_m^2 can be computed by recurrence (Mahmoud (1992) gives an exact formula). Chern and Hwang (2001) proves asymptotic normality in the range $3 \le m \le 26$ by a method of recursive moments. If $m > 26$, the associated urn is no longer an extended urn in the sense of Smythe. Chern and Hwang (2001) argues that no limit distribution for the size of the m-ary tree exists under the central limit scaling because of too large oscillations in all higher moments. What happens after $m \ge 27$ is an area of recent research that was dealt with in Chauvin and Pouyanne (2004) and in Fill and Kapur (2005).

8.1.4 2–3 Trees

To guarantee fast retrieval, m–ary trees are balanced by operations that grow the tree by first accessing an insertion position, and if that causes imbalance, keys are sent up, recursively, and if necessary, they go all the way back, forcing

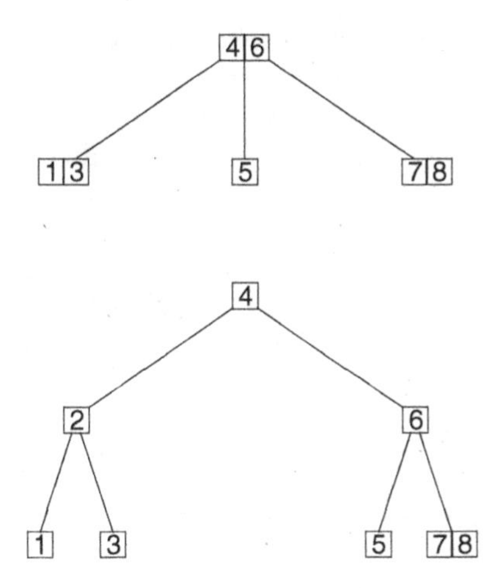

FIGURE 8.5
A 2–3 tree before and after inserting the key 2.

a new root to appear. We illustrate this balancing in 2–3 trees. The nodes here are of two types, type 1 holds one key, and type 2 holds two keys (the corresponding branching is, respectively, 2 and 3). The search phase for the insertion of a new key follows the same general principles as in the m-ary tree. When insertion falls in a leaf node of type 1, it is promoted into type 2. When insertion of a key K overflows a leaf of type 2, already carrying x and y, the median of the three keys x, y, and K is sent up, and the type-2 leaf is split into two type-1 leaves. If the median promotion puts it in a type-1 internal node, the node is expanded into type-2 to accommodate the new key, and the two new type-1 leaves, together with the sibling (possibly an empty node) of the old type-2 leaf are arranged according to order as in the usual ternary search tree. If instead, the median promotion puts it in an already filled type-2 node, the node is split, recursively, etc. Figure 8.5 illustrates the insertion of a key into a 2–3 tree. All the leaves in a 2–3 tree grown from n keys are at the same level, guaranteeing that any search will be done in $O(\log n)$ time, as $n \to \infty$.

The fringe-balancing principle, established in Yao (1978), states that most of the nodes are leaves—the number of leaves in the tree well approximates the size. More precisely, Yao (1978) argues that bounds on the expected size can be found from the expected number of leaves.

THEOREM 8.4
(Yao, 1978, Bagchi and Pal, 1985). Let S_n be the number of nodes after n insertions in an initially empty 2–3 tree. Then,

$$\frac{9}{14}n + o(n) \;\le\; \mathbf{E}[S_n] \;\le\; \frac{6}{7}n + o(n).$$

PROOF We first determine L_n, the number of leaves. Each key in a type-1 leaf defines two insertion gaps; associate two white balls with these gaps. Each pair of keys in a type-2 leaf defines three insertion gaps; associate three blue balls with these gaps. From the evolution rules of the tree, the schema

$$\begin{pmatrix} -2 & 3 \\ 4 & -3 \end{pmatrix}$$

underlies the urn of white and blue balls. By Proposition 3.1, we have

$$\mathbf{E}[W_n] \sim \frac{4}{7}n, \qquad \mathbf{E}[B_n] \sim \frac{3}{7}n.$$

On average, the number of leaves is given by

$$\mathbf{E}[L_n] = \frac{1}{2}\mathbf{E}[W_n] + \frac{1}{3}\mathbf{E}[B_n] \sim \frac{3}{7}n. \tag{8.1}$$

It is well known that the number of leaves in an m-ary tree on n nodes is related to I_n, the number of internal nodes, by

$$L_n = (m-1)I_n + 1.$$

The size of the 2–3 tree with L_n leaves is maximized if all internal nodes are of type-1, in which case the tree is binary with $L_n = I_n + 1$. The size is minimized if all internal nodes are of type-2, in which case the tree is ternary with $L_n = 2I_n + 1$. Thus, the size, $S_n = L_n + I_n$, is sandwiched:

$$L_n + \frac{L_n - 1}{2} \le S_n \le L_n + L_n - 1.$$

The result follows upon taking expectations and plugging in Eq. (8.1). ■

Yao (1978) describes a series of refinements by which the bounds in Theorem 8.4 can be improved. These refinements consider larger and larger shrubs on the fringe of the tree. They can be modeled by urns with more colors.

8.1.5 Paged Binary Trees

Another generalization of binary search trees, suitable for internal memory algorithms working in conjunction with a slow external memory, is the *paged binary trees*, the leaves of which are buckets that can carry up to c keys (see Hoshi and Flajolet, 1992). Each internal node carries one key and has a right and a left child. When a searching algorithm climbs along a path starting at the root, and reaches the leaf at the end of the path, the block of data in the leaf is fetched from external memory. A batch of data is brought into internal memory at one time, where they can be subjected to fast searching. Usually

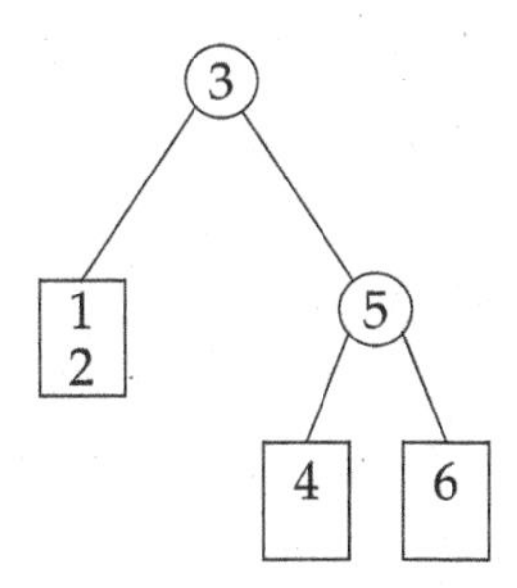

FIGURE 8.6
A paged binary tree with page capacity 2.

the capacity c is chosen to coincide with a "page" of memory, the standard unit of size of a data block in computer memory.

The tree grows from n keys according to a splitting policy. Up to c keys go into the root node. In practice the keys of a node (page) are kept in sorted order to enhance fast local searching. Toward a balanced split, $c = 2b$ is usually an even number. When the $(2b + 1)$st key appears, the tree is restructured: The root bucket is turned into an internal node retaining only the median of the $2b + 1$ keys. Two new pages are created and adjoined as the right and left children of the internal node. Each of the two leaves is half filled. The b keys below the median are placed in the left page, the rest in the right. This branching facilitates later search and insertion. A subsequent key is guided to the left subtree if it is less than the root key, otherwise it goes into the right subtree. Within the appropriate subtree the insertion procedure is applied recursively. Figure 8.6 illustrates a paged tree with page capacity 2, arising from the permutation $6, 3, 1, 4, 2, 5$; the process experiences two splits: First when the key 1 is inserted, the median 3 of the set $\{6, 3, 1\}$ is kept in the root, then again when 5 is inserted, the node containing 4 and 6 splits, keeping the median 5 at the root of a new right subtree. In Figure 8.6 the pages are shown as boxes and the internal nodes as circles. The choice $b = 1$ is unrealistic in practice, it is only chosen here for the purpose of illustration. In a practical implementation a choice like $c = 1024 = 1\text{K}$ may be reasonable.

The associated urn consists of balls of $b + 1$ different colors, and has a $(b+1) \times (b+1)$ schema. Color i corresponds to a "gap" (an insertion position) between two keys of a leaf node carrying $i + b - 1$ keys, for $i = 1, \ldots, b+1$. A key falling in an unfilled leaf with $i + b - 1$ keys ($i \le b$) will increase its keys to $i + b$ and consequently the number of gaps to $i + b + 1$. The corresponding growth rule in the associated urn is to promote i balls of color i into $i+1$ balls of color $i + 1$; this insertion affects colors i and $i + 1$ and no other. The urn rule corresponding to splitting is distinguished. A splitting node contains $2b$ keys with $2b + 1$ gaps of color $b + 1$. The insertion falls in one of these gaps, and an overflow occurs forcing a splitting: Two half-filled leaves appear, each containing b keys, thus having $b + 1$ insertion gaps each; the urn gains $2b + 2$

balls of color 1. The schema is

$$\mathbf{A} = \begin{pmatrix} -(b+1) & b+2 & 0 & 0 & 0 & \dots & 0 & 0 \\ 0 & -(b+2) & b+3 & 0 & 0 & \dots & 0 & 0 \\ 0 & 0 & -(b+3) & b+4 & 0 & \dots & 0 & 0 \\ \vdots & \vdots & & \ddots & & & \vdots & \vdots \\ 0 & 0 & & \dots & & & -2b & (2b+1) \\ 2b+2 & 0 & & \dots & & & 0 & -(2b+1) \end{pmatrix}.$$

The matrix has a banded structure similar to that of the m–ary search tree. The eigenanalysis needed for the convergence theorems is similar, and we only highlight the main points. The characteristic equation $|\mathbf{A}^T - \lambda\mathbf{I}| = 0$ expands into

$$(\lambda + b + 1)(\lambda + b + 2)\dots(\lambda + 2b + 1) = \frac{(2b+2)!}{(b+1)!},$$

by expanding via the first row, for example. The simple roots of this characteristic equation satisfy $\Re\,\lambda_2 < \frac{1}{2}\Re\,\lambda_1 = \frac{1}{2}$, for $b \le 58$. For all $b \ge 59$, $\Re\,\lambda_2 > \frac{1}{2}$. This scheme is an extended urn scheme for b up to 58 (last capacity admitting normality is $c = 116$). The rest of the details of the calculation of the eigenvalues and the corresponding eigenvector are similar to the work done for the m–ary search tree (see Subsection 8.1.3).

THEOREM 8.5

(Chern and Hwang, 2001). Let S_n be the size of a paged binary tree (with page capacity $c = 2b$, with $1 \le b \le 58$), after n insertions. Then

$$S_n^* = \frac{S_n - n/((2b+1)(H_{2b+1} - H_{b+1}))}{\sqrt{n}} \xrightarrow{\mathcal{D}} \mathcal{N}(0, \sigma_b^2),$$

for some effectively computable constant σ_b^2.

The limiting variance σ_b^2 can be obtained with quite a bit of tedious recurrence computation. Chern and Hwang (2001) proves asymptotic normality for $1 \le b \le 58$ by a method of moments (the last capacity for which normality holds is 116). Chern and Hwang (2001) argues that for $b > 58$, the normed random variable S_n^* no longer has a normal limit law.

8.1.6 Bucket Quad Trees

The quad tree, introduced in Finkel and Bentley (1974), is a suitable data structure for d-dimensional data (see Samet (1990) or Yao (1990) for applications).

A point in d dimensions cuts up the space into 2^d quadrants. The first point in a sequence of d-dimensional data defines the 2^d quadrants; other data are grouped according to the quadrant they belong to. The 2^d quadrants are canonically numbered $1, 2, \dots, 2^d$. To expedite work with blocks of data, the nodes are turned into buckets to hold c points. Devroye and Vivas (1998) suggest a balancing policy that associates with each bucket a guiding *index*, which is a dummy point composed of d coordinates, the ith coordinate of which is the median of the c keys' ith coordinates. As median extraction of c numbers is involved in each coordinate, it is therefore customary to take $c = 2b + 1$, an odd number.

The bucket quad tree has 2^d subtrees, numbered $1, \ldots, 2^d$, from left to right. It grows by putting up to c keys in the root node. When a bucket is filled to capacity, its index is computed and stored with the data. A subsequent point belonging to the qth quadrant (according to a partition by the index) goes to the qth subtree, where it is subjected to the same algorithm recursively to be inserted in the subtree. Figure 8.7 shows six points in the unit square, and the corresponding two-dimensional quad tree with bucket capacity 3; the index for the first three point is indicated by the symbol $\times$; the canonical count goes from the bottom left quadrant of the index and proceeds clockwise.

A node holding j keys in this tree has $j + 1$ insertion positions, $j = 1, \ldots, c - 1$. We shall consider an idealized probability distribution on quad trees induced by growing them by choosing an insertion position at random (all insertion positions being equally likely). This probability model is most suited for simulation.

The balls in the associated $(2b+1) \times (2b+1)$ urn scheme have $2b+1$ different colors. Color i accounts for $i - 1$ keys in a leaf node, for $i = 1, \ldots, 2b + 1$. A key falling in an unfilled leaf with $i - 1$ keys will be placed in the leaf and the number of attraction positions increases from i to $i + 1$. The corresponding growth rule in the associated urn is to promote i balls of color i into $i+1$ balls of color $i + 1$; this insertion affects colors i and $i + 1$ and no other. When the $(2b + 1)$st key joins a node containing $2b$ keys, 2^d insertion positions appear

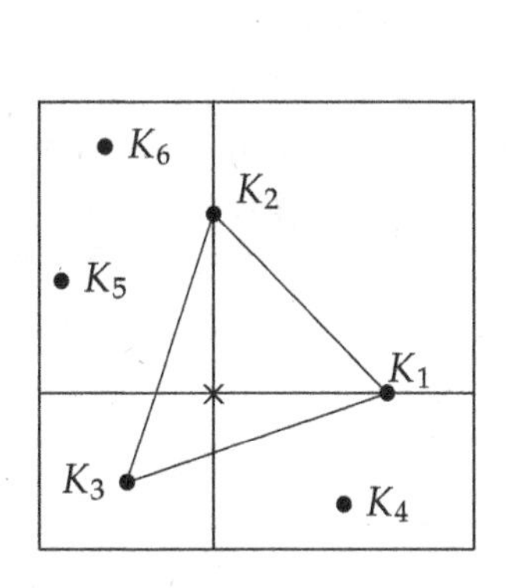

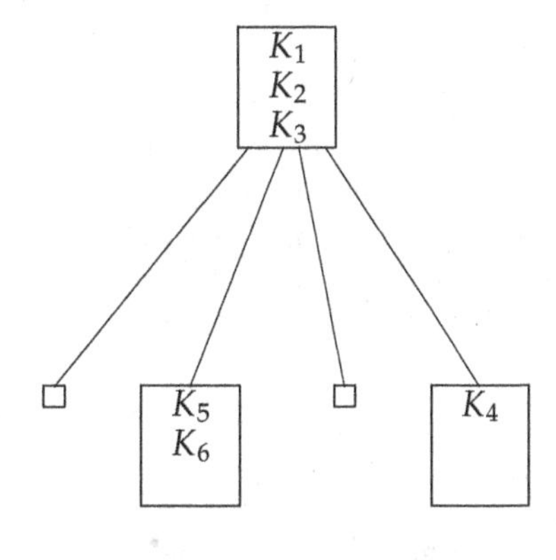

FIGURE 8.7
Points in the unit square and their corresponding quad tree with bucket capacity 3. The symbol $\times$ is the index of K_1, K_2, and K_3.

on the next level. The complete schema is

$$\mathbf{A} = \begin{pmatrix} -1 & 2 & 0 & 0 & 0 & \dots & 0 & 0 \\ 0 & -2 & 3 & 0 & 0 & \dots & 0 & 0 \\ 0 & 0 & -3 & 4 & 0 & \dots & 0 & 0 \\ \vdots & \vdots & & & \ddots & & \vdots & \vdots \\ 0 & 0 & & & \dots & & -2b & (2b+1) \\ 2^d & 0 & & & \dots & & 0 & -(2b+1) \end{pmatrix}.$$

The matrix has a banded structure similar to that of the m–ary search tree. The eigenanalysis needed for the convergence theorems is similar, and we only highlight the main points (see Subsection 8.1.3). The characteristic equation $|\mathbf{A}^T - \lambda\mathbf{I}| = 0$ expands into

$$(\lambda + 1)(\lambda + 2) \dots (\lambda + 2b + 1) = (2b + 1)!\, 2^d,$$

by expanding via the first row, for example. The principal eigenvalue λ_1 is clearly a function of both the capacity and dimension. For instance, with $c = 3$ (i.e., $b = 1$) and $d = 4$, the bucket 4–dimensional quad tree has a principal eigenvalue $\lambda_1 = 2.651649306\dots$, and the corresponding left eigenvector $\mathbf{v} = (0.6030697065\dots, 0.2592928516\dots, 0.1376374422\dots)$. Let $X_n^{(i)}$ be the number of balls of color i, for $i = 1, 2, 3$, after n draws from an initial urn with one ball of color 1. According to a strong analog of Theorem 6.4 (Janson, 2004), the 4–dimensional quad tree has the componentwise asymptotic profile

$$\begin{pmatrix} \mathbf{E}\big[X_n^{(1)}\big] \\ \mathbf{E}\big[X_n^{(2)}\big] \\ \mathbf{E}\big[X_n^{(3)}\big] \end{pmatrix} \xrightarrow{a.s.} \lambda_1 \mathbf{v} = \begin{pmatrix} 1.599129368\dots \\ 0.6875537100\dots \\ 0.3649662279\dots \end{pmatrix}.$$

From these measures one can deduce a strong law for the size by an idea similar to the one we applied for m–ary trees (see Subsection 8.1.3). We first consider the actual leaves and the virtual leaves (balls of color 1) as the external node in an extended 16–way tree. We leave some of the minor details for the reader to complete and get

$$\frac{S_n}{n} \xrightarrow{a.s.} 0.6030697064\dots.$$

8.1.7 Bucket *k–d* Trees

The *k–d* tree was introduced by Bentley in (1975) as a means of data storage amenable to computational geometry algorithms, such as range searching (Samet (1990) surveys applications of this tree). The bucket *k–d* tree is a tree

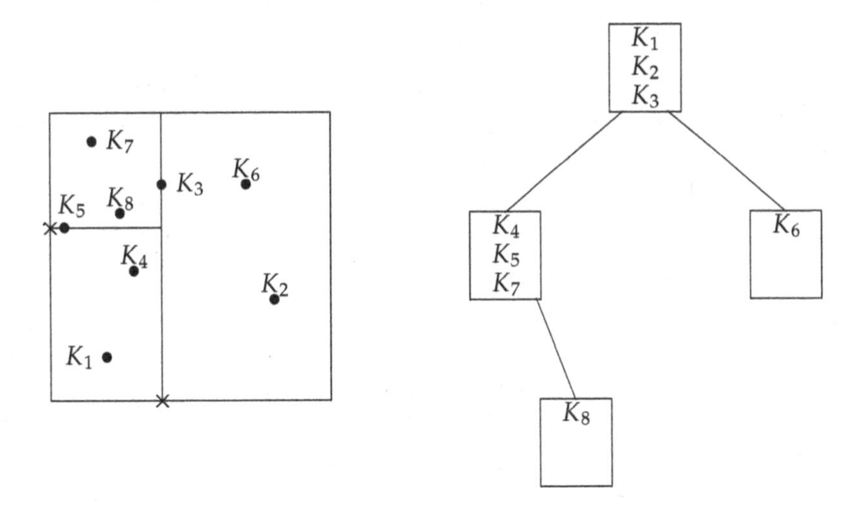

FIGURE 8.8
Points in the unit square and their corresponding 3–2 tree.

for d dimensional data.[1] It is a *binary* tree of bucketed nodes of capacity k each. Up to k keys (usually an odd number, say $2b+1$) go into the root. When $2b + 1$ keys aggregate in the root node, a dummy index is computed and stored with the keys. The index is the median of the first coordinates of the points in the node. A subsequent key goes into the left subtree if the key's first coordinate is less than the root's index; otherwise the key goes into the right subtree. In whichever subtree the key goes, it is subjected recursively to the same insertion algorithm, but at level ℓ in the tree, the $(\ell + 1)$st coordinate of the inserted key is used in a comparison against an index that is the median of the $(\ell+1)$ coordinates of the k keys of a filled node (if the node is not filled, the key adjoins itself to that node). The process cycles through the coordinates with the interpretation that $d+1$ wraps around into the first coordinate. Figure 8.8 shows eight points in the unit square and the corresponding 3–2 tree; the indexes are indicated by the symbol $\times$.

A node holding j keys in this tree has $j + 1$ equally likely insertion positions, $j = 1, \ldots, k-1$. The k–d tree's $(2b+1) \times (2b+1)$ associated urn grows just like that of the quad tree for data in one dimension. The balls in the associated urn have $2b + 1$ different colors. The difference concerns the policy of handling an overflow. An insertion hitting a leaf containing $2b$ keys, fills out the leaf; and two insertion positions appear on the next level. The complete

[1] When $k = 2$ and $d = 3$ the name coincides with that of the 2–3 tree discussed earlier. This is only a historical coincidence, and the two types of trees are different and intended for different purposes.

schema is

$$\begin{pmatrix} -1 & 2 & 0 & 0 & 0 & \dots & 0 & 0 \\ 0 & -2 & 3 & 0 & 0 & \dots & 0 & 0 \\ 0 & 0 & -3 & 4 & 0 & \dots & 0 & 0 \\ \vdots & \vdots & & & \ddots & & \vdots & \vdots \\ 0 & 0 & & & \dots & & -2b & (2b+1) \\ 2 & 0 & & & \dots & & 0 & -(2b+1) \end{pmatrix}.$$

The k–d urn scheme coincides with that of a bucket 1–dimensional quad tree. For instance, with $k = 3$ (that is, $b = 1$), in the random bucket 3–2 tree the number of nodes allocated after n key insertions is

$$\frac{S_n}{n} \xrightarrow{a.s.} 0.3938823094\dots.$$

8.2 Recursive Trees

The random recursive tree is a tree model that underlies Burge's parallel sorting algorithm (Burge, 1958). It is also a combinatorial model for a variety of applications such as contagion, chain letters, philology, etc. (see the survey in Smythe and Mahmoud, 1996 and the 40 plus references therein).

The model has been generalized in a number of directions to take into account the experience of the chain letter holder or to fit operation on computers. These generalizations are sketched in subsections following the analysis of the standard model. The associated results were mostly derived by other methods, such as recurrence and moments methods. Bergeron, Flajolet, and Salvy (1992) give a completely analytic approach to the subject. It has recently become clear that a unifying approach via urn models ties all these results together.

8.2.1 Standard Recursive Trees

Starting out with one root node, *the random recursive tree* grows in discrete time steps. At each step a node in the tree is chosen uniformly at random to be the parent of a new entrant. If the nth node is labeled with n, all root-to-leaf paths will correspond to increasing sequences. There is no restriction on node degrees in recursive trees and they can grow indefinitely. Figure 8.9 shows all the recursive trees of order 4.

Recursive trees have been proposed as a model for chain letters (Gastwirth (1977)), where a company is founded to market a particular item (lottery tickets, a good luck charm, etc.). The initial recruiter looks for a willing participant to buy a copy of the letter. The recruiter and the new letter holder

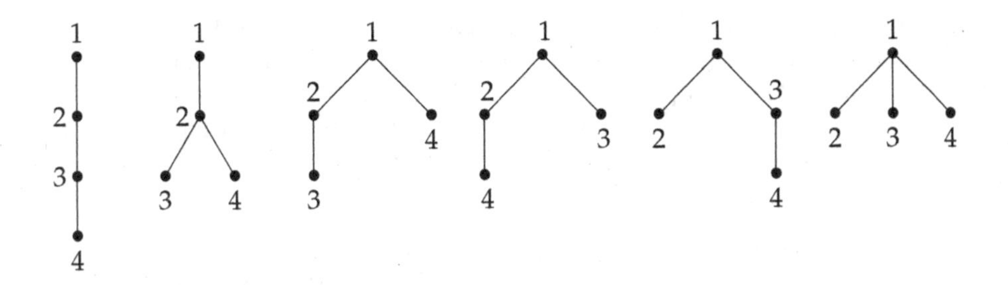

FIGURE 8.9
Recursive trees of order 4.

compete with equal chance to recruit a new participant. The process proceeds in this way, where at each stage all existing participants compete with equal chance to attract the next participant in the scheme.

There is a profit associated with selling a copy of the letter. The promise is that everybody will make profit. Nevertheless, when the number of participants grows to n, it is unavoidable that many letter holders purchased the letter but did not sell it. These are frustrated participants in the scheme (leaves in the tree).

How many frustrated participants are there? This is the question that Najock and Heyde (1982) addresses, and finds out that about half of the participants will be frustrated, on average. Najock and Heyde (1982) characterizes the exact and asymptotic distribution of the number of leaves in a random recursive tree, using recurrence methods; we give a proof based on urns. The proof will need representation in terms of the Eulerian numbers of the first kind; review the material in Subsection 7.2.2.

It is not all that surprising that properties of permutations come into the picture in the analysis of recursive trees, since these trees model, among many other things, sorting algorithms that work on minimizing ascents in a given input.

THEOREM 8.6
(Najock and Heyde, 1982). Let L_n be he number of leaves in a random recursive tree of size n. This random variable has the exact distribution

$$\mathbf{P}(L_n = k) = \frac{1}{(n-1)!} \left\langle {n-1 \atop k} \right\rangle,$$

where $\langle {n \atop k} \rangle$ is the Eulerian number of the first kind, for $k = 1, \dots, n-1$. The number of leaves also exhibits Gaussian tendency:

$$\frac{L_n - \frac{1}{2}n}{\sqrt{n}} = \mathcal{N}\left(0, \frac{1}{12}\right).$$

PROOF Color the leaves of a recursive tree with white, the rest of the nodes with blue. When a white leaf recruits, it is turned into an internal blue node and acquires a new node as a white child. When a blue (internal) node recruits, it remains internal but attracts a white leaf as a child. This is a Friedman's urn with the schema

$$\begin{pmatrix} 0 & 1 \\ 1 & 0 \end{pmatrix},$$

starting out with one white ball. If W_n is the number of white balls after n draws, then $L_n = W_{n-1}$.

With $\alpha = -1, \delta = -1$, and $\tau_n = n+1$, the functional Eq. (3.7) is specialized here to

$$\psi_{n+1}(t) = \frac{\psi_n'(t)}{n+1} = \frac{\psi_{n-1}''(t)}{(n+1)n} = \cdots = \frac{1}{(n+1)!} \times \frac{d^{n+1}}{dt^{n+1}} \psi_0(t).$$

Reverting to the moment generating function $\phi_n(t)$, cf. Eq. (3.6), with the boundary condition $\phi_0(t) = e^t$, we have the solution

$$\phi_n(t) = \frac{1}{n!}(1-e^t)^{n+1} \frac{d^n}{dt^n}\left(\frac{e^t}{1-e^t}\right).$$

We shall show by induction[2] that

$$\frac{d^n}{dt^n}\left(\frac{e^t}{1-e^t}\right) = \frac{1}{(1-e^t)^{n+1}} \sum_{k=1}^{n} \left\langle {n \atop k} \right\rangle e^{kt}. \tag{8.2}$$

The basis at $n = 1$ follows from the boundary conditions. Assuming the assertion to be true for n, we proceed with

$$\begin{aligned}
\frac{d^{n+1}}{dt^{n+1}}\left(\frac{e^t}{1-e^t}\right) &= \frac{d}{dt}\left(\frac{1}{(1-e^t)^{n+1}} \sum_{k=1}^{n} \left\langle {n \atop k} \right\rangle e^{kt}\right) \\
&= \frac{1}{(1-e^t)^{n+2}} \sum_{k=1}^{n} \left(k\left\langle {n \atop k} \right\rangle (1-e^t)e^{kt} + (n+1)\left\langle {n \atop k} \right\rangle e^{(k+1)t}\right) \\
&= \frac{1}{(1-e^t)^{n+2}} \left(\sum_{k=1}^{n} k\left\langle {n \atop k} \right\rangle e^{kt} + \sum_{k=1}^{n} (n+1-k)\left\langle {n \atop k} \right\rangle e^{(k+1)t}\right)
\end{aligned}$$

---

[2] The author thinks this result must be well known, but could not find an explicit reference to it in the literature.

$$= \frac{1}{(1-e^t)^{n+2}} \left(\left\langle {n \atop 1} \right\rangle e^t + \sum_{k=2}^{n} \left(k \left\langle {n \atop k} \right\rangle + (n-k+2) \left\langle {n \atop k-1} \right\rangle \right) e^{kt} + \left\langle {n \atop n} \right\rangle e^{(n+1)t} \right)$$

$$= \frac{1}{(1-e^t)^{n+2}} \left(\left\langle {n+1 \atop 1} \right\rangle e^t + \sum_{k=2}^{n} \left\langle {n+1 \atop k} \right\rangle e^{kt} + \left\langle {n+1 \atop n+1} \right\rangle e^{(n+1)t} \right);$$

the last transition was accomplished via the recurrence Eq. (7.4) and the boundary conditions $\langle {r \atop 1} \rangle = \langle {r \atop r} \rangle = 1$. The induction is complete.

According to the relation Eq. (8.2), the probability generating function for the leaves is

$$\sum_{k=1}^{n-1} \mathbf{P}(L_n = k)\, t^k = \sum_{k=1}^{n-1} \mathbf{P}(W_{n-1} = k)\, t^k = \phi_{n-1}(\ln t) = \frac{1}{(n-1)!} \sum_{k=1}^{n-1} \left\langle {n-1 \atop k} \right\rangle t^k.$$

The central limit tendency is an immediate application of Theorem 3.4. ■

Gastwirth and Bhattacharya (1984) addresses a variation regarding the profit in a chain letter scheme. In some chain letter schemes the success of the recruits of a letter holder adds to his own profit. A letter holder gets a commission whenever anyone in his own subtree recruits. A measure of this success is the size (number of nodes) of the subtree rooted at k.

THEOREM 8.7

(Gastwirth and Bhattacharya, 1984). Let $S_{k,n}$ be the size of the subtree rooted at the kth entrant to a recursive tree. As both k and n increase to infinity in such a way that $k/n \to \rho$, the distribution of this size approaches a geometric law:

$$S_{k,n} \xrightarrow{\mathcal{D}} \text{Geo}(\rho).$$

PROOF Right after the kth node joins the tree, color the entrant k with white, all else ($k-1$ nodes) with blue. Whenever a node recruits, paint its child with the same color. The Pólya-Eggenberger urn

$$\begin{pmatrix} 1 & 0 \\ 0 & 1 \end{pmatrix}$$

underlies the growth after the tree size has reached k; the urn starts out with the initial conditions $W_0 = 1$ and $B_0 = k-1$. So, $S_{k,n} = W_{n-k}$, which has the exact Pólya-Eggenberger distribution. Let $\tilde{W}_n$ be the number of white ball drawings in n draws (i.e., $W_n = \tilde{W}_n + 1$). Write the distribution of W_{n-k}

(see Theorem 3.1) for n large enough as

$$\mathbf{P}(\tilde{W}_{n-k} = j) = \frac{j! \times (k-1)k \dots (n-j-2)}{k(k+1)\dots(n-1)} \binom{n-k}{j}$$
$$= \frac{(k-1)\,\Gamma(n-j-1)\,\Gamma(n-k+1)}{\Gamma(n)\,\Gamma(n-k-j+1)}.$$

Passage to the limit, via Stirling's approximation to the Gamma function, as $k, n \to \infty$ in such a way that $k/n \to \rho$, gives

$$\mathbf{P}(S_{k,n} - 1 = j) \sim \frac{k}{n^{j+1}}(n-k)^j \to (1-\rho)^j \rho. \quad \blacksquare$$

Meir and Moon (1988) brought the notion of the number of nodes of a certain outdegree in a recursive tree into the spotlight. They use an analytic method for simply-generated trees and related families. They calculated the means of the number of nodes of outdegree 0 (the leaves), 1, and 2, and came up with a partial computation of the variance-covariance matrix. Mahmoud and Smythe (1992) approached the subject via urn modeling and obtained a joint central limit. Janson (2005b) extends the result to the joint distribution of nodes of outdegree up to arbitrary k (fixed).

THEOREM 8.8

(Janson, 2005b). Let $X_n^{(j)}$ be the number of nodes of outdegree j (for $j = 0, 1, \dots, k$) in a random recursive tree. The vector $\mathbf{X}_n = (X_n^{(1)}, X_n^{(2)}, \dots X_n^{(k)})^T$ converges in distribution to a multivariate normal vector:

$$\frac{1}{\sqrt{n}}(\mathbf{X}_n - n\boldsymbol{\mu}_k) \xrightarrow{\mathcal{D}} \mathcal{N}_k(\mathbf{0}_k, \Sigma_\mathbf{k}).$$

where $\boldsymbol{\mu}_k = (\frac{1}{2}, \frac{1}{4}, \dots, \frac{1}{2^{k+1}})^T$, $\mathbf{0}_k$ is a vector of k zeros, and the covariance matrix is given by coefficients of a generating function.

PROOF Suppose the leaves (nodes of outdegree 0) are colored with color 1, nodes of outdegree 1 are colored with color 2, and so on, so that nodes of outdegree k are colored with color $k+1$. All nodes of higher outdegrees are colored with the color $k+2$. By the same combinatorial arguments used before to connect the tree to an urn, one reasons that when a node of outdegree j recruits, its ball in the urn is removed and replaced with a ball of the next color up as if its outdegree is grown by 1, also a leaf (ball of color 1) is added. The exceptions are nodes of color 1 and nodes of high outdegree (larger than k). When nodes of color 1 recruit, we add a node of color 2, and when nodes of outdegree higher than k recruit, only a leaf (ball of color 1) is added:

$$\begin{pmatrix} 0 & 1 & 0 & 0 & \dots & 0 \\ 1 & -1 & 1 & 0 & \dots & 0 \\ 1 & 0 & -1 & 1 & \dots & 0 \\ \vdots & \vdots & \vdots & & \dots & 0 \\ & & & \ddots & & \\ 1 & 0 & 0 & \dots & 0 & 0 \end{pmatrix}.$$

This is an extended urn scheme, so the rest of the proof follows from Theorem 6.5 and the required eigenvalue analysis. ■

Janson (2005b) gives a formula for the variance-covariance matrix that involves the extraction of coefficients from a multivariate generating function.

There is interest also in the structure of the branches or the subtrees rooted at node k. While the size of the subtree rooted at k measures the success of the kth entrant in a chain letter scheme, the number of leaves is a measure of his legal liability. The leaves in the subtree are unsuccessful participants in the scheme, who purchased the letter but were not able to sell it. These leaves represent the potential number of complaints against the kth participant.

THEOREM 8.9

(Mahmoud and Smythe, 1991). Let $L_n^{(k)}$ be the number of leaves in the subtree rooted at k in a random recursive tree. For k fixed,

$$\frac{L_n^{(k)}}{n} \xrightarrow{\mathcal{D}} \frac{1}{2}\beta(1, k-1),$$

as $n \to \infty$.

PROOF A color code renders the colored nodes balls in a Pólya-Eggenberger urn. As in the proof of Theorem 8.7, we wait until the kth entrant joins the scheme then color it with white, and color all the preceding nodes with blue. After $n-k$ additional draws, the size $S_{k,n}$ of the kth branch (the subtree rooted at k) is W_{n-k}, the number of white balls in the urn after $n-k$ draws. According to Theorem 3.2, the size of the branch follows the convergence

$$\frac{S_{k,n}}{n-k} \xrightarrow{\mathcal{D}} \beta(1, k-1).$$

The number of leaves L_n in a recursive tree is about half its size (refer to Theorem 8.6). Formally, in the kth branch there are $L_n^{(k)} = L_{S_{k,n}}$ leaves, and as $S_{k,n} \xrightarrow{a.s.} \infty$,

$$\frac{L_{S_{k,n}}}{S_{k,n}} \xrightarrow{P} \frac{1}{2}.$$

By Slutsky's theorem, the multiplication of the last two relations together with $(n-k)/n \to 1$ yields the result. ■

It is worth noting that in the original proof of Mahmoud and Smythe (1991), the (scaled) difference between the number of white and blue balls is shown to be a mixture of normals, the mixing density is that of $\beta(1, k-1)$.

8.2.2 Pyramids

There is no restriction on the outdegree of the nodes of a recursive tree. Some chain letter schemes put a restriction on the number of copies a letter holder can sell. If a letter holder wishes to continue after he makes a certain number of sales, he must reenter (purchase a new copy). A reentry is a new participant (a new node in the tree). We illustrate the urn and associated type of result in the binary case. The tree modeling in this scheme is a *pyramid*, a recursive tree where outdegrees are restricted to at most 2. A binary pyramid grows out of a root. When a node attracts its second child, it is saturated, and not allowed to recruit. The unsaturated nodes are equally likely recruiters. Only the first five trees from the left in Figure 8.9 are binary pyramids.

Color the leaves with white, the rest of the unsaturated nodes with blue. Leave the saturated nodes uncolored. When a white leaf recruits, it remains unsaturated (but turns blue as it is now of outdegree 1), and acquires a leaf under it; when a blue node recruits, it becomes saturated (colorless), but also acquires a leaf under it. From the point of view of the urn, colorless balls are tossed out. The schema is

$$\begin{pmatrix} 0 & 1 \\ 1 & -1 \end{pmatrix}.$$

The eigenvalues of this matrix are the golden ratio ϕ and its conjugate ϕ^* which are namely

$$\phi := \frac{\sqrt{5}-1}{2}, \qquad \phi^* := -\frac{\sqrt{5}+1}{2}.$$

Let W_n be the number of white balls in the urn after n draws. Then $L_n = W_{n-1}$. By the almost sure variants of Theorem 6.2 that were referred to at the end of Chapter 6, we recover the following strong law.

THEOREM 8.10

(Gastwirth and Bhattacharya, 1984). Let L_n be the number of leaves in a binary pyramid of size n. Then, as $n \to \infty$,

$$\frac{L_n}{n} \xrightarrow{a.s.} \frac{1}{2}(3-\sqrt{5}).$$

The original proof of Theorem 8.10 is established via diffusion theory, where additionally a central limit theorem is obtained.

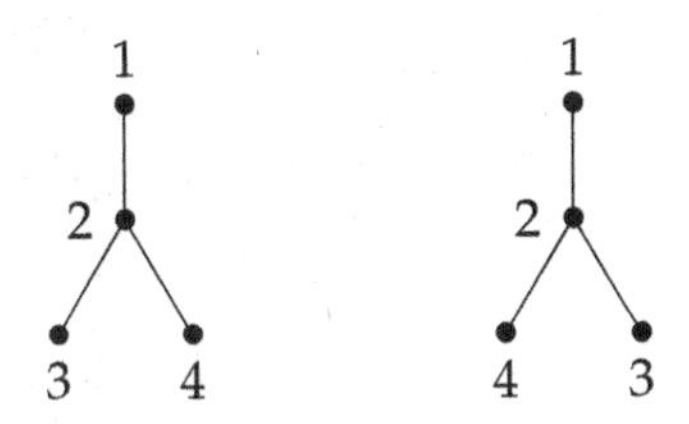

FIGURE 8.10
Two different plane-oriented trees.

8.2.3 Plane-Oriented Recursive Trees

Orientation in the plane was not taken into account in the definition of recursive trees. The two labeled trees in Figure 8.10 are only two drawings of the *same* recursive tree. If different orientations are taken to represent different structures, we arrive at a definition of a *plane-oriented recursive tree.* In such a tree, if a node has outdegree Δ; there are Δ children under it, with $\Delta + 1$ "gaps." The leftmost and rightmost insertion positions are gaps, too. If these gaps are represented by external nodes, we obtain an extended plane-oriented recursive tree. Figure 8.11 shows one of the plane-oriented recursive trees of Figure 8.10 after it has been extended; the external nodes are shown as colored squares in Figure 8.11. The blank squares correspond to the white external nodes and the darkened squares correspond to the blue external nodes in a white-blue coloring scheme that will be discussed shortly.

The plane-oriented recursive tree grows by choosing one of the gaps at random, all gaps in the tree being equally likely. This uniform distribution on gaps gives rise to a uniform distribution on plane-oriented recursive trees.

Mahmoud, Smythe, and Szymański (1993) characterized the exact and limits distribution of the number of leaves in a random plane-oriented recursive tree. The exact distribution involves Eulerian numbers of the *second* kind—contrasted with the exact distribution of the number of leaves in standard recursive trees, which involves Eulerian numbers of the first kind.

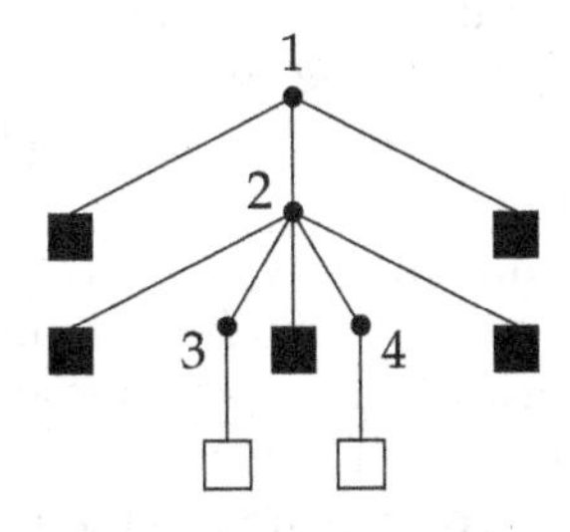

FIGURE 8.11
An extended plane-oriented recursive tree.

THEOREM 8.11

(Mahmoud, Smythe, and Szymański, 1993). In a plane-oriented recursive tree of size n, the number of leaves, L_n, exhibits the Gaussian tendency

$$\frac{L_n - \frac{2}{3}n}{\sqrt{n}} \xrightarrow{\mathcal{D}} \mathcal{N}\left(0, \frac{1}{9}\right).$$

PROOF The equally likely objects in this model are the insertion positions or gaps (external nodes in the extended tree). Color each gap underneath a leaf with white, the rest of the gaps with blue. Figure 8.11 gives an instance of this coloring. When a leaf recruits (insertion hits a white gap), the leaf is turned into an internal node and acquires a new leaf as a child, with a white external node under it. The white gap that got hit is eliminated, but two blue gaps appear underneath the old leaf, as right and left siblings of the new leaf. When insertion hits a blue gap, it turns into a leaf, with one white gap under it; two blue gaps appear as siblings of the new leaf (net gain of only one blue gap). This is an urn process with the Bagchi-Pal schema

$$\begin{pmatrix} 0 & 2 \\ 1 & 1 \end{pmatrix}.$$

The central limit tendency is an immediate application of Theorem 3.5. ■

8.2.4 Bucket Recursive Trees

Bucket recursive trees were introduced in Mahmoud and Smythe (1995) to generalize the ordinary recursive tree into one with nodes carrying multiple labels.

Every node in a bucket recursive tree has capacity b. The tree grows by adding labels (recruiting agents) in $\{1, \dots, n\}$; at the jth step j is inserted. The affinities of all the recruiters are equal. The agents work in offices of capacity b. The first agent goes into the root and recruits a second agent for the same office, then the two compete with equal chance to recruit a third, and so on. Up to b recruiters go into the root, filling it to full capacity. The b agents of the root office compete with equal chance to recruit the $(b+1)$st agent, who starts a new subordinate office (node attached as a child of the root). For each new entrant, the existing members of the tree compete with equal probabilities (adaptive in time) to recruit the entrant. If an agent belonging to an unfilled office succeeds in attracting the new entrant, the entrant joins the recruiter's office. If the recruiter's office is full, the new entrant starts a new subordinate office attached as a child of the recruiter's office. Figure 8.12 shows a bucket recursive tree with bucket capacity 2, after 8 recruits join the system.

After n agents are in the system, the number of offices is S_n, the size of the tree. An urn of balls of b colors corresponds to the recruiting process. The affinity of an office with i agents in it is i and is represented in the urn by i balls of color i. When an office of capacity $i < b$ recruits, the urn loses i balls of color i and gains $i + 1$ balls of color $i + 1$. When someone in a full office

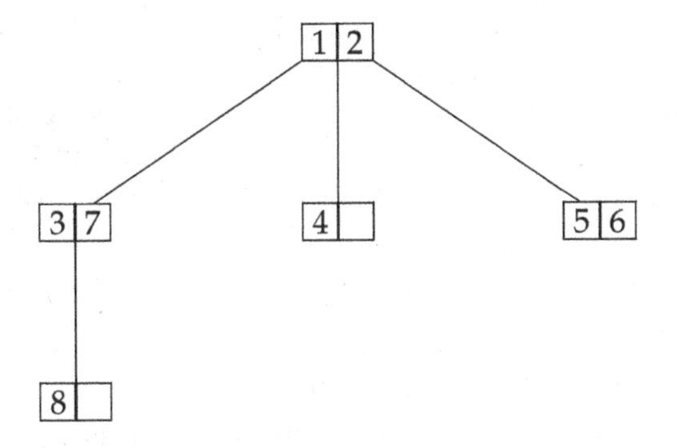

FIGURE 8.12
A bucket recursive tree with bucket capacity 2.

recruits, we only add a ball of color 1 to the urn. The schema is

$$\mathbf{A} = \begin{pmatrix} -1 & 2 & 0 & 0 & 0 & \dots & 0 & 0 \\ 0 & -2 & 3 & 0 & 0 & \dots & 0 & 0 \\ 0 & 0 & -3 & 4 & 0 & \dots & 0 & 0 \\ \vdots & \vdots & & & \ddots & & \vdots & \vdots \\ 0 & 0 & & & \dots & & -(b-1) & b \\ 1 & 0 & & & \dots & & 0 & 0 \end{pmatrix}.$$

The matrix has a banded structure similar to that of the m–ary search tree. The eigenanalysis needed for the convergence theorems is similar, and we only highlight the main points (see the discussion in Subsection 8.1.3). The characteristic equation $|\mathbf{A}^T - \lambda\mathbf{I}| = 0$ expands into

$$(\lambda+1)(\lambda+2)\dots(\lambda+b) = b!,$$

by expanding via the first row of $\mathbf{A}^T - \lambda\mathbf{I}$, for example. Up to $b = 26$, this is an extended urn scheme, the eigenvalues of which have the property that $\Re\,\lambda_2 < \frac{1}{2}\lambda_1 = \frac{1}{2}$. For $2 \le b \le 26$, one obtains a multivariate normal law among the number of balls of different kinds. One obtains the average size of the tree from the profile of colors.

THEOREM 8.12

(Mahmoud and Smythe, 1995). Let S_n be the size of a bucket recursive tree after n insertions. If the node capacity b satisfies $2 \le b \le 26$, then

$$\frac{S_n - n/H_b}{\sqrt{n}} \xrightarrow{\mathcal{D}} \mathcal{N}(0, \sigma_b^2),$$

where σ_b^2 is an effectively computable constant.

The limiting variance σ_b^2 requires some rather lengthy computation. If $b \geq 27$, the associated urn no longer satisfies the sufficient condition for asymptotic normality in an extended urn in the sense of Smythe. It is argued in Mahmoud and Smythe (1995) that $(S_n - n/H_b)/\sqrt{n}$ does not converge to a normal limit, if $b \geq 27$. It is shown instead that under a different scale, a nonnormal limit exists.

8.2.5 Sprouts

A *random recursive sprout* is a random tree that grows just like the ordinary recursive tree, except that the recruiter attracts a random number of nodes instead of just 1. If the recruiter is internal, $Y \geq 1$ leaves are adjoined to it; and if a leaf is chosen as parent, $X \geq 1$ leaves are attached to it as children, and the parent leaf turns into an internal node. The random variables X and Y are of course discrete on positive integers, and may generally be different. The standard recursive tree is a case where deterministically $X \equiv Y \equiv 1$.

The random sprout extends the domain of application of recursive trees. For example, it may model a chain letter scheme, where the company starts out with one recruiter who seeks buyers for a copy of a letter he is holding. At some point the recruiter will sell a copy of his letter to X persons whom he may meet socially. The initial founder of the company remains in contention and all $X + 1$ letter holders go competing for the next batch of buyers with equal chance. The chain letter scheme propagates in society, however, whenever someone who sold copies of the letter before finds buyers he sells according to the distribution Y (presumably stochastically larger than X). The model takes experience into account. Those who have the experience of selling before can generally sell more than new inexperienced letter holders.

The definition of a sprout embedded in real time derives its motivation from the growth of the Internet, and similar types of virtual networks. In reality, nodes are added into the net at random times, and not necessarily at discrete time ticks. Also, the growth of a network like the Internet is accelerating as society has increasing demand for on-line services and nations are getting into the era of e-government. Sprouts provide a realistic and flexible model that adds these features. It also adds flexibility in the tree growth to allow several simultaneous entries, instead of adding one node at a time. For example, a new computer center may be established in a university, ministry, or hospital, requesting for efficiency in all its departments to have direct connections to the Internet. Managers of new sites may generally commit a large number of connections to well-established computer centers deeming them more reliable. We know from poissonization that urn scheme (and hence the trees they model) embedded in real time have exponential average rate of growth, which might be a good reflection of the reality of the Internet in which we witness very rapid acceleration of increase in size. Discrete trees always remain linear or sublinear in size with respect to discrete time.

A *random sprout embedded in continuous time* is a random tree that grows in real time as follows. Initially, the sprout starts out as a root node. After a random period of time t_1, a random number X of leaves is added as children adjoined to the root, which is turned into an *internal* node. The growth process continues by having a batch of nodes joining at time t_2, by choosing a node at random from the existing structure as a parent for the entire batch of children. If the node chosen is internal, $Y \ge 1$ leaves are adjoined to it; and if a leaf is chosen as parent, $X \ge 1$ leaves are attached to it as children, and the parent leaf turns into an internal node. The process is perpetuated with the joining of additional batches of random size at times $t_3, t_4, \dots$, etc., always choosing a parent at random from the tree existing at any of these renewal times, and always adding an independent realization of X, if the chosen parent is a leaf, and an independent realization of Y, if the chosen parent is an internal node. The random variables X and Y are of course discrete on positive integers. However, the two random variables may generally have different distributions. The standard recursive tree is a case where deterministically $X \equiv Y \equiv 1$, and the interarrival times are $t_k - t_{k-1} \equiv 1$.

Suppose the leaves of the sprout are colored white and the internal nodes are colored blue, and think of these nodes as balls in an urn. The progress of the tree corresponds to the growth of an urn with the generator

$$\mathbf{A} = \begin{pmatrix} X-1 & 1 \\ Y & 0 \end{pmatrix}.$$

For standard embedding with independent Exp(1) interarrival times, let $S(t)$ be the size of the sprout, $W(t)$ be the number of leaves (white balls) and $B(t)$ be the number of internal nodes (blue balls) at time t. Let $\mu_X := \mathbf{E}[X]$, and $\mu_Y := \mathbf{E}[Y]$. Here, $(W(0), B(0)) = (1, 0)$, and the transposed average generator is

$$\mathbf{B} = \begin{pmatrix} \mu_X - 1 & \mu_Y \\ 1 & 0 \end{pmatrix},$$

with eigenvalues

$$\lambda_{1,2} = \frac{(\mu_X - 1) \pm \sqrt{(\mu_X - 1)^2 + 4\mu_Y}}{2}.$$

As both $\mu_X \ge 1$ and $\mu_Y \ge 1$, the two eigenvalues are real and distinct, and λ_2 is negative. We represent the exponential matrix in terms of idempotents:

$$e^{\mathbf{B}t} = \mathcal{E}_1 e^{\lambda_1 t} + \mathcal{E}_2 e^{\lambda_2 t},$$

and we have

$$\begin{aligned} \mathcal{E}_1 &= \frac{1}{\lambda_1 - \lambda_2}(\mathbf{B} - \lambda_2 \mathbf{I}) \\ &= \frac{1}{\lambda_1 - \lambda_2} \begin{pmatrix} \mu_X - 1 - \lambda_2 & \mu_Y \\ 1 & -\lambda_2 \end{pmatrix}. \end{aligned}$$

It follows that

$$\mathcal{E}_1 \begin{pmatrix} W(0) \\ B(0) \end{pmatrix} = \frac{1}{\lambda_1 - \lambda_2} \begin{pmatrix} \mu_X - 1 - \lambda_2 \\ 1 \end{pmatrix} \neq \mathbf{0};$$

similarly

$$\mathcal{E}_2 \begin{pmatrix} W(0) \\ B(0) \end{pmatrix} = \frac{1}{\lambda_2 - \lambda_1} \begin{pmatrix} \mu_X - 1 - \lambda_1 \\ 1 \end{pmatrix}.$$

By Theorem 4.1, we have

$$\begin{pmatrix} \mathbf{E}[W(t)] \\ \mathbf{E}[B(t)] \end{pmatrix} = \frac{1}{\lambda_1 - \lambda_2} \begin{pmatrix} \mu_X - 1 - \lambda_2 \\ 1 \end{pmatrix} e^{\lambda_1 t} + \frac{1}{\lambda_2 - \lambda_1} \begin{pmatrix} \mu_X - 1 - \lambda_1 \\ 1 \end{pmatrix} e^{\lambda_2 t}.$$

Adding up the two components, we arrive at the average size of a random sprout

$$\mathbf{E}[S(t)] = \frac{\mu_X - \lambda_2}{\lambda_1 - \lambda_2} e^{\lambda_1 t} + \frac{\mu_X - \lambda_1}{\lambda_2 - \lambda_1} e^{\lambda_2 t}.$$

For instance, if $X = 1 + \text{Poi}(2)$, $Y = 1 + \text{Poi}(7)$, the schema of the sprout is

$$\mathbf{A} = \begin{pmatrix} \text{Poi}(2) & 1 \\ 1 + \text{Poi}(7) & 0 \end{pmatrix},$$

and we have

$$\begin{pmatrix} \mathbf{E}[W(t)] \\ \mathbf{E}[B(t)] \end{pmatrix} = \frac{1}{6} \begin{pmatrix} 4 \\ 1 \end{pmatrix} e^{4t} + \frac{1}{6} \begin{pmatrix} 2 \\ -1 \end{pmatrix} e^{-2t}.$$

At time t, there will be an average of about $\frac{4}{6}e^{4t}$ leaves, and $\frac{1}{6}e^{4t}$ internal nodes, and the size of the tree is about $\frac{5}{6}e^{4t}$ on average.

Exercises

8.1 (Quintas and Szymański, 1992). The nonuniform recursive tree with bounded degrees was introduced as a models for the growth of certain chemical compounds. The binary model introduced in Quintas and Szymański (1992) coincides with binary search trees. Let us consider a ternary model. A large polymer grows in the form of an unrooted ternary tree. The nodes attract the next molecule according to weights or affinities that are determined by their degrees. A node of degree i has affinity proportional to $4 - i$. Thus, nodes of degree 4 are saturated and a new molecule entering the polymer can only be adjoined to nodes of degree 1, 2, or 3. We can consider the polymer to grow out of a tree of size 2. In this model nodes with low degree are "hungry" for chemical bonding and are more likely to attract the next molecule, and nodes of higher degree are less hungry. What is the asymptotic proportion of molecules that are hungriest (leaves) in this tree (in probability)?

8.2 A frugal binary search tree for real numbers (reside in four bytes) gets rid of all nill pointers to empty subtrees at the fringe. This is accomplished by having two types of nodes: Internal with at least one nonempty subtree and leaves, and the economy is in the latter type. Each key is stored in four bytes, and each pointer is also stored in four bytes. A leaf carries only one key and contains no pointers. An internal node carries one key and two pointers. Derive a weak law for the physical size (in bytes) of such a tree.

8.3 A binary search tree is embedded in real time. The keys, elements of a random permutation of $\{1, \ldots, n\}$, appear at times with interarrival periods that are independent of each other, and each is distributed like Exp(1), the exponential random variable with mean 1. When a key appears, it is placed in the tree in accordance with the binary search trees insertion algorithm. Let $S(t)$ be the virtual size of the tree by time t (the number of nodes in it).

(a) What is the distribution of $S(t)$?

(b) Show that, as $t \to \infty$,

$$\frac{S(t)}{e^t} \xrightarrow{\mathcal{D}} \text{Exp}(1).$$

8.4 A recursive tree grows in real time t. At interarrival times that are independent of each other and each is distributed like Exp(1) nodes appear; when a new node appears, it chooses a parent in the existing tree at random. What is the asymptotic average number of leaves in the recursive tree at time t?

8.5 A binary pyramid grows in real time t. At interarrival times that are independent and each is distributed like Exp(1) nodes appear; when a new node appears, it chooses as parent an unsaturated node in the existing pyramid at random. What is the average number of leaves in the pyramid at time t?

9

Urn Schemes in Bioscience

In this chapter we discuss some Pólya urns and some new Pólya-like schemes that find uses and applications in various biosciences, such as evolution of species and health science. Hoppe's urn scheme captures the essence of the popular Wright-Fisher model for a Mendelian theory of evolution, where big changes occur abruptly via mutation, rather than gradually as in the Darwinian theory. Pólya urn models like the standard ones we encountered in previous chapters, and variations thereupon, find sound applications in topics of bio and health sciences such as epidemiology, ecology, and clinical trials.

9.1 Evolution of Species

The pioneering work of Ewens in mathematical models for population genetics opened the door for new Pólya-like urns. The work led to a remarkable formula known as Ewens' sampling formula. In the standard Pólya urn schemes there is a *fixed* set of admissible colors that can appear in the urn. Models for evolution of species represent animals from a given species by balls of the same color. In view of evolution, a new species (new color) may appear at some point. In such a model we must allow the number of colors to increase with time, and consider an *infinite* set of admissible colors. The models are also deemed as a sensible representation for genetic alleles, where once in a while a new allelic type (new color) appears as a result of mutation.

Ewens arrived at his sampling formula via a heuristic argument, which itself can be presented by urn models. The ultimate mixing is a result of continual evolution, and after a long time stationary models come into the picture. We shall discuss the genetic urn models leading to Ewens' sampling formula, which in itself is captured by Hoppe's urn scheme.

9.1.1 Wright-Fisher Allelic Urn Models

In a non-Darwinian model there is no change in the genes on average. The hereditary process can be represented by balls in an urn. The balls undergo a rebirth process and produce balls in a subsequent urn. Let us start with two colors, representing say two genes. Suppose there are m balls in an urn, of which i are white (of certain genetic characteristics) and $m - i$ are blue (of different genetic characteristics). Balls are sampled with replacement, m times, to give a chance for each ball to appear once on average. If a sampled ball is white, we deposit a white ball in a new urn, and if the sampled ball is blue we put a blue ball in the new urn. The number of white balls that appear in the size-m sample is $\text{Bin}(m, \frac{i}{m})$. This is the number of white balls we deposit in the new urn. On average, there are $m \times \frac{i}{m} = i$ white balls in the new urn, and on average there is no change in the proportion of the white genes. These models appeared in the work of Fisher as far back as 1922, and later in the work of Wright in the 1930s. In modern literature they are referred to as *simple Wright-Fisher models*. The one we outlined is one of the simplest genetic models for a fixed population of alleles (of two types) without selection, mutation, or sexes.

In a simple Wright-Fisher model, after a new urn is filled, the process is repeated to produce a third urn, and so forth. It is clear that there is a positive probability that a new urn becomes monochromatic. Once this uniformity is attained, the urns remain the same for ever. This can be seen from a Markov model. We consider a very small example.

Example 9.1

Suppose we have one white ball and one blue in the initial urn. We represent the evolution by the states of a Markov chain with three states, where a state is described by the number of white balls in it. We choose to start in State 1. From State 1, we can move into state 0, if no white balls appear in the sample, remain in State 1 if only one white ball appears in the sample, or progress to State 2, if two white balls appear in the sample. The corresponding probabilities are $\frac{1}{4}$, $\frac{1}{2}$, and $\frac{1}{4}$. These numbers are $\mathbf{P}(X = k)$, for $k = 0, 1, 2$, for a binomial random variable X distributed like $\text{Bin}(2, \frac{1}{2})$. These numbers constitute row 1 of the Markov chain transition matrix. Likewise, we can compute the numbers across rows 0 and 2, and obtain the full transition matrix

$$\mathbf{M} = \begin{pmatrix} 1 & 0 & 0 \\ \frac{1}{4} & \frac{1}{2} & \frac{1}{4} \\ 0 & 0 & 1 \end{pmatrix}.$$

It is clear from a straightforward induction that the n–step transition matrix is

$$\mathbf{A}^n = \begin{pmatrix} 1 & 0 & 0 \\ \frac{2^n-1}{2^{n+1}} & \frac{2}{2^{n+1}} & \frac{2^n-1}{2^{n+1}} \\ 0 & 0 & 1 \end{pmatrix}.$$

It follows that after n steps the probability state vector is

$$\begin{aligned}\pi_n &= (0 \quad 1 \quad 0)\mathbf{A}^n \\ &= \begin{pmatrix} \dfrac{2^n - 1}{2^{n+1}} & \dfrac{2}{2^{n+1}} & \dfrac{2^n - 1}{2^{n+1}} \end{pmatrix} \\ &= \begin{pmatrix} \dfrac{1}{2} & 0 & \dfrac{1}{2} \end{pmatrix} + \begin{pmatrix} -\dfrac{1}{2^{n+1}} & \dfrac{1}{2^n} & -\dfrac{1}{2^{n+1}} \end{pmatrix},\end{aligned}$$

and we see that this state vector converges componentwise to $(\frac{1}{2} \quad 0 \quad \frac{1}{2})$. The uniformity comes rapidly at an exponential rate. Rather quickly the urn is either in a state of all white or all blue. Ultimately, there is zero probability of a split, and probability 1 of total uniformity. There is probability $\frac{1}{2}$ of an ultimate all-white urn, and there is probability $\frac{1}{2}$ of an ultimate all-blue urn.

It is also plain to show that $(\, p \quad 0 \quad 1 - p\,)$, for any $p \in [0, 1]$ is a stationary distribution, as this row vector is a solution to $\boldsymbol{\pi}\mathbf{M} = \boldsymbol{\pi}$, for $\boldsymbol{\pi} = (\,\pi_0 \quad \pi_1 \quad \pi_2\,)$, with $\pi_0 + \pi_1 + \pi_2 = 1$. ◇

Example 9.1 can be easily generalized to m total number of balls of which i are white, and we almost surely get an ultimate monochromatic urn, at a rapid exponential rate. The ultimate probability of an all white urn is i/m, the initial proportion of white balls in the urn, and the $(m + 1)$-component row vector $(\, p \quad 0 \quad 0 \quad \dots \quad 0 \quad 0 \quad 1 - p\,)$ is a stationary distribution.

9.1.2 Gene Miscopying

In the simple Wright-Fisher model of Subsection 9.1.1, m balls of two colors exist in an urn and m balls are sampled with replacement. In the meanwhile, a second urn is filled with an exact copy of the sample. Whenever a ball appears in the sample, a ball of the same color is placed in the new urn. When the new urn is filled the process is repeated to fill yet another urn. The balls represent genes. In reality copying gene properties is not a perfect operation and there are chances of miscopying. In a simplified model, we allow color miscopying. When a ball is sampled there is a small chance of "crossing over" to the other color. When a white ball appears in the sample, we place a white ball in the new urn with probability $1 - \alpha$, otherwise we place a blue ball in the new urn with probability α (small). And vice versa, When a blue ball appears in the sample, we place a blue ball in the new urn with probability $1 - \beta$, otherwise we place a white ball in the new urn with probability β (small). The operation is then repeated with future generations of newer urns.

Let us revisit Example 9.1 and admit miscopying.

Example 9.2

We have one white ball and one blue in the initial urn and the process of creating new urns is perpetuated. The underlying Markov chain has three states labeled by the number of white balls in the urn, there are no white balls. The sample (with replacement) will show two blue balls. They can both give blue children (with probability $(1-\beta)^2$), both children can flip color into white (with probability β^2), or one gives a white child and one gives blue (with probability $2\beta(1-\beta)$). By symmetry, if we are in State 2, we have two white balls and we can move into states 0, 1, 2 with probabilities α^2, $2\alpha(1-\alpha)$, $(1-\alpha)^2$, respectively. In State 1, we have one white and one blue ball. We can move into a state with two blue (zero white) balls in four different ways, depending on the sequence of colors in the sample, and on whether they are miscopied or not. We represent the white and blue colors in the sample with W and B, and the miscopying operation with Yes (Y) or No (N), and observe that we can move from State 1 into State 0, if we have one of the four events

$$WWYY, \qquad WBYN, \qquad BWNY, \qquad BBNN,$$

which, respectively, occur with probability

$$\frac{1}{4}\alpha^2, \qquad \frac{1}{4}\alpha(1-\beta), \qquad \frac{1}{4}(1-\beta)\alpha, \qquad \frac{1}{4}(1-\beta)^2.$$

The sum of these four probabilities is the entry $M_{1,0}$ in the transition matrix $\mathbf{M}$. The entry $M_{1,2}$ is symmetrical to $M_{1,0}$, and the middle entry $M_{1,1}$ is $1 - M_{1,0} - M_{1,2}$. The complete transition matrix is

$$\mathbf{M} = \begin{pmatrix} (1-\beta)^2 & 2\beta(1-\beta) & \beta^2 \\ \frac{1}{4}(1+\alpha-\beta)^2 & \frac{1}{2}-\frac{1}{2}(\alpha-\beta)^2 & \frac{1}{4}(1-\alpha+\beta)^2 \\ \alpha^2 & 2\alpha(1-\alpha) & (1-\alpha)^2 \end{pmatrix}.$$

The stationary distribution state vector is the solution of $\boldsymbol{\pi}\mathbf{M} = \boldsymbol{\pi}$, with components adding up to 1. Let $\mathcal{W}$ be a random variable with the stationary distribution. Then from the solution we find

$$\mathbf{P}(\mathcal{W}=0) = \frac{\alpha((\alpha-\beta)(\alpha+\beta-2)-1)}{(\alpha+\beta)((\alpha+\beta)(\alpha+\beta-2)-1)},$$

$$\mathbf{P}(\mathcal{W}=1) = \frac{\beta((\beta-\alpha)(\alpha+\beta-2)-1)}{(\alpha+\beta)((\alpha+\beta)(\alpha+\beta-2)-1)},$$

$$\mathbf{P}(\mathcal{W}=2) = \frac{4\alpha\beta(\alpha+\beta-2)}{(\alpha+\beta)((\alpha+\beta)(\alpha+\beta-2)-1)}.$$

The average number of balls in the steady state is

$$\mathbf{E}[\mathcal{W}] = \frac{2\beta}{\alpha+\beta}. \quad \diamond$$

The arguments in Example 9.2 can be generalized to bigger examples with m balls in the starting urn, but the transition matrix becomes rather complicated. However, we can find the steady state average from a direct consideration. In the steady state each ball appears once on average in the sample, a white ball contributes a white ball in the next urn with probability $(1-\alpha)$ and a blue ball contributes a white ball upon miscopying with probability β, and the average solves the equation

$$\mathbf{E}[\mathcal{W}] = (1 - \alpha)\mathbf{E}[\mathcal{W}] + \beta(m - \mathbf{E}[\mathcal{W}]),$$

giving

$$\mathbf{E}[\mathcal{W}] = \frac{m\beta}{\alpha + \beta}.$$

Typical values of the miscopying probabilities α and β are in the range of 10^{-5}. If they are of comparable values, a decent proportion of each color appears in the stationary distribution. For example, with $\alpha = 10^{-5}$ and $\beta = 3 \times 10^{-5}$, on average $\frac{3}{4}$ of the balls are white, and the rest are blue.

9.1.3 Mutation

The operation of miscopying genes often results in mutations or gene types not seen before. In terms of urns, new colors appear and old ones disappear. In fact, at the appropriate rates, any color eventually disappears. In this setting there is no stationarity in the conventional sense. Alternatively, one can think of stationarity of *partitions* among the existing and ever changing colors. A partition of m species (genes) describes how many colors there are that are represented by i balls, for various values of i. For example, if the number of balls is $m = 5$, and initially there are three yellow balls, one black and one red, the partition is $(2, 0, 1, 0, 0, 0, \dots)$, specifying that there are two species each represented by one color (gene) each, and there is one species represented by three identical genes. (There are no species represented by 2, 4, 5, or higher numbers of animals). It is not feasible for a population of total size m to have any species represented by $(m + 1)$ or more animals. We therefore can use a truncated m–tuple to specify the partition. In the example with $m = 5$, we can unambiguously use the truncated partition $(2, 0, 1, 0, 0)$ to simplify the notation.

If after a long period of evolution one observes that the urn has drifted into one with three white balls, one green and one pink, one sees again the partition $(2, 0, 1, 0, 0)$ and surmises that this partition potentially has a stationary probability.

We are drawing m balls with replacement from an urn, and are filling a new urn with balls of colors determined by the sample. If a ball of a certain color appears in the sample, with some positive probability a mutation occurs (a ball of an entirely new color, or a gene of new properties, is placed in the new urn). The probability of a mutation is α. Otherwise, no mutation occurs and perfect gene copying takes place in the new generation—a ball of the same color is placed in the new urn (with probability $1-\alpha$). For instance, the balls at the beginning may be of colors 1,1,2,3,3. The sequence of colors 1,2,1,3,3 may appear in the sample. We may have a perfect copying in the first three draws, and balls of colors 1,2,1 are placed in a new urn. We may then experience two mutations for the next two draws and balls of color 4 and 5 are placed in the new urn which now carries the genes 1,1,2,4,5.

We proceed under a stationarity assumption, which is consistent with the ecological surroundings. Consider the stationary probability Q_2 of drawing two balls of the same color. The two balls may have the same parent: A ball was chosen twice from the parent urn (also in the stationary distribution) and no mutation occurred. The probability of choosing the common parent is $\sum_{k=1}^{m} \frac{1}{m^2} = \frac{1}{m}$. Or, the two balls my be children of two different balls of the same color (chosen with probability Q_2 from the stationary parent urn), and no mutation occurred. The probability of choosing two different balls is $1 - \frac{1}{m}$. Note that two balls of the same color in a stationary urn cannot be mutations from the parent urn, because necessarily all mutations are of different colors. The stationary probability satisfies the equation

$$Q_2 = \frac{1}{m}(1-\alpha)^2 + \left(1 - \frac{1}{m}\right) Q_2(1-\alpha)^2,$$

with solution

$$Q_2 = \frac{(1-\alpha)^2}{(1-\alpha)^2 + m\alpha(2-\alpha)}.$$

With the typical order of 10^{-5} for α, the stationary probability is well approximated by

$$Q_2 \approx \frac{1}{1+2m\alpha} := \frac{1}{1+\theta}.$$

(The factor $2m\alpha$ is usually called θ.)

We can argue hierarchically stationary probabilities for withdrawing three balls of the same color (or Q_3), for withdrawing four balls of the same color (or Q_4), and so on. For Q_3 the argument comes from assembling probabilities of three mutually exclusive events: The three balls are nonmutated replicas of one, two, or three parent balls of the same color. The probability of picking the same ball three times is $p' = \sum_{k=1}^{m} \frac{1}{m^3} = \frac{1}{m^2}$. The probability of picking three different balls is $p''' = 3! \sum_{1 \le i < j < k \le m} \frac{1}{m^3} = \frac{6}{m^3}\binom{m}{3}$. Of course, the probability of picking the same ball twice, and a different ball once is $1 - p' - p'''$, and we get

$$Q_3 = (1-\alpha)^3 \left(\frac{1}{m^2} + \frac{3}{m}\left(1 - \frac{1}{m}\right) Q_2 + \left(1 - \frac{1}{m}\right)\left(1 - \frac{2}{m}\right) Q_3\right).$$

If we ignore α relative to constants, the approximate solution is

$$Q_3 \approx \frac{2}{(\theta+1)(\theta+2)}.$$

The next stationary probability is

$$Q_4 \approx \frac{6}{(\theta+1)(\theta+2)(\theta+3)}.$$

Carrying on in this manner, we have

$$Q_i \approx \frac{(i-1)!}{(\theta+1)(\theta+2)\dots(\theta+i-1)}.$$

We can also consider the probability of a multicolor sample in the steady state. For example, we can ask about the probability that a sample of two balls contains a pair of balls of different colors. Note that the question is concerned with the partition in the sample, and does not address which particular colors are in it. By a natural extension of the notation introduced, one calls such steady-state probability $Q_{1,1}$, for which we have the approximation

$$Q_{1,1} = 1 - Q_2 \approx \frac{\theta}{\theta+1} = \frac{2!\,\theta^2}{2!\,\theta(\theta+1)}.$$

These are the heuristic arguments behind Ewens' sampling formula.

9.1.4 Hoppe's Urn Scheme

Ewens' sampling formula can be obtained conceptually from a simple Pólya-like scheme, which was first discussed in Hoppe (1984). In this scheme one color is considered "special," and every time a ball of this color is picked, a ball of a new color (a new species) appears. The special balls are generators of other species. The scheme progresses as follows. At the beginning there are $\theta \ge 1$ special balls, of a distinguished color, say black. This color does not account by itself for any species. The evolution of Hoppe's urn progresses in discrete time steps. At any step a ball is sampled at random from the urn (all balls being equally likely). If the ball is black we put it back in the urn together with a ball of a new color (new species). The colors corresponding to the species are numbered $1, 2, 3, \dots$ as the need arises. If a nonspecial ball appears in the sample, the ball is returned to the urn together with a ball of the same color. Figure 9.1 illustrates the sample paths arising from sampling from a Hoppe's urn starting with one black ball. The links in the diagram are labeled with the probability—a link coming down from one configuration into another is labeled with the probability of making this particular transition.

In this sampling process the selection of a special ball generates a new species, animals of a new kind are born. In the context of alleles, the sampling of a special ball is a mutation, and induces a new allelic type that did not exist before. The selection of a numbered ball (an animal from a species) gives a ball of the same color, with the interpretation that the species is multiplying within itself.

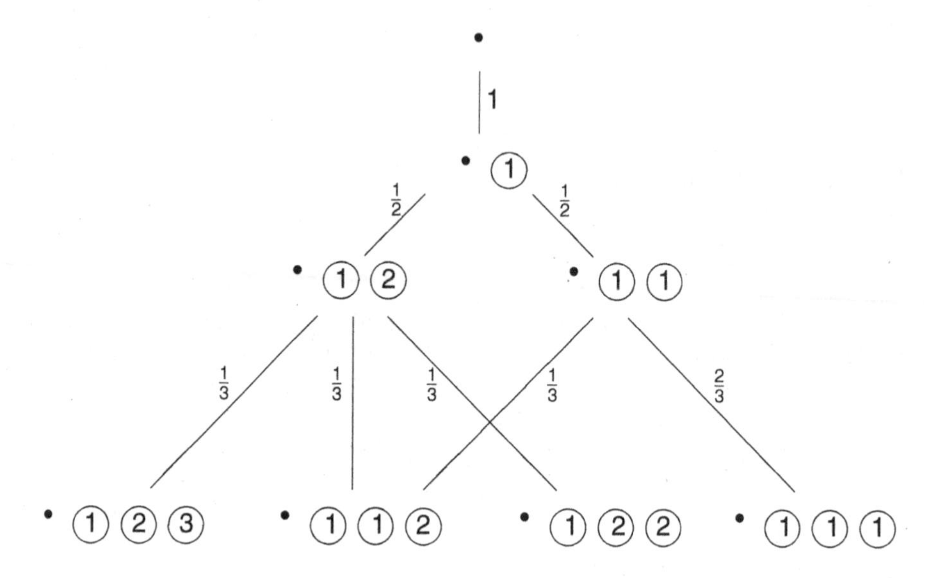

FIGURE 9.1
Sample paths for three draws from Hoppe's urn.

Natural questions to ask are concerned about the number of species and the size of the population within each. Let us take up the number of species first. This random variable has a distribution related to the number of cycles in a random permutation, and the limit distribution of a suitably normalized version has a Gaussian distribution and has a close connection to the theory of records. We first develop the mean and variance. These results are given by formulas that are most compact in terms of the nth generalized harmonic number of order k:

$$\mathcal{H}_n^{(k)}(x) = \frac{1}{x^k} + \frac{1}{(x+1)^k} + \cdots + \frac{1}{(x+n-1)^k};$$

we drop the superscript when it is 1. The usual harmonic number H_n is $\mathcal{H}_n(1)$. The generalized harmonic numbers $\mathcal{H}_n(x)$ have asymptotics similar to the standard ones—given any $x \ge 1$, the generalized harmonic number $\mathcal{H}_n(x)$ is sandwiched between $\mathcal{H}_n(\lceil x \rceil) = H_{\lceil x \rceil + n - 1} - H_{\lceil x \rceil - 1}$ and $\mathcal{H}_n(\lfloor x \rfloor) = H_{\lfloor x \rfloor + n - 1} - H_{\lfloor x \rfloor - 1}$. And so, for fixed $x \ge 1$ the asymptotics of $\mathcal{H}_n(x)$ are the same as those of the ordinary harmonic numbers H_n, because

$$\begin{aligned}\ln\left(n + \lceil x \rceil - 1\right) + O(1) &= H_{\lceil x \rceil + n - 1} - H_{\lceil x \rceil - 1} \\ &\le \mathcal{H}_n(x) \\ &\le H_{\lfloor x \rfloor + n - 1} - H_{\lfloor x \rfloor - 1} \\ &= \ln\left(n + \lfloor x \rfloor - 1\right) + O(1),\end{aligned}$$

that is, for fixed $x \geq 1$,

$$\mathcal{H}_n(x) \sim \ln n, \qquad \text{as } n \to \infty.$$

Similarly, for fixed $k \geq 2$, as $n \to \infty$,

$$O(1) = H^{(k)}_{\lceil x \rceil + n - 1} - H^{(k)}_{\lceil x \rceil - 1} \leq \mathcal{H}^{(k)}_n(x) \leq H^{(k)}_{\lfloor x \rfloor + n - 1} - H^{(k)}_{\lfloor x \rfloor - 1} = O(1).$$

The presentation of results for Hoppe's urn and Ewens' sampling formula go via Stirling's numbers of the first kind. We recall from previous chapters Pochhammer's symbol for the rising factorial:

$$\langle x \rangle_n = x(x+1)\dots(x+n-1).$$

If we expand the right-hand side, we get a polynomial of degree n. In classical combinatorics this polynomial is viewed as a generating function for a sequence of numbers known as the signless Stirling's numbers of the first kind. That is, the coefficients of x^k, for $k = 1, \dots, n$, in the expansion are the nth signless Stirling numbers of the first kind, which count the number of permutations of n elements with k disjoint cycles. The kth of these numbers is denoted by $\left[{n \atop k}\right]$.[1] For instance,

$$\langle x \rangle_3 = x(x+1)(x+2) = 2x + 3x^2 + x^3,$$

and

$$\left[{3 \atop 1}\right] = 2, \qquad \left[{3 \atop 2}\right] = 3, \qquad \left[{3 \atop 3}\right] = 1.$$

Let K_n be the number of nonspecial colors (species) in a Hoppe's urn after n draws. Results for mean and variance, as well as the distribution, can be developed easily from a representation of K_n as a sum of independent random variables. Each draw places an extra ball in the urn, and after i draws from the urn, there is a total of $\tau_i = \theta + i$ balls. The probability of picking a special ball in the ith draw is θ/τ_{i-1}. Let Y_i be the indicator that assumes value 1 if the ball in the ith draw is special, and 0 otherwise. Thus, $Y_i = \text{Ber}(\frac{\theta}{\tau_{i-1}})$. These Bernoulli random variables are independent, with $\mathbf{E}[Y_i] = \frac{\theta}{\tau_{i-1}}$ and variance $\mathbf{Var}[Y_i] = \frac{\theta}{\tau_{i-1}}(1 - \frac{\theta}{\tau_{i-1}})$. We have the representation

$$K_n = Y_1 + Y_2 + \dots + Y_n. \tag{9.1}$$

[1] The large square brackets notation was introduced by Karamata (see Karamata, 1935) and promoted by Knuth (see Knuth, 1992) to parallel the binomial coefficient symbolism.

PROPOSITION 9.1

(Ewens, 1972). Let K_n be the number of nonspecial colors in a Hoppe's urn after n draws, starting with θ black balls.[2] *Then*

$$\mathbf{E}[K_n] = \theta\mathcal{H}_n(\theta) \sim \theta \ln n,$$

and

$$\mathbf{Var}[K_n] = \theta\mathcal{H}_n(\theta) - \theta^2\mathcal{H}_n^{(2)}(\theta) \sim \theta \ln n.$$

PROOF Taking expectations of Eq. (9.1),

$$\mathbf{E}[K_n] = \frac{\theta}{\theta} + \frac{\theta}{\theta+1} + \cdots + \frac{\theta}{\theta+n-1} = \theta\mathcal{H}_n(\theta).$$

By independence, we also have

$$\begin{aligned}\mathbf{Var}[K_n] &= \mathbf{Var}[Y_1] + \mathbf{Var}[Y_2] + \cdots + \mathbf{Var}[Y_n] \\ &= \sum_{i=0}^{n-1} \frac{\theta}{\theta+i}\left(1 - \frac{\theta}{\theta+i}\right) \\ &= \theta\mathcal{H}_n(\theta) - \theta^2\mathcal{H}_n^{(2)}(\theta).\end{aligned}$$

The asymptotic equivalents of the mean and variance follow from the discussion preceding the theorem in which it was shown that the generalized harmonic numbers have the same asymptotics as the ordinary harmonic numbers. ■

The rates of growth in the mean and variance tell us that K_n is highly concentrated around its mean value. We state this remark formally.

COROLLARY 9.1

$$\frac{K_n}{\ln n} \xrightarrow{P} \theta.$$

PROOF By Chebyshev's inequality, for any fixed $\varepsilon > 0$,

$$\mathbf{P}(|K_n - \mathbf{E}[K_n]| > \varepsilon) \le \frac{\mathbf{Var}[K_n]}{\varepsilon^2}.$$

Replace ε by $\varepsilon\mathbf{E}[K_n]$, to get

$$\mathbf{P}\left(\left|\frac{K_n}{\mathbf{E}[K_n]} - 1\right| > \varepsilon\right) \le \frac{\mathbf{Var}[K_n]}{\varepsilon^2\mathbf{E}^2[K_n]} = O\left(\frac{1}{\ln n}\right) \to 0.$$

[2] This number is also the number of species in Ewens' sampling after n steps of regeneration.

Therefore,

$$\frac{K_n}{\mathbf{E}[K_n]} \xrightarrow{P} 1.$$

Multiply this relation by the convergence relation $\mathbf{E}[K_n]/\ln n \to \theta$ to get

$$\frac{K_n}{\ln n} \xrightarrow{P} \theta,$$

in accordance with Slutsky's theorem (cf. Billingsley, 1995). ■

THEOREM 9.1
(Ewens, 1972). Let K_n be the number of nonspecial colors in Hoppe's urn after n draws.[3] *Then, for $k = 1, 2, \ldots, n$,*

$$\mathbf{P}(K_n = k) = \frac{\theta^k}{\langle\theta\rangle_n} \left[{n \atop k} \right].$$

PROOF Let $\psi_X(z)$ be the probability generating function of a generic nonnegative integer random variable X, that is,

$$\psi_X(z) = \mathbf{E}[z^X] = \sum_{k=0}^{\infty} \mathbf{P}(X = k)\, z^k.$$

Starting with the representation in Eq. (9.1) we write the probability generating function of K_n in terms of the probability generating function of the sum of the Bernoulli random variables Y_i and decompose it by independence:

$$\begin{aligned}
\psi_{K_n}(z) &= \psi_{Y_1+Y_2+\cdots+Y_n}(z) \\
&= \psi_{Y_1}(z)\, \psi_{Y_2}(z) \ldots \psi_{Y_n}(z) \\
&= \frac{\theta z}{\theta} \left(\left(1 - \frac{\theta}{\theta+1}\right) + \frac{\theta z}{\theta+1} \right) \ldots \left(\left(1 - \frac{\theta}{\theta+n-1}\right) + \frac{\theta z}{\theta+n-1} \right) \\
&= \frac{\theta z}{\theta} \left(\frac{\theta z + 1}{\theta+1} \right) \ldots \left(\frac{\theta z + n - 1}{\theta + n - 1} \right) \\
&= \frac{1}{\langle\theta\rangle_n} \sum_{k=1}^{n} \left[{n \atop k} \right] (\theta z)^k.
\end{aligned}$$

We extract the coefficient of z^k to establish the theorem. ■

The probabilities in Theorem 9.1 may not be easy to compute for large n. One needs to simplify the calculation via asymptotic equivalents in terms of

[3] This is also the number of different allelic types in Ewens' sampling after n regeneration steps.

standard functions. There are known asymptotic equivalents for the Stirling numbers ${n \brack k}$. However, they are not very convenient for all ranges of k. We shall furnish an approximation to the exact probabilities in Ewens' theorem via asymptotic normal distributions. To establish a central limit theorem for K_n, we can appeal to one of the classical central limit theorems. A convenient form is Lyapunov's central limit theorem, as it only uses up to third moments of a sum of independent random variables. The sum in K_n is that of Bernoulli random variables, and moment calculations are simple. We state without proof the central limit theorem here for reference. A reader interested in a proof of this classical result can look up Billingsley (1995).

THEOREM 9.2

(Lyapunov's central limit theorem). Suppose $X_1, X_2, \ldots$ are independent, with finite absolute third moments ($\mathbf{E}|X_i^3| < \infty$, for all i). Let $\mathbf{E}[X_i] = \mu_i$, and $\mathbf{Var}[X_i] = \sigma_i^2$. If the family $\{X_i\}_{i=1}^{\infty}$ satisfies Lyapunov's condition:

$$\lim_{n\to\infty} \frac{\sum_{i=1}^n \mathbf{E}|X_i - \mu_i|^{\delta+2}}{\left(\sum_{k=1}^n \sigma_k^2\right)^{\frac{1}{2}\delta+1}} = 0,$$

for some $\delta > 0$, then

$$\frac{\sum_{i=1}^n X_i - \sum_{i=1}^n \mu_i}{\left(\sum_{k=1}^n \sigma_k^2\right)^{\frac{1}{2}}} \xrightarrow{\mathcal{D}} \mathcal{N}(0,1).$$

COROLLARY 9.2

Let K_n be the number of nonspecial colors in Hoppe's urn after n draws. Then

$$\frac{K_n - \theta \ln n}{\sqrt{\ln n}} \xrightarrow{\mathcal{D}} \mathcal{N}(0,\theta).$$

PROOF We verify Lyapunov' s condition with $\delta = 1$. Set

$$p_i = \frac{\theta}{\theta + i - 1},$$

which is the probability of success of the ith Bernoulli random variable in the representation in Eq. (9.1). Then, $\mu_i = \mathbf{E}[Y_i] = p_i$, $\sigma_i^2 = \mathbf{Var}[Y_i] = p_i(1 - p_i)$, and

$$\begin{aligned}
\mathbf{E}|Y_i - \mu_i|^3 &= (1 - p_i)\,|0 - p_i|^3 + p_i\,|1 - p_i|^3 \\
&= \left(1 - \frac{\theta}{\theta + i - 1}\right)\left(\frac{\theta}{\theta + i - 1}\right)^3 + \frac{\theta}{\theta + i - 1}\left(1 - \frac{\theta}{\theta + i - 1}\right)^3 \\
&= \frac{\theta}{\theta + i - 1} - \frac{3\theta^2}{(\theta + i - 1)^2} + \frac{4\theta^3}{(\theta + i - 1)^3} - \frac{2\theta^4}{(\theta + i - 1)^4}.
\end{aligned}$$

The sum in the numerator of Lyapunov' s condition is

$$\sum_{i=1}^{n} \mathbf{E}|Y_i - \mu_i|^3 = \theta\mathcal{H}_n(\theta) - 3\theta^2\mathcal{H}_n^{(2)}(\theta) + 4\theta^3\mathcal{H}_n^{(3)}(\theta) - 2\theta^4\mathcal{H}_n^{(4)}(\theta)$$
$$\sim \theta \ln n, \qquad \text{as } n \to \infty.$$

By independence, the sum of the variances in the denominator of Lyapunov's condition is simply $\mathbf{Var}[K_n]$, for which Proposition 9.1 provides the asymptotic equivalent $\theta \ln n$. Thus,

$$\frac{\sum_{i=1}^{n} \mathbf{E}|Y_i - \mu_i|^3}{\left(\sum_{k=1}^{n} \sigma_k^2\right)^{\frac{3}{2}}} \sim \frac{\theta \ln n}{(\theta \ln n)^{\frac{3}{2}}} \to 0,$$

and Lyapunov's condition has been verified. With K_n being $\sum_{i=1}^{n} Y_i$, the technical statement of Lyapunov's condition yields

$$\frac{K_n - \mathbf{E}[K_n]}{\sqrt{\mathbf{Var}[K_n]}} \xrightarrow{\mathcal{D}} \mathcal{N}(0, 1).$$

To put the result in the stated form we need adjustments via the multiplicative and additive versions of Slutsky's theorem. Namely, the rates found in Eq. (9.1) give

$$\frac{\sqrt{\mathbf{Var}[K_n]}}{\sqrt{\theta \ln n}} = \frac{\sqrt{\theta\mathcal{H}_n(\theta) - \theta^2\mathcal{H}_n^{(2)}(\theta)}}{\sqrt{\theta \ln n}} \to 1,$$

so, a multiplication gives

$$\frac{K_n - \mathbf{E}[K_n]}{\sqrt{\theta \ln n}} \xrightarrow{\mathcal{D}} \mathcal{N}(0, 1).$$

We then add

$$\frac{\mathbf{E}[K_n] - \theta \ln n}{\sqrt{\theta \ln n}} = \frac{\theta\mathcal{H}_n(\theta) - \theta \ln n}{\sqrt{\theta \ln n}} = \frac{O(1)}{\sqrt{\theta \ln n}} \to 0. \qquad \blacksquare$$

9.1.5 Ewens' Sampling Formula

Hoppe's urn appears as a stationary model for the Wright-Fisher non-Darwinian evolution where big changes occur abruptly via mutation as in the Mendelian theory. The variety of the species and their sizes are important issues in population genetics. A description of the profile of species can be formulated in a number of equivalent ways. An interesting question is how many species are of a given size. The multivariate view of this question is known as Ewens' sampling formula. Suppose there are A_1 species represented by one animal each, A_2 species represented by two animals each, and so on. After n steps of evolution there can be up to n species, and Ewens' sampling formula specifies the probability

$$\mathbf{P}(A_1 = a_1, A_2 = a_2, \dots, A_n = a_n)$$

for any n-tuple $(a_1, \dots, a_n)$. Of course, this probability is 0 if the tuple is not feasible—a feasible tuple must satisfy

$$\sum_{i=1}^{n} i a_i = n.$$

As an instance, refer to the third generation in Figure 9.1. The leftmost configuration corresponds to the partition $a_1 = 3, a_2 = 0, a_3 = 0$, and thus

$$\mathbf{P}(A_1 = 3, A_2 = 0, A_3 = 0) = \frac{1}{6}.$$

The rightmost configuration corresponds to the partition $a_1 = 0, a_2 = 0, a_3 = 1$, and thus

$$\mathbf{P}(A_1 = 0, A_2 = 0, A_3 = 1) = \frac{2}{6}.$$

The two middle configurations, though they are different in details, both correspond to the partition $a_1 = 1, a_2 = 1, a_3 = 0$. Among these two configurations the one to the left can be reached via two different histories, with probabilities $\frac{1}{6}$ each. The one to the right also has probability $\frac{1}{6}$, whence

$$\mathbf{P}(A_1 = 1, A_2 = 1, A_3 = 0) = \frac{3}{6}.$$

A *configuration* is specified with the number of animals in species i, for $i = 1, 2, \dots$, or more compactly by K_n and the number of animals in species i, for $i = 1, 2, \dots, K_n$. A configuration gives rise to a partition of n in a number of species of size 1, a number of species of size 2, etc., up to size n. Of course, some of the numbers in the partition may be 0. A configuration can be reached recursively from certain configurations in the previous generations. For instance, the configuration in Figure 9.2 (a) induces the partition $A_1 = 3, A_2 = 0, A_3 = 2$. This configuration may be reached from any of the configurations in Figure 9.2 (b). The bottom configuration in 9.2 (b) evolves into that in Figure 9.2 (a), if the special black ball is sampled, and the middle one in Figure 9.2 (b) evolves into that in Figure 9.2 (a), if a ball of color 1 is sampled, whilst the top configuration in Figure 9.2 (b) evolves into that in Figure 9.2 (a), if a ball of color 3 is sampled.

It is convenient to describe partitions as row vectors, for which we use the boldface notation. For a random partition we write $\mathbf{A}_n$ for the row vector with random components $A_1, \dots, A_n$ after n draws. The notation $\mathbf{a}_n$ is for a row vector with n fixed components. In general, after $n + 1$ draws, if we are given a feasible partition

$$\mathbf{A}_{n+1} = (a_1, \dots, a_{n+1}),$$

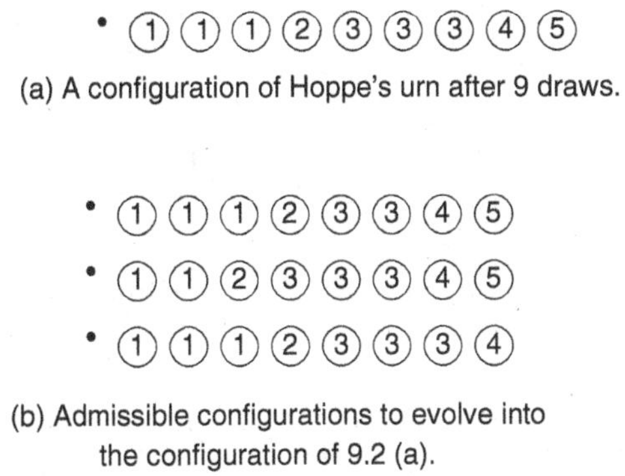

• (1) (1) (1) (2) (3) (3) (3) (4) (5)

(a) A configuration of Hoppe's urn after 9 draws.

• (1) (1) (1) (2) (3) (3) (4) (5)

• (1) (1) (2) (3) (3) (3) (4) (5)

• (1) (1) (1) (2) (3) (3) (3) (4)

(b) Admissible configurations to evolve into the configuration of 9.2 (a).

FIGURE 9.2
Evolution of configurations in Hoppe's urn.

with $a_{n+1} = 0$, it can be realized from $\mathbf{b}_n = (b_1, \ldots, b_n)$, a previous partition (after n draws), if

$$\mathbf{b}_n = (a_1, \ldots, a_{i-1}, a_i + 1, a_{i+1} - 1, a_{i+2}, \ldots, a_n), \tag{9.2}$$

and a ball from a species of size i is drawn in the nth sample (with probability $i(a_i+1)/(\theta+n)$. Here we put that ball back, and add one ball of the same color. This perturbs the partition in only two places, the number of species with i animals is reduced by one species, and the number of species with i members grows by one species, for $i = 1, \ldots, n-1$. Another special boundary case is when the number of species of size one increases, owing to a drawing of one of the special balls. This case arises with probability $\theta/(\theta+n)$.

THEOREM 9.3
(Ewens, 1972, Karlin and McGregor, 1972). In a Hoppe's urn starting with θ special balls, let $\mathbf{A}_n = (A_1, \ldots, A_n)$ be the partition of the total population of all nonspecial balls in such a way that A_i is the number of colors represented by i balls.[4] *Let $\mathbf{a}_n = (a_1, \ldots, a_n)$ be a fixed feasible partition. Then,*

$$\mathbf{P}(\mathbf{A}_n = \mathbf{a}_n) = \frac{n!\, \theta^{a_1+\cdots+a_n}}{1^{a_1}\, 2^{a_2} \ldots n^{a_n}\, a_1!\, a_2! \ldots a_n!\, \langle\theta\rangle_n}.$$

PROOF In figuring out the probability for a partition, we separate the boundary case $a_n = 1$, which follows directly—in this case the partition after n draws must be $\mathbf{a}_n = (a_1, \ldots, a_n) = (0, 0, \ldots, 0, 1)$ with one species of size n. This can only happen by perpetuating $n-1$ draws after the first from balls of

[4] In Ewens' sampling formula A_i is the number of species of population size i.

color 1. This event occurs with probability $1 \times \frac{1}{\theta+1} \times \frac{2}{\theta+2} \times \frac{3}{\theta+3} \times \cdots \times \frac{n-1}{\theta+n-1}$, and so

$$\mathbf{P}(\mathbf{A}_n = \mathbf{a}_n) = \frac{(n-1)!}{\langle \theta + 1 \rangle_{n-1}} = \frac{n!\, \theta^{0+\cdots+0+1}}{1^0\, 2^0 \ldots (n-1)^0 n^1\, 0!\, 0! \ldots 0!\, 1!\, \langle \theta \rangle_n}.$$

We prove Ewens' sampling formula by induction on n. For the basis at $n = 1$, the only feasible partition after one draw is $a_1 = 1$, and we have

$$\mathbf{P}(\mathbf{A}_1 = \mathbf{a}_1) = \mathbf{P}(A_1 = 1) = 1 = \frac{\theta}{\theta} = \frac{1!\, \theta^{a_1}}{1^{a_1} a_1!\, \langle \theta \rangle_1}.$$

Assume the formula holds for n. In what follows we take $a_{n+1} = 0$ (having directly proved the sampling formula for $a_{n+1} = 1$ above). We condition the calculation of $\mathbf{P}(\mathbf{A}_{n+1} = \mathbf{a}_{n+1})$ on $\mathbf{b}_n = (b_1, \ldots, b_n)$, which are feasible partitions at the nth stage, that is, partitions satisfying $\sum_{i=1}^{n} i b_i = n$. As in the discussion preceding the theorem, only certain ones among these can progress into $\mathbf{a}_{n+1} = (a_1, \ldots, a_{n+1})$. Subsequently, we have

$$\begin{aligned}\mathbf{P}(\mathbf{A}_{n+1} = \mathbf{a}_{n+1}) &= \sum_{\mathbf{b}_n} \mathbf{P}(\mathbf{A}_{n+1} = \mathbf{a}_{n+1} \mid \mathbf{A}_n = \mathbf{b}_n)\, \mathbf{P}(\mathbf{A}_n = \mathbf{b}_n) \\ &= 0 + \sum_{\tilde{\mathbf{b}}_n} \mathbf{P}(\mathbf{A}_{n+1} = \mathbf{a}_{n+1} \mid \mathbf{A}_n = \tilde{\mathbf{b}}_n)\, \mathbf{P}(\mathbf{A}_n = \tilde{\mathbf{b}}_n),\end{aligned}$$

where $\tilde{\mathbf{b}}_n = (\tilde{b}_1, \ldots, \tilde{b}_n)$ are partitions that can progress into $\mathbf{a}_{n+1}$. If $\tilde{\mathbf{b}}_n$ is of the form in Eq. (9.2), by the induction hypothesis, the probability $\mathbf{P}(\mathbf{A}_n = \tilde{\mathbf{b}}_n)$ is

$$\begin{aligned}&\frac{n!\, \theta^{\tilde{b}_1+\cdots+\tilde{b}_n}}{1^{\tilde{b}_1}\, 2^{\tilde{b}_2} \ldots n^{\tilde{b}_n}\, \tilde{b}_1!\, \tilde{b}_2! \ldots \tilde{b}_n!\, \langle \theta \rangle_n} \\ &\quad = \frac{n!\, \theta^{a_1+a_2+\cdots+a_{i-1}+(a_i+1)+(a_{i+1}-1)+a_{i+2}+\cdots+a_n+0}}{1^{a_1}\, 2^{a_2} \ldots (i-1)^{a_{i-1}} i^{a_i+1} (i+1)^{a_{i+1}-1} (i+2)^{a_{i+2}} \ldots n^{a_n} (n+1)^0} \\ &\quad\quad \times \frac{1}{a_1!\, a_2! \ldots a_{i-1}!\, (a_i+1)!\, (a_{i+1}-1)!\, a_{i+2}! \ldots a_n!\, 0!\, \langle \theta \rangle_n} \\ &\quad = \frac{n!\, \theta^{a_1+\cdots+a_n+a_{n+1}}}{1^{a_1}\, 2^{a_2} \ldots (n+1)^{a_{n+1}}\, a_1!\, a_2! \ldots a_{n+1}!\, \langle \theta \rangle_n} \times \frac{(i+1)a_{i+1}}{i(a_i+1)},\end{aligned}$$

and the conditional probability of progressing into $\mathbf{a}_{n+1}$ is $i(a_i+1)/(\theta+n)$.

On the other hand, if $\tilde{\mathbf{b}}_n$ is the partition $(a_1 - 1, a_2, \ldots, a_n)$ it can progress into the partition $\mathbf{a}_{n+1} = (a_1, a_2, \ldots, a_n, 0)$ if any of the special balls is sampled, which adds a new species (of size 1). The conditional probability of this progress is $\theta/(\theta+n)$. By the induction hypothesis, the probability of such a boundary case is

$$\frac{n!\,\theta^{a_1-1+a_2+\cdots+a_n}}{1^{a_1-1}\,2^{a_2}\ldots n^{a_n}\,(a_1-1)!\,a_2!\,\ldots a_n!\,\langle\theta\rangle_n}$$
$$= \frac{n!\,a_1\theta^{a_1+a_2+\cdots+a_n+0-1}}{1^{a_1}\,2^{a_2}\ldots n^{a_n}(n+1)^0\,a_1!\,a_2!\,\ldots a_n!\,0!\,\langle\theta\rangle_n}.$$

We thus have

$$\mathbf{P}(\mathbf{A}_{n+1}=\mathbf{a}_{n+1}) = \sum_{i=1}^{n}\frac{n!\,\theta^{a_1+\cdots+a_{n+1}}}{1^{a_1}\,2^{a_2}\ldots(n+1)^{a_{n+1}}\,a_1!\,a_2!\,\ldots a_{n+1}!\,\langle\theta\rangle_n}$$
$$\times\frac{(i+1)a_{i+1}}{i(a_i+1)}\times\frac{i(a_i+1)}{\theta+n}$$
$$+\frac{n!\,a_1\theta^{a_1+a_2+\cdots+a_n+a_{n+1}-1}}{1^{a_1}\,2^{a_2}\ldots(n+1)^{a_{n+1}}\,a_1!\,a_2!\,\ldots a_{n+1}!\,\langle\theta\rangle_n}\times\frac{\theta}{\theta+n}$$
$$=\frac{n!\,\theta^{a_1+a_2+\cdots+a_{n+1}}}{1^{a_1}\,2^{a_2}\ldots(n+1)^{a_{n+1}}\,a_1!\,a_2!\,\ldots,a_{n+1}!\,\langle\theta\rangle_{n+1}}\times\sum_{i=0}^{n}(i+1)a_{i+1}.$$

We complete the induction by the feasibility requirement—after $n+1$ draws, $\mathbf{a}_{n+1}$ is a feasible partition with $\sum_{i=0}^{n}(i+1)a_{i+1}=n+1$. ∎

There are a number of equivalent formulas to that of Ewens' sampling formula of Theorem 9.3, some of which are suitable for other types of applications. For example, Donnelly and Tavaré (1986) give a formula for the joint distribution of the number of species, and the number of animals in each. Let X_j be the number of animals in the jth species. Recalling the notation K_n for the number of colors (species) after n draws, the Donnelly-Tavaré formula is

$$\mathbf{P}(K_n=k;\,X_1=x_1,\,X_2=x_2,\ldots,X_k=x_k)$$
$$=\frac{(n-1)!\,\theta^k}{\langle\theta\rangle_n x_k(x_k+x_{k-1})\ldots(x_k+x_{k-1}+\cdots+x_2)},$$

for feasible $(x_1,\ldots,x_k)$.

In his original paper, Ewens (1972) also gives a formula in which the species are "delabeled" in the sense that the original sequencing is randomized. The K_n species are labeled at random (by a random permutation of $\{1,\ldots,K_n\}$). If we let Y_i be the number of animals in the ith species according to this randomized labeling, Ewens (1972) gives the conditional formula

$$\mathbf{P}(Y_1=y_1,\ldots,Y_k=y_k\mid K_n=k)=\frac{n!}{k!\,n_1\ldots n_k\left[{n\atop k}\right]}.$$

The latter conditional formula is used in the statistical testing of the non-Darwinian theory of evolution as the distribution of the null hypothesis of that theory. What is appealing about this conditional formula is that it does not depend on θ, which in turn relies on many unknown parameters, such as true mutation rates that led to the steady state.

9.1.6 Phylogenetic Trees

Family trees have been used at least since the Middle Ages to represent genealogy. Certain forms of trees have been used as models for phylogeny (see Aldous, 1995). An explicit form was suggested in McKenzie and Steele (2000) as a model for evolutionary relations allowing simple testing of the similarity between actual and simulated trees.

The phylogenetic tree is a tree of unlabeled ternary internal nodes and labeled leaves that represent species on the hereditary scale. The edges connected to the leaves are called *pendants*. The tree grows under some probability model. One appealing model is the process in which a species modeled by this stochastic system evolves by having a uniformly chosen pendant "split." Recall from Subsection 8.1.1 the extended binary search tree (under the random permutation model), the external nodes of which are equally likely insertion positions. With the exception of the root, each node in such a tree is either an external node (of degree 1), or a node with two children (of degree 3). Except for the root, the influence of which disappears asymptotically, the phylogenetic tree is essentially an extended binary search tree; Theorem 8.1 applies.

Other flavors of phylogenetic trees are discussed in McKenzie and Steele (2000), such as the unrooted version of the model discussed above, and rooted and unrooted versions under a uniform probability model (all combinatorial trees of a particular size are equally likely). Similar results are derived, also with the aid of urns.

Figure 9.3 shows a phylogenetic tree on four species (leaves) and one on five species evolving from it by the splitting of an edge (pointed to by a down arrow). A pair of leaves at distance 2 (two leaves that are adjacent to a common internal node) are called a *cherry*. The number of cherries represents the number of species that are "cousins" or most similar on the hereditary scale.

THEOREM 9.4

(McKenzie and Steele, 2000). Let C_n be he number of cherries in a rooted uniform phylogenetic tree on n species. Then

$$\frac{C_n - \frac{n}{4}}{\sqrt{n}} \xrightarrow{\mathcal{D}} \mathcal{N}\Big(0, \frac{1}{16}\Big).$$

PROOF What is equally likely in this phylogenetic tree model are the edges. Color the edges in a cherry with white, and all the other pendants with blue. Nonpendant edges are colored with red. When a white pendant splits, the number of cherries remains the same, with two associated white pendants as before the splitting. But the new edge connected to the common internal

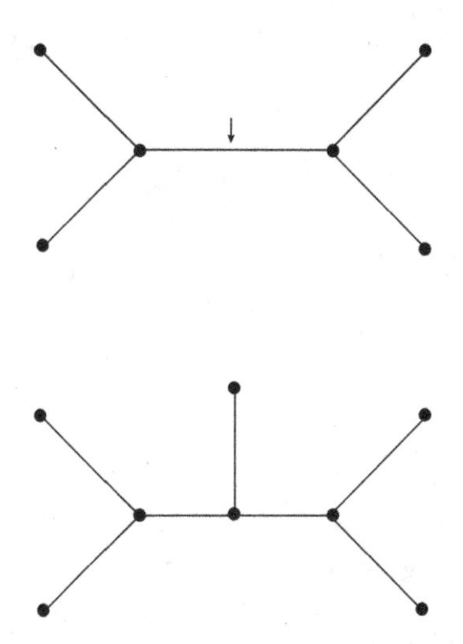

FIGURE 9.3
A phylogenetic tree on four species, and a tree on five species arising from it by the splitting of the edge pointed to.

vertex of the cherry is not pendant; the number of red pendants increases by one. The old white sibling pendant of the splitting edge is no longer involved in a cherry, and must be turned blue. When a blue pendant splits, a new cherry appears, with two associated white edges, the remaining part of the splitting edge becomes a nonpendant red edge. When a red edge splits, it is partitioned into two red edges (net increase of one red edge), and a blue pendant is attached.

The colored edges evolve as the balls of an extended urn with the scheme

$$\mathbf{A} = \begin{pmatrix} 0 & 1 & 1 \\ 2 & -1 & 1 \\ 0 & 1 & 1 \end{pmatrix};$$

the rows and columns of the scheme are in the order white, blue, red. An eigenvalue analysis of $\mathbf{A}^T$, as required in Theorem 6.5, gives the three eigenvalues $\lambda_1 = 2, \lambda_2 = 0, \lambda_3 = -2$. The principal eigenvalue is $\lambda_1 = 2$, with corresponding eigenvector

$$\mathbf{v} = \begin{pmatrix} v_1 \\ v_2 \\ v_3 \end{pmatrix} = \frac{1}{4} \begin{pmatrix} 1 \\ 1 \\ 2 \end{pmatrix}.$$

By Theorem 6.4, if W_n is the number of white balls after n draws,

$$\frac{W_n}{n} \xrightarrow{P} \lambda_1 v_1 = \frac{1}{2},$$

and by Theorem 6.5 the marginal distribution for the number of white pendants satisfies

$$\frac{W_n - \frac{n}{2}}{\sqrt{n}} \xrightarrow{\mathcal{D}} \mathcal{N}\Big(0, \frac{1}{4}\Big);$$

the variance is computed by recurrence methods. But then $W_n = 2C_n$. ■

9.2 Competitive Exclusion

An urn model for the competitive and exclusive evolution is proposed in Bailey (1972). The underlying ecological scenario is as follows. The liveable habitat comprises "niches," such as lakes, trees, forests, soil, etc. Various species have their own preference of niche or niches, and in many cases the choice is dictated by biological structure. For example, while fish live in the water, amphibian animals, such as frogs, can live in water and over ground, and certain types of bacteria may be able to exist in all the given niches.

A simplified model proposed for the relative frequency of species follows an urn scheme. Say there are n urns (niches) and a population of creatures categorized in s species. Each species has its own size. Bailey's model is concerned with case $s \le n$. Each species is associated with balls of a certain color. We have s colors, which we can number $1, 2, \ldots, s$. The species can coexist in a niche—balls of different colors may share an urn. Assume the urns are distinguishable and are numbered $1, 2, \ldots, n$. Each species chooses a number of niches to live in. The ecological rules are such that there are no empty urns, and each species has size at least one (there is at least one ball of color m, for each $m = 1, \ldots, s$). The idea behind these restrictions is to guarantee that there are no "phantom" urns or species. If there are any phantom niches or species, such as an urn that is not used, we can simply throw it out at the beginning, and redefine our n and s without changing the essence of the model. Cohen (1966) suggests to endow species i with a "quota" of i niches, as a way or reflecting *Gause's axiom*, or the ecological principle of strong competitive exclusion, according to which different species can coexist in niches that are not sufficiently distinct, but the competition ultimately drives most of them to extinction. In the Cohen (1966) model balls hit the urns in sufficient numbers so that each species meets its quota.

For definiteness, let us go over an illustrative example. Suppose $n = 5$, and $s = 4$. The first species immediately meets its quota (of one niche) by choosing one niche, say niche 4. A ball of color 1 is deposited in urn number 4. The second species, of quota two, comes along and chooses niches 3 and 5, and thus satisfies its quota requirement. One ball of color 2 is deposited in urn 3 and another one is deposited in urn 5. The members of the third species (balls of color 3) may be placed successively in niches (urns) 3, 4, 4, 3, 5. Note that the third ball of this color hit an urn already chosen by the second ball of this

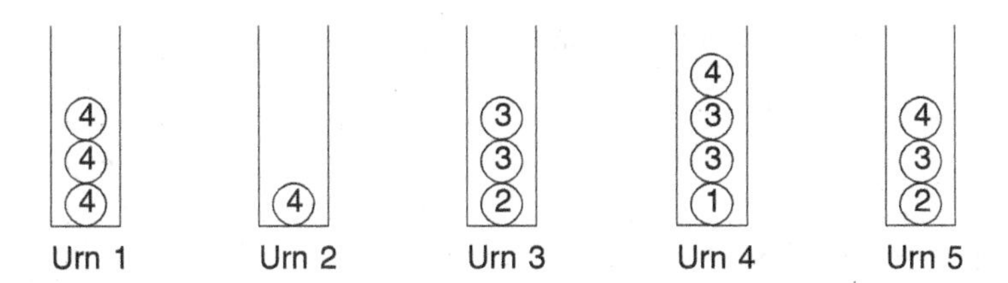

FIGURE 9.4
An urn model for competitive exclusion.

color, only two urns are occupied with balls of color 3 after three hits—the hits must then continue because the quota of 3 urns is not yet met, and it happens again that the next ball hits a previously hit urn. It is the fifth ball of color 3 that chooses a third new urn and completes the quota for this color. Lastly, the members of the fourth species (balls of color 4) are placed in niches (urns) 1, 2, 1, 1, 5, 4 and meet the quota of urns for this color; see Figure 9.4.

Let the total number of balls of color m dropped in all the urns be $X_{m,n}$. In the example discussed above and illustrated in Figure 9.4, $X_{1,5} = 1$, $X_{2,5} = 2$, $X_{3,5} = 5$, and $X_{4,5} = 6$. The number of balls of color m in the system is the population size of the mth species. It is clear that the $X_{m,n}$' s are independent. We can represent $X_{m,n}$ as a sum of independent geometric random variables. The first member in the mth species chooses an urn, all urns being equally likely. To get a second urn for the species, other members come by and try, they may hit a new urn with probability $(n-1)/n$, or may hit the already chosen one with probability $1/n$. The species keeps trying, until a second urn is secured. The number of attempts to get the second urn is $\text{Geo}((n-1)/n)$. This process goes on—after $j < m$ urns have been obtained for the mth species, there are $n - j$ urns free of balls of color m; the hits continue until one of them is chosen as the next urn required toward the quota, and the probability of each hit to succeed is $(n - j)/n$, requiring a $\text{Geo}((n - j)/n)$ number of attempts. So,

$$X_{m,n} = 1 + \text{Geo}\left(\frac{n-1}{n}\right) + \text{Geo}\left(\frac{n-2}{n}\right) + \cdots + \text{Geo}\left(\frac{n-m+1}{n}\right).$$

Note the similarity of this problem to coupon collection, where a collector is determined to get m different coupons from a publicly available set of n coupons. The collector acquires the coupons in stages. After collecting $j < m$ coupons, the collector keeps trying to get the next one according to a geometric random variable $\text{Geo}((n - j)/n)$. The typical challenge in the coupon collection problem is the acquisition of all n coupons in the complete set. So, $X_{n,n}$ is important in the context of coupon collection. The completive exclusion problem with all its lingo and results can be entirely stated in terms of coupon collection, but we shall stay within the ecological framework in all our phraseology in the sequel.

This convolution has interesting asymptotic properties (as $n \to \infty$) with multiple phases as m spans a large range. The average behavior instantiates the phases. Recall that Geo(p) has mean $1/p$. Hence

$$\mathbf{E}[X_{m,n}] = \frac{n}{n} + \frac{n}{n-1} + \cdots + \frac{n}{n-m+1} = n(H_n - H_{n-m}),$$

where H_j is the jth harmonic number. Employing the asymptotic properties of the harmonic numbers, we have

$$\begin{aligned}\mathbf{E}[X_{m,n}] &= n\left[\left(\ln n + \gamma + O\left(\frac{1}{n}\right)\right) - \left(\ln(n-m) + \gamma + O\left(\frac{1}{n-m}\right)\right)\right] \\ &= n\left[\left(\ln n + O\left(\frac{1}{n}\right)\right) - \left(\ln n + \ln\left(1 - \frac{m}{n}\right) + O\left(\frac{1}{n-m}\right)\right)\right] \\ &= n\left[-\ln\left(1 - \frac{m}{n}\right) + O\left(\frac{1}{n}\right) + O\left(\frac{1}{n-m}\right)\right]. \qquad (9.3)\end{aligned}$$

Expand the logarithm to get

$$\mathbf{E}[X_{m,n}] = \left[n\left[\frac{m}{n} + \frac{m^2}{2n^2} + \frac{m^3}{3n^3} + \cdots\right] + O(1) + O\left(\frac{n}{n-m}\right)\right].$$

The series can be stopped after the first term and the error is $O(m^2/n^2)$. So, if m is of a lower order of magnitude than n, then

$$\mathbf{E}[X_{m,n}] = m + o\left(\frac{m^2}{n}\right) + O(1) = m + O(1) + o(m).$$

For example, a species appearing at stage $5\lfloor \ln n \rfloor$ will have size asymptotic to $5 \ln n$.

On the other hand, if m is of the exact order n, such as αn, for some $\alpha \in (0, 1)$, from Eq. (9.3) we alternatively have

$$\frac{1}{n}\mathbf{E}[X_{m,n}] \to \ln\left(\frac{1}{1-\alpha}\right).$$

If we allow s to reach n (or be asymptotic to n), species with indexes as high as values asymptotic to n appear late, and for $m \sim n$, $\mathbf{E}[X_{m,n}]$ is asymptotic to $n \ln n$.

Phases in the variance follow suit. Recall that the variance of Geo(p) is q/p^2, with $q = 1 - p$. By independence, the variance of $X_{m,n}$ is the sum of the variances of the independent geometric random variables in it:

$$\begin{aligned}\mathbf{Var}[X_{m,n}] &= \sum_{j=1}^{m} \frac{n(j-1)}{(n-j+1)^2} \\ &= \sum_{k=n-m+1}^{n} \frac{n(n-k)}{k^2} \\ &= n^2(H_n^{(2)} - H_{n-m}^{(2)}) - n(H_n - H_{n-m}),\end{aligned}$$

with variance magnitude ranging from $O(1)$ for very small m (such as any fixed m), to $\Omega(n^2)$ for large values of m (such as $m = n - o(n)$).[5]

Like the mean and variance, the distribution of $X_{m,n} - m$ has multiple hidden phase changes, in fact an infinite number of them. We shall next look into the Poisson distributions hidden in it. Let $\psi_{m,n}(t)$ be the moment generating function of $X_{m,n} - m$. Recall that the moment generating function of Geo(p) is $pe^t/(1 - qe^t)$. The independence of the geometric random variables in $X_{m,n}$ gives

$$\begin{aligned}
\psi_{m,n}(t) &= e^{-mt}\mathbf{E}\left[\exp\left(\left(1 + \text{Geo}\left(\frac{n-1}{n}\right) + \cdots + \text{Geo}\left(\frac{n-m+1}{n}\right)\right)t\right)\right] \\
&= e^{-mt}\prod_{k=1}^{m}\mathbf{E}\left[e^{\text{Geo}(\frac{n-k+1}{n})t}\right] \\
&= \prod_{k=1}^{m}\frac{n-k+1}{n-(k-1)e^t}. \qquad (9.4)
\end{aligned}$$

Several results ensue. Let us start with the exact distribution in terms of the Stirling numbers $\{{r \atop j}\}$ of the second kind; review the material in Subsection 7.2.4. Stirling numbers of the second kind have several useful well known identities and generating function representation, of which the pertinent expansion is

$$\frac{x^j}{(1-x)(1-2x)\dots(1-jx)} = \sum_{k=1}^{\infty}\left\{{k \atop j}\right\}x^k. \qquad (9.5)$$

THEOREM 9.5

Let $X_{m,n}$ be the population size of the mth species in a competitive exclusion model with n niches (urns) and $s \le n$ species (ball colors). Then,

$$\mathbf{P}(X_{m,n} = k) = \frac{m!}{n^k}\binom{n}{m}\left\{{k-1 \atop m-1}\right\}.$$

PROOF It is more direct here to deal with the probability generating function

$$Q_{m,n}(z) = \sum_{k=1}^{\infty}\mathbf{P}(X_{m,n} = k)\, z^k = \phi_{m,n}(\ln z),$$

where $\phi_{m,n}(.)$ is the moment generating function of $X_{m,n}$. In terms of the falling factorial, in Pochhammer's notation $(n)_m = n(n-1)\dots(n-m+1)$, we can write the probability generating function $Q_{m,n}$ from the product

[5] A function $g(n)$ is $\Omega(h(n))$, if there is a positive constant C and a positive integer n_0, such that $|g(n)| \ge C|h(n)|$, for all $n \ge n_0$, or in other words, if $h(n) = O(g(n))$.

representation in Eq. (9.4) as follows:

$$Q_{m,n}(z) = \frac{(n)_m\, z^m}{n(n-z)(n-2z)\dots(n-(m-1)z)}$$
$$= \frac{(n)_m\, z\left(\frac{z}{n}\right)^{m-1}}{n\left(1-\frac{z}{n}\right)\left(1-\frac{2z}{n}\right)\dots\left(1-\frac{(m-1)z}{n}\right)}.$$

By the expansion in Eq. (9.5), we have

$$Q_{m,n}(z) = \frac{(n)_m\, z}{n}\sum_{k=1}^{\infty}\left\{ {k \atop m-1} \right\}\left(\frac{z}{n}\right)^k.$$

The probability as stated in the theorem follows by extracting the coefficient of z^k on both sides. ■

Toward asymptotics, we represent this exact distribution as a convolution of Poisson random variables.

THEOREM 9.6
Let $X_{m,n}$ be the population size of the mth species in a competitive exclusion model with n niches (urns) and $s \le n$ species (ball colors). Then,

$$X_{m,n} - m \overset{\mathcal{D}}{=} \sum_{j=1}^{\infty} j\,\text{Poi}\left(\frac{1}{j}\sum_{k=1}^{m}\left(\frac{k-1}{n}\right)^j\right),$$

PROOF Let

$$b_{m,n}(t) = \prod_{k=1}^{m}(n-(k-1)e^t),$$

and take its logarithm:

$$\ln b_{m,n}(t) = \sum_{k=1}^{m}\ln\left(n\left(1-\frac{k-1}{n}e^t\right)\right)$$
$$= m\ln n + \sum_{k=1}^{m}\ln\left(1-\frac{k-1}{n}e^t\right) \tag{9.6}$$
$$= m\ln n - \sum_{k=1}^{m}\sum_{j=1}^{\infty}\left(\frac{k-1}{n}\right)^j\frac{e^{jt}}{j}.$$

We can put the moment generating function $\psi_{m,n}$ (of $X_{m,n} - m$, cf. Eq. (9.4)) in the form

$$\psi_{m,n}(t) = \frac{b_{m,n}(0)}{b_{m,n}(t)} = \exp\left(\sum_{k=1}^{m}\sum_{j=1}^{\infty}\frac{1}{j}\left(\frac{k-1}{n}\right)^{j}(e^{jt}-1)\right).$$

Recall that the moment generating function of $\text{Poi}(\lambda)$ is $\exp(\lambda(e^{t}-1))$, and so $\exp(\lambda(e^{rt}-1))$ is the moment generating function of $r\text{Poi}(\lambda)$. If we define

$$\lambda_j(m) = \frac{1}{j}\sum_{k=1}^{m}\left(\frac{k-1}{n}\right)^{j},$$

we produce the convolution form

$$\psi_{m,n}(t) = \exp\left(\sum_{j=1}^{\infty}\lambda_j(m)(e^{jt}-1)\right),$$

which is the moment generating function of a sum of independent $j\text{Poi}(\lambda_j(m))$ random variables. ■

The cocktail of Poisson random variables in Theorem 9.6 gives rise to multiple asymptotics, depending on the value of m relative to n. Roughly speaking for any fixed j, the random variable $j\text{Poi}(j^{-1}\sum_{k=1}^{m}((k-1)/n)^{j}) = j\text{Poi}$ $(O(m^{j+1}/n^{j}))$ is "very small" with high probability, if $m = o(n^{j/(j+1)})$, as the Poisson parameter is too small, when n is large. If m is in the range starting at $\Omega(n^{(j-1)/j})$ and reaching out to $o(n^{j/(j+1)})$, the first $j-1$ Poisson random variable "dominate," while the others are asymptotically small. For instance, with $m = \lceil n^{0.7}\rceil$, the Poisson parameters involved with $j = 1, 2, 3, 4\ldots$ are of the orders $n^{0.4}, n^{0.1}, n^{-0.2}, n^{-0.5}, \ldots$, and only the first two have a nonnegligible asymptotic effect. For each positive integer j, the orders of magnitude lying between $\Omega(n^{(j-1)/j})$ and $o(n^{j/(j+1)})$ define a new range, with $\Omega(n^{(j-1)/j})$ being its lower *boundary*. Every time m crosses the lower boundary over into a new range, one extra Poisson random variable is released into the picture as an effective nonnegligible contributor in the convolution.

Before we study the various ranges, let us add rigor to the above argument by justifying the asymptotic negligibility in the tail. For this we need asymptotic equivalents to the moment generating function. The negative of the logarithm in Eq. (9.6) can be written as

$$-\ln\left(1-\frac{k-1}{n}e^{t}\right) = \sum_{j=1}^{r-1}\left(\frac{k-1}{n}\right)^{j}\frac{e^{jt}}{j} + O\left(\left(\frac{k}{n}\right)^{r}\right),$$

and the sum of the negative logarithms is

$$-\sum_{k=1}^{m} \ln\left(1 - \frac{k-1}{n}e^t\right) = \sum_{k=1}^{m}\sum_{j=1}^{r-1}\left(\frac{k-1}{n}\right)^j \frac{e^{jt}}{j} + O\left(\frac{m^{r+1}}{n^r}\right).$$

Therefore,

$$\psi_{m,n}(t) = \exp\left(\sum_{j=1}^{r-1}\sum_{k=1}^{m}\left(\frac{k-1}{n}\right)^j \frac{e^{jt}-1}{j} + O\left(\frac{m^{r+1}}{n^r}\right)\right). \tag{9.7}$$

The asymptotic behavior appears piecemeal. There are four essential ranges, and an infinite number of hidden rates of convergence.

COROLLARY 9.3
(Baum and Billingsley, 1965). If $m = o(\sqrt{n})$, as $n \to \infty$, then

$$X_{m,n} - m \xrightarrow{P} 0.$$

PROOF Take $r = 1$ in Eq. (9.7), to get

$$\psi_{m,n}(t) = \exp\left(O\left(\frac{m^2}{n}\right)\right) \to 1 = \mathbf{E}[e^{0\times t}].$$

The right-hand side is the moment generating function of 0, and $X_{m,n} - m$ converges in distribution (and consequently) in probability to 0. ∎

Corollary 9.3 can also be proved by comparing the relative rate of growth of the mean and variance via Chebyshev's inequality, as was done in a number of instances (cf. Corollaries 3.2 and 9.1, for example).

COROLLARY 9.4
(Baum and Billingsley, 1965). If $m \sim \lambda\sqrt{n}$, as $n \to \infty$, then

$$X_{m,n} - m \stackrel{\mathcal{D}}{=} \mathrm{Poi}\left(\frac{\lambda^2}{2}\right).$$

PROOF Take $r = 2$ in Eq. (9.7), to get

$$\begin{aligned}\psi_{m,n}(t) &= \exp\left(\sum_{k=1}^{m}\left(\frac{k-1}{n}\right)(e^t - 1) + O\left(\frac{m^3}{n^2}\right)\right)\\ &= \exp\left(\frac{m(m-1)}{2n}(e^t-1) + O\left(\frac{1}{\sqrt{n}}\right)\right)\\ &\to e^{\frac{1}{2}\lambda^2(e^t-1)}.\end{aligned}$$

Hence $X_{m,n} - m$ converges in distribution to a $\mathrm{Poi}(\frac{1}{2}\lambda^2)$ random variable. ∎

Corollary 9.4 applies to a growth of m as $\lambda\sqrt{n}$, for any fixed λ. This covers cases like $m = \lfloor 5\sqrt{n}\rfloor$, $m = 5\lfloor 17n^{\frac{1}{4}} + 6\sqrt{n}\rfloor$. There is behavior of m growing at the exact rate of $\sqrt{n}$ for which Corollary 9.4 does not apply. These are the anomalous cases where $m \sim \lambda_n\sqrt{n}$, and $\lambda_n = O(1)$ but is not convergent to a limit $\lambda > 0$. This includes, for example, cases where λ_n contains oscillations, such as the case $n = 2\lceil(7+2\cos n)\sqrt{n}\,\rceil$. In these anomalous cases the asymptotic moment generating function oscillates, and $X_{m-n} - m$ does not converge in distribution at all to any limiting random variable.

For ranges of m above $\sqrt{n}$, we get normal distributions, so long as $n-m$ is not $O(1)$. It is interesting to see hidden Poisson rates of convergence. For instance if m is both $\Omega(\sqrt{n})$ and $o(n^{\frac{2}{3}})$ the asymptotic moment generating function is $\exp(\frac{m^2}{2n}(e^t - 1))$, which is that of $\mathrm{Poi}(\frac{m^2}{2n})$. For example, if $m = \lceil n^{0.6}\rceil$, $X_{m,n}$ behaves asymptotically like $\lceil n^{0.6}\rceil + \mathrm{Poi}(\frac{1}{2}n^{0.2})$. If m is both $\Omega(n^{\frac{2}{3}})$ and $o(n^{\frac{3}{4}})$ the asymptotic moment generating function is $\exp(\frac{m^2}{2n}(e^t - 1)) + \frac{m^3}{6n^2}(e^{2t} - 1))$, which is the convolution of $\mathrm{Poi}(\frac{m^2}{2n}) + 2\mathrm{Poi}(\frac{m^2}{6n^2})$. This goes on for ranges of m, where m is both $\Omega(n^{(j-1)/j})$ and $o(n^{j/(j+1)})$, for $j = 2, 3, \ldots$, where the distribution of $X_{m,n} - m$ can be approximated by $j - 1$ Poisson random variables. In all these cases with appropriate normalization a central limit theorem holds, as we shall see. The asymptotic normal equivalent is one that captures the behavior of the first asymptotic Poisson random variable. The existence of lower-order Poisson random variables affects the rate of convergence. The inclusion of these lower-order random variables provides a better approximation for small values of n. Table 9.1 compares three approximations to $\mathbf{P}(X_{m,n} \le m)$ for a sequence of increasing n, and $m = \lceil n^{\frac{2}{3}}\rceil$. For $n = 8$, the asymptotic normal distribution is far from the exact, but approximation with one Poisson random variables is reasonably close, and approximation by two Poisson random variables is in error by about only 1%. Even at n as high has 2^9, the normal approximation is off by about 0.7 percentage points.

For the range $m \sim \alpha n, 0 < \alpha < 1$, we shall prove a central limit theorem. This was shown in Rényi (1962) and reproved in Baum and Billingsley (1965) by a proof that focused on a careful analysis of the asymptotics of the characteristic function. Hurley and Mahmoud (1994) gives a proof, presented below, that appeals to standard central limit theorems (which was deemed in Baum and Billingsley to be "prohibitive.") The proof we present employs the following standard theorem (which can be found in many textbooks).

TABLE 9.1

Comparison of Three Approximations to $P(X_{\lceil n^{2/3}\rceil,n} \le \lceil n^{2/3}\rceil)$ with the Exact Probability

n	Exact	Normal	One Poisson	Two Poissons
2^3	0.4102	0.1950	0.4724	0.4234
2^6	0.1290	0.0859	0.1534	0.1318
2^9	0.0164	0.0237	0.0195	0.0166
2^{12}	2.92×10^{-4}	2.38×10^{-4}	3.4×10^{-4}	2.93×10^{-4}

THEOREM 9.7
(Lindeberg's central limit theorem) Suppose $X_1, X_2, \dots$ are independent, with corresponding distribution functions $F_1, F_2, \dots$. Let X_i have mean μ_i and finite variance σ_i^2, and set

$$s_n^2 = \sum_{i=1}^{n} \sigma_i^2.$$

If the family $\{X_i\}$ satisfies Lindeberg's condition

$$\lim_{n\to\infty} \frac{1}{s_n^2} \sum_{i=1}^{n} \int_{|x-\mu_i|\ge \varepsilon s_n} (x-\mu_i)^2 \, dF_i = 0,$$

for every $\varepsilon > 0$, then

$$\frac{\sum_{i=1}^{n} X_i - \sum_{i=1}^{n} \mu_i}{s_n} \xrightarrow{\mathcal{D}} \mathcal{N}(0,1).$$

We have already developed an asymptotic equivalent for the mean in this case:

$$\mathbf{E}[X_{m,n}] = n(H_n - H_{n-m}) \sim n \ln\left(\frac{1}{1-\alpha}\right).$$

This will be a shift factor. The standard deviation is also needed as scale factor. We have the asymptotic equivalent of the variance from simple approximations to the harmonic numbers:

$$\begin{aligned}
\sigma_{m,n}^2 &= \mathbf{Var}[X_{m,n}] \\
&= n^2\big(H_n^{(2)} - H_{n-m}^{(2)}\big) - n(H_n - H_{n-m}) \\
&\sim n^2 \left[\left(\frac{\pi^2}{6} - \frac{1}{n} + O\left(\frac{1}{n^2}\right)\right) - \left(\frac{\pi^2}{6} - \frac{1}{n-m} + O\left(\frac{1}{(n-m)^2}\right)\right)\right] \\
&\quad - n \ln\left(\frac{1}{1-\alpha}\right) \\
&\sim \left(\frac{\alpha n}{1-\alpha}\right) - n \ln\left(\frac{1}{1-\alpha}\right).
\end{aligned}$$

THEOREM 9.8
(Rényi, 1962, Baum and Billingsley, 1965). Let $X_{m,n}$ be the population size of the mth species in a competitive exclusion model with n niches (urns) and $m \sim \alpha n$, $0 < \alpha < 1$, species (ball colors). Then,

$$\frac{X_{m,n} - n \ln\left(\frac{1}{1-\alpha}\right)}{\sqrt{\left(\frac{\alpha}{1-\alpha} - \ln\left(\frac{1}{1-\alpha}\right)\right) n}} \xrightarrow{\mathcal{D}} \mathcal{N}(0,1),$$

PROOF Let G_k denote the kth geometric random variable $\text{Geo}((n-k+1)/n)$ in the convolution

$$X_{m,n} = 1 + \text{Geo}\left(\frac{n-1}{n}\right) + \text{Geo}\left(\frac{n-2}{n}\right) + \cdots + \text{Geo}\left(\frac{n-m+1}{n}\right).$$

This random variable has success rate $p_k = (n-k+1)/n$, and failure rate $q_k = 1 - p_k$. Let $\sigma^2_{m,n} = \mathbf{Var}[X_{m,n}]$, and $\mu_k = \mu_k(n) = \mathbf{E}[G_k]$. Fix $\varepsilon > 0$, and define *Lindeberg's quantity*

$$L_{m,n}(\varepsilon) = \frac{1}{\sigma^2_{m,n}} \sum_{k=1}^{m} \sum_{|x-\mu_k| \ge \varepsilon\sigma_{m,n}} (x-\mu_k)^2 \, \mathbf{P}(G_k = x).$$

We verify that

$$\lim_{n\to\infty} L_{m,n}(\varepsilon) = 0.$$

As

$$1 < \mu_k = \frac{n}{n-k+1} < \frac{1}{1-\alpha} + o(1),$$

while $\sigma_{m,n} \to \infty$, for n sufficiently large, $\{x : x \le \mu_k - \varepsilon\sigma_{m,n}\}$ is a set of negative integers, and it is impossible for G_k to fall in it. Therefore,

$$\begin{aligned} L_{m,n}(\varepsilon) &= \frac{1}{\sigma^2_{m,n}} \sum_{k=1}^{m} \sum_{x \ge \mu_k + \varepsilon\sigma_{m,n}} (x-\mu_k)^2 \, \mathbf{P}(G_k = x) \\ &= \frac{1}{\sigma^2_{m,n}} \sum_{k=1}^{m} \sum_{y \in \mathcal{Y}_k(m,n)} y^2 p_k q_k^{y+\mu_k-1}, \end{aligned}$$

where

$$\mathcal{Y}_k(m,n) = \{y = j - \mu_k \ : \ j \ge \mu_k + \varepsilon\sigma_{m,n}, \text{ and } j \text{ is a positive integer}\}.$$

Thus,

$$\begin{aligned} L_{m,n}(\varepsilon) &= \frac{1}{\sigma^2_{m,n}} \sum_{k=1}^{m} p_k q_k^{\mu_k+1} \sum_{y \in \mathcal{Y}_k(m,n)} y^2 q_k^{y-2} \\ &\le \frac{1}{\sigma^2_{m,n}} \sum_{k=1}^{m} p_k q_k^{\mu_k+1} \sum_{\mathcal{Y}_k(m,n) \ni y \ge v_k(m,n)} y^2 q_k^{y-2}, \end{aligned}$$

where $v_k = v_k(m,n) = \lceil \min_y \mathcal{Y}(m,n) \rceil = \lceil \mu_k + \varepsilon\sigma_{m,n} \rceil$, which is at least 2 for large n. The inner series is a variant of the second derivative of a geometric series. Summing this geometric-like series we obtain the bound

$$\begin{aligned}
L_{m,n}(\varepsilon) &\le \frac{1}{\sigma^2_{m,n}} \sum_{k=1}^{m} p_k q_k^{\mu_k+1} \Big[\big(v_k^2 - (2v_k^2 - 2v_k - 1)q_k\big)\, q_k^{v_k-2} + q_k^{v_k}(v_k-1)^2\Big]\, p_k^{-3} \\
&< \frac{1}{\sigma^2_{m,n}} \sum_{k=1}^{m} q_k^{\mu_k+1} \Big[\big(v_k^2 - (v_k^2 - 2 + (v_k-1)^2)\, q_k\big)\, q_k^{v_k-2} + q_k^{v_k} v_k^2\Big]\, p_k^{-2} \\
&< \frac{1}{\sigma^2_{m,n}} \sum_{k=1}^{m} q_k^{\mu_k+1} \frac{q_k^{v_k-2}}{p_k^2} v_k^2 \left(1 + q_k^2\right);
\end{aligned}$$

we only dropped negative terms. As $1 \le \mu_k \le 1/(1-\alpha)+1$, for large n and each $1 \le k \le m$,

$$L_{m,n}(\varepsilon) < \frac{2}{\sigma^2_{m,n}} \sum_{k=1}^{m} \left(\frac{k-1}{n}\right)^{v_k+1} \left(\frac{n}{n-k+1}\right)^2 v_k^2.$$

The sequence μ_k, hence v_k, is increasing, and we can bound the right-hand side by using the least power v_1 for q_k uniformly, and replacing v_k^2 by v_m^2: Recalling that $m \sim \alpha n$, for some $0 < C < 1$ and large n, $1 \le m \le Cn$, and

$$\begin{aligned}
L_{m,n}(\varepsilon) &< \frac{2v_m^2 n^2}{\sigma^2_{m,n}(n-m+1)^2} \sum_{k=1}^{m} \left(\frac{k}{n}\right)^{v_1+1} \\
&< \frac{2v_m^2 n^2}{\sigma^2_{m,n}(n-Cn+1)^2 n^{v_1+1}} \sum_{k=1}^{m} k^{v_1+1} \\
&\le \frac{2v_m^2 n^2}{\sigma^2_{m,n}(n-Cn)^2 n^{v_1+1}} \sum_{k=1}^{m} m^{v_1+1} \\
&< \frac{2v_m^2}{\sigma^2_{m,n}(1-C)^2 n^{v_1+1}} m^{v_1+2} \\
&< \frac{2\lceil \mu_m + \varepsilon\sigma_{m,n}\rceil^2}{\sigma^2_{m,n}(1-C)^2 n^{v_1+1}} (Cn)^{v_1+2} \\
&< O(n) \times C^{\varepsilon\sigma_{m,n}}.
\end{aligned}$$

However, $0 < C < 1$, and $\sigma_{m,n} \to \infty$ (at a linear rate), as $n \to \infty$, and it follows that $\lim_{n\to\infty} L_{m,n}(\varepsilon) = 0$; Lindeberg's condition has been verified for the sequence $X^*_{m,n} = \frac{X_{m,n} - \mathbf{E}[X_{m,n}]}{\sqrt{\mathbf{Var}[X_{m,n}}}$, and

$$X^*_{m,n} \xrightarrow{\mathcal{D}} \mathcal{N}(0,1).$$

We can replace the mean and the variance by their asymptotic equivalents via Slutsky's theorems. ■

The last range, when m comes very close to n (only differing from it by $O(1)$), exhibits a curious behavior, where the logarithm of an appropriate chi-squared distribution appears as a limit for a scaled and shifted version of $X_{m,n}$. This is the range where $m = n - b_n$, with $b_n = O(1)$, as is the case in coupon collection to obtain a complete set (the case $m = n$). Erdős and Rényi (1961) proved the result in the context of coupon collection with $m = n$, but Baum and Billingsley (1965) expanded the range to $m = n - O(1)$. The proof for this range runs along the lines for ranges already discussed, therefore, it will only be sketched. It is based on a careful analysis of the moment generating function (or characteristic function), then identifying the nature of the limiting random variable.

THEOREM 9.9

(Erdős and Rényi, 1961, Baum and Billingsley, 1965). Let $X_{m,n}$ be the population size of the mth species in a competitive exclusion model with n niches (urns) and $m = n - b_n$ species (ball colors), with $b_n \to b$ (a constant). Then,

$$2ne^{-\frac{X_{m,n}}{n}} \xrightarrow{\mathcal{D}} \chi^2(2b+2).$$

PROOF *(sketch).* Let $\mu_{m,n} = \mathbf{E}[X_{m,n}] = n(H_n - H_{b_n})$. Let us write the moment generating function for $X_{m,n}/n$. It is the same as that of the raw variable $X_{m,n}$, but with the dummy variable of the moment generating function scaled by n:

$$\phi_{X_{m,n}/n}(t) = e^{mt/n} \prod_{k=1}^{m} \frac{n-k+1}{n-(k-1)e^{t/n}}.$$

We then approximate $e^{t/n}$ with $1 + t/n$, which is the Taylor series truncated after two terms. For large n, and for the given range of m we have

$$\phi_{(X_{m,n}-\mu_{m,n})/n}(t) \sim e^{-t\sum_{k=b_n+1}^{n}\frac{1}{k}} \prod_{k=1}^{m} \frac{n-k+1}{n-(k-1)(1+t/n)} \sim \prod_{k=b_n+1}^{n} \frac{e^{-\frac{1}{k}t}}{1-t/k}.$$

Recalling that the $\text{Exp}(\lambda)$ random variable has moment generating function $1/(1-\lambda t)$, we recognize the latter product as the moment generating function of the convolution

$$W_{b_n,n} = \sum_{k=b_n+1}^{n} \left(Y_k - \frac{1}{k}\right),$$

with $Y_k = \text{Exp}(1/k)$, and the Y_k's are independent. If b_n converges to a limit $b, (X_{m,n} - \mu_{m,n})/n$ will converge to the random variable $W_{b,\infty}$.[6]

Inductively, we can show that $W_{b_n,n}$ has the density

$$f_n(x) = \begin{cases} n\binom{n-1}{b_n} e^{-(b_n+1)(x+A_{b_n+1})}(1 - e^{-(x+A_{b_n+1})})^{n-b_n-1}, & \text{if } x > 0, \\ 0, & \text{elsewhere,} \end{cases}$$

where $A_{b_n+1} = \gamma - H_{b_n}$. If $b_n \to b$, the density $f_n(x)$ converges to

$$\frac{1}{b!} e^{-(b+1)(x+A_{b+1})} e^{-e^{(x+A_{b+1})}}.$$

A change of variables shows that the limiting density function has the form $x^b e^{-x/2}/(2^{b+1}\Gamma(b+1))$, which is that of a chi-squared random variable with $2b+2$ degrees of freedom. Since $(X_{m,n} - \mu_{m,n})/n$ converges in law to $W_{b,\infty}$, $\exp(-X_{m,n}/n + \ln(2n))$ converges in distribution to $\chi^2(2b+2)$. ■

In the Theorem 9.9 it was assumed that $m = n - b_n$, for $b_n \to b$. If b_n does not converge, such as in a case like $m = n - b_n = n - \lceil 3 + \sin n \rceil)$, the proof indicates that $\chi^2(2\lceil 3 + \sin n \rceil + 2)$ will be a good approximation, for large n. But of course, as the asymptotic chi-squared random variables change with n, in these anomalous cases there will be no convergence in distribution to a unique law.

9.3 Epidemiology: Kriz' Polytonomous Urn Scheme

The dichromatic Pólya urn with the scheme

$$\mathbf{A} = \begin{pmatrix} 1 & 0 \\ 0 & 1 \end{pmatrix}$$

is an old classic model for contagion that goes back to Eggenberger and Pólya (1923). It is presumed to be a model for the spreading of a disease in a society, where a newcomer's first contact is with an infected person and the newcomer becomes infected, or with an immune person and the newcomer becomes immune. Many natural generalizations lead to multivariate Pólya distributions with epidemiologic interpretation. The diagonal elements need not be 1 (as in the case of the arrival of married couples), and the number of colors need not be two. We have seen a multicolor hyper Pólya-Eggenberger

[6] This random variable is itself defined as an infinite series, which can be shown to converge almost surely, in view of Kolmogorov's criterion of convergence of the corresponding variance series.

urn scheme in Exercise 5.2. We recall the result here to build the Kriz model around it. We have an urn of up to k colors, with the hyper Pólya-Eggenberger schema

$$\mathbf{A} = \begin{pmatrix} s & 0 & 0 & \dots & 0 \\ 0 & s & 0 & \dots & 0 \\ \vdots & \vdots & & \ddots & \vdots \\ 0 & 0 & & \dots & s \end{pmatrix}. \tag{9.8}$$

Initially there are τ_i balls of color i, and a total of $\tau = \sum_{i=1}^{k} \tau_i$ balls in the urn. Let $X_i^{(n)}$ be the number of times balls of color i have been withdrawn by the time of the nth draw. Then,

$$\mathbf{P}\left(X_1^{(n)} = x_1, \dots, X_k^{(n)} = x_k\right) = \binom{n}{x_1, \dots, x_k} \frac{\prod_{j=1}^{k} \tau_j(\tau_j + s)\dots(\tau_i + (x_j - 1)s)}{\tau(\tau + s)\dots(\tau + (n-1)s))},$$

for a feasible partition of n, that is, $\sum_{i=1}^{k} x_i = n$, for nonnegative integers $x_1, \dots, x_k$, and of course the probability is 0, if $x_1, \dots, x_k$ are not a feasible partition of n.

A real society may have multiple infectious diseases spreading in it, and multiple societies may be evolving simultaneously such as the different villages in a region or cities in a state. The Kriz (1972) polytonomous model deals with this extension, and is useful in comparative epidemiological studies.

In the Kriz model we let m Pólya-Eggenberger multicolor urns progress in parallel, with the ith urn having up to k_i different ball colors, and the hyper Pólya-Egenberger schema $\mathbf{A}_i$, which is a $k_i \times k_i$ version of Eq. (9.8), for $i = 1, \dots, m$. The ball colors in each urn are assumed to be distinct and all urns start with the same total number of balls τ. For instance, with $m = 2$, $k_1 = 3, k_2 = 2$, and $s = 1$ we have two Pólya-Eggenberger urns: One has a 3×3 schema, and the other has a 2×2 schema, and we can think of the first urn to be a scheme of white, blue, and green balls, while the second is a scheme on the set of red and black colors. Simultaneously, from each urn we draw a ball at random, and we replace it in the urn together with another ball of the same color.

Let $X_{(i,j)}^{(n)}$ be the number of times a ball of the jth color in the ith urn appears in the sampling up to the nth draw. As the urns develop independently of each other, the multivariate distribution in question is that of m independent multivariate Pólya-Eggenberger random variables:

$$\mathbf{P}(\{X_{(i,j)}^{(n)} = x_{i,j} \ : \ i = 1, \dots, m, \ j = 1, \dots, k_i\})$$
$$= \prod_{i=1}^{m} \binom{n}{x_{i,1}, \dots, x_{i,k_i}} \frac{\prod_{j=1}^{k_i} \prod_{r=1}^{x_{i,j}} (\tau_{i,j} + (r-1)s)}{(\tau(\tau + s)\dots(\tau + (n-1)s)},$$

where $\tau_{i,j}$ is the initial number of balls of the jth color in the ith urn.

9.4 Clinical Trials

Urn schemes have been proposed as models for the discriminatory monitoring of alternative medical treatments. They were introduced in Zelen (1969), who cites an origin of the formulation in the context of gambling strategies in Robbins (1952). To differentiate between two clinical treatments (surgeries, laboratory tests, procedures or the administration of drugs) one sets up an urn. Suppose the two treatments are $\mathcal{A}$ and $\mathcal{B}$. It is assumed that it is the nature of the treatment that the result is instantaneously obtained. The urn contains white balls corresponding to the degree of success in Treatment $\mathcal{A}$, and blue balls corresponding to the degree of success in Treatment $\mathcal{B}$. Let W_n and B_n be, respectively, the number of white and blue balls in the urn after n draws, and let $\tilde{W}_n$ and $\tilde{B}_n$ be, respectively, the number of times a white or a blue ball is drawn among the first n draws. In the absence of prior information we start with an equal number of balls of each color: $W_0 = B_0$. Let the total starting number of balls be $\tau_0 = 2W_0$.

When Treatment $\mathcal{A}$ is administered, it may succeed with a certain hidden probability $p_\mathcal{A}$, or fail with probability $q_\mathcal{A} = 1 - p_\mathcal{A}$. Similarly, Treatment $\mathcal{B}$ succeeds and fails, respectively, with probability $p_\mathcal{B}$ and $q_\mathcal{B} = 1 - p_\mathcal{B}$.

9.4.1 Play-the-Winner Schemes

To select a treatment for the next patient, the clinician picks at random a ball from the urn. If the ball is white, it is replaced in the urn, Treatment $\mathcal{A}$ is given and the result is observed. If it is observed that the treatment succeeds, we add one white ball so that the scale is tilted in favor of Treatment $\mathcal{A}$ in the future. Alternatively, if Treatment $\mathcal{A}$ is given and fails, we add one blue ball to favor Treatment $\mathcal{B}$ in the future. We do the opposite if a blue ball is withdrawn. In this case, we replace the ball and give Treatment $\mathcal{B}$. If the treatment succeeds, we add a blue ball to the urn to favor $\mathcal{B}$ in the future, and if the treatment fails we add a white ball to favor $\mathcal{A}$ in the future. The scheme works in favor of success, and is called *Play-the-Winner* scheme. Let $X_\mathcal{A}$ and $X_\mathcal{B}$ be, respectively, $\mathrm{Ber}(p_\mathcal{A})$ and $\mathrm{Ber}(p_\mathcal{B})$ random variables. The urn scheme is then given by the replacement matrix

$$\begin{pmatrix} X_\mathcal{A} & 1 - X_\mathcal{A} \\ 1 - X_\mathcal{B} & X_\mathcal{B} \end{pmatrix};$$

as usual white precedes blue in indexing both rows and columns. It is ethically unacceptable to try treatments with rates of success that are not at least believed to be favorable (with probability of success at least $\frac{1}{2}$). This will be assumed for the rest of this discussion—both $p_\mathcal{A}$ and $p_\mathcal{B}$ are at least $\frac{1}{2}$.

A question of central importance in this urn setup is to see that after running the scheme for a sufficiently long time, it gravitates toward chances of success that are higher than schemes that operate in the absence of information gained throughout. For instance, a scheme that does not "learn" from the accumulating information might use a *randomized clinical trial*. In this case, each time a patient needs a treatment, the choice is randomized: A coin is flipped and with probability $\frac{1}{2}$ Treatment $\mathcal{A}$ is selected, and with probability $\frac{1}{2}$ Treatment $\mathcal{B}$ is selected. Let Y be the indicator of success in an individual patient. Then

$$Y = \begin{cases} X_{\mathcal{A}}, & \text{with probability } \frac{1}{2}; \\ X_{\mathcal{B}}, & \text{with probability } \frac{1}{2}. \end{cases}$$

By conditioning on the event $\mathcal{H}$ controlling the randomization (Heads and the complement Tails, for example, in coin flipping) the randomized treatment then has the average rate of success

$$\begin{aligned} \mathbf{E}[Y] &= \mathbf{E}[X \mid \mathcal{H}]\,\mathbf{P}(\mathcal{H}) + \mathbf{E}[X \mid \mathcal{H}^c]\,\mathbf{P}(\mathcal{H}^c) \\ &= \frac{1}{2}\mathbf{E}[X_{\mathcal{A}}] + \frac{1}{2}\mathbf{E}[X_{\mathcal{B}}] \\ &= \frac{p_{\mathcal{A}} + p_{\mathcal{B}}}{2}. \end{aligned}$$

Over n treatments, each succeeding according to Y_i, an independent copy of Y, we have a total number of successes give by

$$S_n = Y_1 + \cdots + Y_n.$$

According to the strong law of large numbers, the proportion of times there is success is then

$$\frac{S_n}{n} \xrightarrow{a.s.} \mathbf{E}[Y] = \frac{p_{\mathcal{A}} + p_{\mathcal{B}}}{2}.$$

A randomized clinical trial merely attains the average of the two rates of success in Treatments $\mathcal{A}$ and $\mathcal{B}$.

Does the Play-the-Winner scheme do better? Does it learn well from the accumulated experience? The eigenvalues of the Play-the-Winner average generator $\mathbf{E}[\mathbf{A}]$ are $\lambda_1 = 1$ and $\lambda_2 = p_{\mathcal{A}} + p_{\mathcal{B}} - 1 \in [0, 1]$. The principal eigenvalue of $\mathbf{E}[\mathbf{A}^T]$ is 1 with a corresponding eigenvector

$$\begin{pmatrix} v_1 \\ v_2 \end{pmatrix} = \frac{1}{2 - p_{\mathcal{A}} - p_{\mathcal{B}}} \begin{pmatrix} 1 - p_{\mathcal{B}} \\ 1 - p_{\mathcal{A}} \end{pmatrix}.$$

Each treatment adds one ball to the urn. After $n-1$ treatments, we have $n+\tau_0-1$ balls in the urn. We choose the next treatment to be $\mathcal{A}$, with probability $W_{n-1}/(n+\tau_0-1)$ and we choose it to be $\mathcal{B}$, with probability $B_{n-1}/(n+\tau_0-1)$. The indicator of success with the nth patient is

$$\tilde{Y}_n = \begin{cases} X_{\mathcal{A}}^{(n)}, & \text{with probability } \dfrac{W_{n-1}}{n+\tau_0-1}; \\ X_{\mathcal{B}}^{(n)}, & \text{with probability } \dfrac{B_{n-1}}{n+\tau_0-1}, \end{cases}$$

with $X_A^{(n)}$ and $X_B^{(n)}$ being identically distributed independent copies of X_A and X_B. By Theorem 6.4, the proportion of white balls converges in probability to $\lambda_1 v_1 = v_1$, and the proportion of blue balls converges in probability to $\lambda_1 v_2 = v_2$.

Let $\tilde{S}_n$ be the number of successes after n treatments under Play-the-Winner rules. Then,

$$\tilde{S}_n = \tilde{Y}_1 + \cdots + \tilde{Y}_n.$$

Of these n (dependent) random variables there are $\hat{W}_n$ behaving like independent copies of X_A, and $\hat{B}_n$ behaving like independent copies of X_B. We can represent the number of successes as

$$\tilde{S}_n = \left(X_A^{(1)} + \cdots + X_A^{(\tilde{W}_n)}\right) + \left(X_B^{(1)} + \cdots + X_B^{(\tilde{B}_n)}\right).$$

Recall that $\tilde{W}_n$ and $\tilde{B}_n$ are, respectively, the number of times a white or a blue ball appears in the sample after n draws. In view of the latter partition, we have the convergence relations for each part:

$$\frac{\tilde{W}_n}{n} \xrightarrow{P} v_1 = \frac{1 - p_B}{2 - p_A - p_B}, \qquad \tilde{W}_n \xrightarrow{a.s.} \infty, \qquad \frac{X_A^{(1)} + \cdots + X_A^{(\tilde{W}_n)}}{\tilde{W}_n} \xrightarrow{a.s.} p_A.$$

Hence,

$$\frac{X_A^{(1)} + \cdots + X_A^{(\tilde{W}_n)}}{n} = \frac{X_A^{(1)} + \cdots + X_A^{(\tilde{W}_n)}}{\tilde{W}_n} \times \frac{\tilde{W}_n}{n} \xrightarrow{P} p_A\left(\frac{1 - p_B}{2 - p_A - p_B}\right).$$

Likewise,

$$\frac{X_B^{(1)} + \cdots + X_B^{(\tilde{B}_n)}}{n} \xrightarrow{P} p_B\left(\frac{1 - p_A}{2 - p_A - p_B}\right).$$

These convergence relations give a success proportion

$$\begin{aligned}\frac{\tilde{S}_n}{n} &= \frac{\left(X_A^{(1)} + \cdots + X_A^{(\tilde{W}_n)}\right)}{n} + \frac{\left(X_B^{(1)} + \cdots + X_B^{(\tilde{B}_n)}\right)}{n} \\ &\xrightarrow{a.s.} \frac{p_A(1 - p_B) + p_B(1 - p_A)}{2 - p_A - p_B}.\end{aligned}$$

The function

$$h(x, y) = \frac{x(1 - y) + y(1 - x)}{2 - x - y},$$

compares favorably relative to $\frac{1}{2}(x + y)$. One always has

$$h(x, y) \geq \frac{1}{2}(x + y).$$

We can see this from the comparison

$$g(x, y) := h(x, y) - \frac{1}{2}(x + y) = \frac{(x - y)^2}{2(2 - x - y)},$$

which is always nonnegative for the entire range of interest (when both $x, y \in (0, 1)$). The difference function $g(p_A, p_B)$ is 0 when p_A and p_A are the same, when Play-the-Winner cannot top randomized clinical trials. But this difference increases as one gets away from the point of equality. For instance, $h(0.7, 0.7) = 0.7 = \frac{1}{2}(0.7 + 0.7)$, whereas $h(0.7, 0.93) \approx 0.886$, and the average of the two success probabilities is 0.815. Therefore, a Play-the-Winner scheme with $p_A \neq p_B$, offers an improvement over randomized clinical trials, and when the two probabilities are the same the Play-the-Winner scheme is no worse.

Several urns for alternate strategies such as Play-the-Winner with immigration and Drop-the-Loser are discussed in the recent literature. The reader is referred to Ivanova (2003, 2006), Ivanova, Rosenberger, Durham, and Flournoy (2000), and several contributions by Wei (1977–1979), and Wie with coauthors (1978, 1988).

Exercises

9.1 A Hoppe's urn starts with θ special balls. Let X_n be the number of balls of color 1 after n draws. (In the biological interpretation, X_n is the size of the population of the oldest species.) What are the mean and variance of X_n? (Hint:

$$X_n = X_{n-1} + \text{Ber}\left(\frac{X_{n-1}}{n + \theta - 1}\right).)$$

9.2 In the competitive exclusion problem, let the total number of balls of color m dropped in all the urns be $X_{m,n}$. In principle, one could compute the exact distribution of $X_{m,n}$. Explore this possibility by finding the exact distribution of $X_{3,n}$.

9.3 In a phylogenetic tree following the edge-splitting model, what proportion (in probability) of nodes are leaves, what proportion of leaves are not in cherries, and what proportion of nodes are internal (not leaves)?

10

Urns Evolving by Multiple Drawing

The sampling procedure considered in all the previous chapters was concerned with drawing one ball and acting according to some predesignated rules. There may be occasion to consider a sampling procedure that involves drawing a set of balls of a fixed or random size, with ball addition rules according to what is drawn in the sample.

Instead of picking one ball at random, consider picking a number of balls. Depending on the multiset that comes out, we replace the drawn multiset in the urn and add a corresponding number of balls of each color. In the case of k colors and multisets of fixed size $m \ge 2$, the ball addition scheme will thus be a nonsquare matrix, with $\binom{k+m-1}{m}$ rows indexed with multisets of size m, and k columns indexed by the colors. The entries of the matrix are numbers a_{ij}, which can be positive or negative, or even random in the most general case.

We make the illustration on the class of urns with $k = m = 2$, which will give us clues on the flavor of this nonclassic urn. We confine our attention to the balanced case where the number of balls added of each color is fixed, with a fixed total number of balls in each addition.

10.1 Urns Evolving by Drawing Pairs of Balls

We have delineated the scope of the illustration to an urn with balls of two colors, where one draws two balls at a time. We assume our two colors to be white W and blue B. The possible outcome multisets of any draw are $\{W, W\}$, $\{W, B\}$, or $\{B, B\}$. If the pair $\{W, W\}$ is drawn, we replace the pair into the urn and add a white balls and b blue balls. Likewise, we add, respectively, c white and d blue balls upon drawing $\{W, B\}$, and we add, respectively, e white and f blue balls upon drawing $\{B, B\}$. Organized as a matrix, the rows of which are labeled with the multisets drawn, and the columns of which are labeled with the color of balls added, we have the associated 3×2 rectangular urn schema

$$\begin{array}{c} \\ WW \\ WB \\ BB \end{array} \begin{array}{c} \begin{array}{cc} W & B \end{array} \\ \begin{pmatrix} a & b \\ c & d \\ e & f \end{pmatrix} \end{array}.$$

As mentioned, we focus our attention on the balanced case where the total number of balls added of all colors is a fixed number $K \geq 1$ (i.e., constant row sum).

The concept of tenability extends to rectangular urn schemes. A rectangular urn scheme that progresses upon the repeated drawing of two balls is *tenable* if it is possible to draw pairs from it and apply the rules indefinitely. If all the entries in the scheme are nonnegative, no further constraints are necessary for tenability. However, if some entries are negative, the starting numbers of balls of the two colors have to satisfy some divisibility conditions, or else the urn may "get stuck." That is, there may come a time when drawing balls according to the rules is no longer feasible. For example, the scheme $\begin{pmatrix} -1 & 2 \\ 0 & 1 \\ 1 & 0 \end{pmatrix}$ is clearly tenable no matter how we start with $\tau_0 \geq 2$ balls, whereas the scheme $\begin{pmatrix} -3 & 5 \\ 0 & 2 \\ 1 & 1 \end{pmatrix}$ may get stuck if $W_0 = 8$, when drawing $\{W, W\}$ persists three times. We shall assume we are dealing only with tenable urns.

Suppose we perpetuate draws of pairs of balls from a tenable urn with a 3×2 rectangular schema, with constant row sum K, that starts out with $\tau_0 \geq 2$ balls in total. The total number of balls after n draws is obviously

$$\tau_n = Kn + \tau_0.$$

Let W_n and B_n be the number of white and blue balls, respectively, after n random draws (a *random draw* refers to one step of drawing two balls without replacement, where all *distinct* pairs of balls are equally likely).

Three indicator random variables influence the evolution. These are I_n^{WW}, I_n^{WB}, and I_n^{BB}, where I_n^{XY} generally indicates a draw at the nth step of the multiset $\{X, Y\}$, that is, I_n^{XY} assumes the value 1 when the multiset $\{X, Y\}$ is drawn at the nth stage, and assumes the value 0 otherwise. Of course, the three indicators I_n^{WW}, I_n^{WB}, and I_n^{BB} are mutually exclusive, only one of them must be 1, the rest are 0. In view of these definitions we have the stochastic recurrence

$$W_n = W_{n-1} + a\, I_n^{WW} + c\, I_n^{WB} + e\, I_n^{BB}. \tag{10.1}$$

Conditioning the last relation on W_{n-1}, we obtain

$$\mathbf{E}[W_n \mid W_{n-1}] = W_{n-1} + a\,\mathbf{E}\big[I_n^{WW} \mid W_{n-1}\big] + c\,\mathbf{E}\big[I_n^{WB} \mid W_{n-1}\big] + e\,\mathbf{E}\big[I_n^{BB} \mid W_{n-1}\big]. \tag{10.2}$$

According to the drawing rules, the conditional expectation of the indicators is

$$\mathbf{E}\big[I_n^{WW} \mid W_{n-1}\big] = \frac{\binom{W_{n-1}}{2}}{\binom{\tau_{n-1}}{2}}; \tag{10.3}$$

$$\mathbf{E}\big[I_n^{WB} \mid W_{n-1}\big] = \frac{\binom{W_{n-1}}{1}\binom{B_{n-1}}{1}}{\binom{\tau_{n-1}}{2}}; \tag{10.4}$$

$$\mathbf{E}\big[I_n^{BB} \mid W_{n-1}\big] = \frac{\binom{B_{n-1}}{2}}{\binom{\tau_{n-1}}{2}}. \tag{10.5}$$

To develop an intuition in the cases where "equilibrium" is attained (that is, when the number of balls of each color reaches a steady proportion over time), we consider the case where $W_n/(Kn)$ converges in some sense to some constant value, say $w \neq 0$. Surely, the average sense is a goal and, if possible, a probabilistic sense (such as convergence in probability) is preferable. If a state of equilibrium is attained, then $B_n/(Kn)$ also approaches a limit (say b) in the same sense, and necessarily

$$w + b = 1. \tag{10.6}$$

This intuitive approach helps us identify cases with normal behavior, an intuition that will be rigorized later on.

Plugging the conditional expectations Eq. ((10.3)–(10.5)) into Eq. (10.2), then taking expectation, one obtains

$$\begin{aligned}\mathbf{E}[W_n] = \mathbf{E}[W_{n-1}] &+ \frac{a}{\tau_{n-1}(\tau_{n-1}-1)}\mathbf{E}[W_{n-1}(W_{n-1}-1)]\\ &+ \frac{2c}{\tau_{n-1}(\tau_{n-1}-1)}\mathbf{E}[W_{n-1}B_{n-1}] + \frac{e}{\tau_{n-1}(\tau_{n-1}-1)}\mathbf{E}[B_{n-1}(B_{n-1}-1)].\end{aligned} \tag{10.7}$$

If a state of equilibrium is attained, one can interpret this recurrence as:

$$wKn \approx wK(n-1) + aw^2 + 2cwb + eb^2,$$

which in view of Eq. (10.6) simplifies in the limit to the quadratic equation

$$(a - 2c + e)w^2 + (2c - 2e - K)w + e = 0,$$

the solutions of which are

$$w = \frac{K + 2(e - c) \pm \sqrt{K^2 + 4\left[K(e - c) + c^2 - ae\right]}}{2(a - 2c + e)},$$

if $a - 2c + e \neq 0$. We call this quadratic equation the *indicial equation*. It clearly suggests that an urn with the linearization effect

$$a - 2c + e = 0$$

is "easier" to handle; its indicial equation is only linear, with the unique solution

$$w = \frac{e}{K + e - a},$$

when this number is nonzero and is well defined. The case $e = 0$, and $K + e - a = 0$ introduces an interesting degeneracy. This latter case is the subject of investigation in Chen and Wei (2005). Take a note that when w is well defined, it must be in $[0, 1]$, because the total number of balls grows linearly, and that is a natural bound on the number of white balls. In the case $w = 0$, the number of white balls grows sublinearly. We look at these cases in the exercises.

Indeed, we shall see in a rigorous way that the balanced case with $w \neq 0$ possesses a linear recurrence equation for average and is approachable with the theory of martingales. A complete asymptotic theory will be derived for this balanced case. The case of a fully quadratic indicial equation (a reference to the case $a - 2c + e \neq 0$) appears to be significantly harder to approach with these tools. Nevertheless, we shall be able to say a few words about a heuristic analysis of the behavior of the urn on average, but not about its asymptotic distribution.

We call an urn scheme with a linear indicial equation a *linear urn scheme*. Using the relation $B_n = \tau_n - W_n$, the conditional recurrence (Eq. (10.7)) becomes

$$\mathbf{E}[W_n \mid W_{n-1}] = \frac{a - 2c + e}{\tau_{n-1}(\tau_{n-1} - 1)}\mathbf{E}\left[W_{n-1}^2 \mid W_{n-1}\right] + g_n W_{n-1} + e,$$

where

$$g_n = \frac{Kn + \tau_0 - K - (e - a)}{Kn + \tau_0 - K}.$$

In a linear urn scheme, the recurrence becomes purely linear. It takes the form

$$\mathbf{E}[W_n \mid W_{n-1}] = g_n W_{n-1} + e, \qquad (10.8)$$

and W_n is almost a martingale, but not quite so. We shall consider in Section 10.3 a martingalization procedure.

10.2 The Chen-Wei Urn

Chen and Wei (2005) considers an interesting class of balanced urn schemes, their urn has the schema

$$\begin{pmatrix} 2C & 0 \\ C & C \\ 0 & 2C \end{pmatrix}.$$

Here is a degenerate case where $e = 0$, $K = a = 2C$, and w is not well defined. The recurrence (Eq. (10.8)) for this degenerate case simplifies to

$$\mathbf{E}[W_n \mid W_{n-1}] = \frac{2Cn + \tau_0}{2Cn + \tau_0 - 2C} W_{n-1}.$$

The double expectation gives the recurrence

$$\mathbf{E}[W_n] = \frac{2Cn + \tau_0}{2Cn + \tau_0 - 2C} \mathbf{E}[W_{n-1}].$$

This recurrence unwinds easily into

$$\begin{aligned} \mathbf{E}[W_n] &= \frac{2Cn + \tau_0}{2Cn + \tau_0 - 2C} \times \frac{2Cn + \tau_0 - 2C}{2Cn + \tau_0 - 4C} \\ &\quad \times \frac{2Cn + \tau_0 - 4C}{2Cn + \tau_0 - 6C} \times \cdots \times \frac{2C + \tau_0}{\tau_0} \mathbf{E}[W_0] \\ &= \frac{2C\,W_0}{\tau_0} n + W_0. \end{aligned}$$

Curiously, the average behavior of this urn asymptotically depends on the initial conditions; recall a similar phenomenon in the Pólya-Eggenberger urn.

10.3 Linear Rectangular Urns

Generally, for the scheme $\begin{pmatrix} a & b \\ c & d \\ e & f \end{pmatrix}$ when $e = 0$, the urn is either untenable or the number of white balls, does not possess a linear rate of growth, or the coefficients of the leading terms depend on the initial conditions. Let us take up next nondegenerate cases ($e \neq 0$) with positive $w = e/(K + e - a)$, where the initial conditions are washed away in the long run. The difference $\theta := e - a$ will appear often. We assume throughout that $\theta \neq 0$. (The case $\theta = 0$ is of no interest; it involves no randomness because $\theta = 0$ means that $a = c = e$, and $W_n \equiv an + W_0$.) To sustain tenability, when $\theta \geq 1$, e cannot be negative, and when $\theta \leq 1$, a cannot be negative. When $0 < w < 1$, the average number of both colors grows linearly, as we shall see, and that is why we refer to such a class of urn schemes as linear.

PROPOSITION 10.1

In a tenable linear urn scheme with the balanced schema $\begin{pmatrix} a & b \\ c & d \\ e & f \end{pmatrix}$ *that has constant row sum* $K \ge 1$, *the exact average number of white balls after* n *draws is*

$$\mathbf{E}[W_n] = \frac{eK}{K+\theta}\, n + \frac{e\tau_0}{K+\theta} + \Big(W_0 - \frac{e\tau_0}{K+\theta}\Big) \frac{\big\langle \frac{\tau_0-\theta}{K} \big\rangle_n}{\big\langle \frac{\tau_0}{K} \big\rangle_n},$$

for $n \ge 0$, *where* $\langle x \rangle_n$ *is the rising factorial* $x(x+1)\ldots(x+n-1)$.

PROOF We show this exact formula by induction on n. In a linear urn scheme, $K+\theta \ne 0$. Under the usual interpretation $\langle x \rangle_0 = 1$, the basis is established at $n = 0$. Assuming the formula holds for some $n \ge 0$, we go back to the mean recurrence (Eq. (10.8)) and write

$$\begin{aligned}
\mathbf{E}[W_{n+1}] &= g_{n+1}\mathbf{E}[W_n] + e \\
&= \frac{Kn+\tau_0-\theta}{Kn+\tau_0}\left(\frac{eK}{K+\theta} n + \frac{e\tau_0}{K+\theta} + \Big(W_0 - \frac{e\tau_0}{K+\theta}\Big)\frac{\big\langle \frac{\tau_0-\theta}{K} \big\rangle_n}{\big\langle \frac{\tau_0}{K} \big\rangle_n}\right) + e \\
&= \frac{Kn+\tau_0-\theta}{Kn+\tau_0}\Big(\frac{eK}{K+\theta} n + \frac{e\tau_0}{K+\theta}\Big) + e \\
&\quad + \Big(\Big(W_0 - \frac{e\tau_0}{K+\theta}\Big) \frac{\big(n + \frac{\tau_0-\theta}{K}\big)\big\langle \frac{\tau_0-\theta}{K} \big\rangle_n}{\big(n+\frac{\tau_0}{K}\big)\big\langle \frac{\tau_0}{K} \big\rangle_n} \\
&= \frac{eK}{K+\theta}(n+1) + \frac{e\tau_0}{K+\theta} + \Big(W_0 - \frac{e\tau_0}{K+\theta}\Big)\frac{\big\langle \frac{\tau_0-\theta}{K} \big\rangle_{n+1}}{\big\langle \frac{\tau_0}{K} \big\rangle_{n+1}},
\end{aligned}$$

which completes the induction. ∎

The formula in Proposition 10.1 is greatly simplified, when $(\tau_0 - \theta)/K$ is a negative integer, say ℓ. The rising factorial passes through 0 and we get simply

$$\mathbf{E}[W_n] = \frac{eK}{K+\theta}\, n + \frac{e\tau_0}{K+\theta},$$

for $n \ge \ell + 1$. Such a simplification is not available for other starting conditions. However, we still get a simplified asymptotic formula for all starting conditions.

COROLLARY 10.1
In a linear tenable urn the average number of white balls after n draws is

$$\mathbf{E}[W_n] = \frac{eK}{K+\theta}\,n + \frac{e\tau_0}{K+\theta} + O\left(\frac{1}{n^{\theta/K}}\right),$$

as $n \to \infty$.

PROOF Write the rising factorials in the exact formula in terms of gamma functions:

$$\mathbf{E}[W_n] = \frac{eK}{K+\theta}\,n + \frac{e\tau_0}{K+\theta} + \left(W_0 - \frac{e\tau_0}{K+\theta}\right)\frac{\Gamma\left(n+\frac{\tau_0-\theta}{K}\right)}{\Gamma\left(n+\frac{\tau_0}{K}\right)} \times \frac{\Gamma\left(\frac{\tau_0}{K}\right)}{\Gamma\left(\frac{\tau_0-\theta}{K}\right)},$$

The statement of the corollary follows from Stirling's approximation to the Gamma function. ■

We could sharpen our asymptotic estimates at will, by including more terms from the known asymptotic expansion of the ratio of Gamma functions.

Variance calculation in random structures tends to be very lengthy. Here we only sketch this calculation. Squaring the stochastic recurrence in Eq. (10.1) we obtain a recurrence for the conditional second moments:

$$\begin{aligned}\mathbf{E}\left[W_n^2 \mid W_{n-1}\right] &= W_{n-1}^2 + (a^2 + 2a\,W_{n-1})\,\mathbf{E}\left[I_n^{WW} \mid W_{n-1}\right] \\ &\quad + (c^2 + 2c\,W_{n-1})\,\mathbf{E}\left[I_n^{WB} \mid W_{n-1}\right] + (e^2 + 2e\,W_{n-1})\,\mathbf{E}\left[I_n^{BB} \mid W_{n-1}\right].\end{aligned}$$

When we plug in the conditional expectations of the three indicators as in Eqs. (10.3)–(10.5), we see that the third-order W_{n-1}^3 vanishes and the coefficient of the second-order term W_{n-1}^2 is $(a-2c+e)/(\tau_{n-1}(\tau_{n-1}-1))$, which cancels out in view of the linearization effect. The rest of the calculation is only tedious algebra to put the equation in a wieldy form. One obtains

$$\begin{aligned}\mathbf{E}\left[W_n^2 \mid W_{n-1}\right] &= \frac{2\tau_{n-1}^2 - 2(1+2\theta)\tau_{n-1} + \theta(\theta+4)}{2\tau_{n-1}(\tau_{n-1}-1)}\,W_{n-1}^2 \\ &\quad + \left(2e + \frac{2(c^2-e^2)}{\tau_{n-1}-1} + \frac{e^2-a^2}{\tau_{n-1}(\tau_{n-1}-1)}\right)W_{n-1} + e^2.\end{aligned}$$

Taking expectations, we put the unconditional second moment recurrence in the form

$$\mathbf{E}\left[W_n^2\right] = \alpha_n \mathbf{E}\left[W_{n-1}^2\right] + \beta_n \tag{10.9}$$

(valid for $n \ge 1$), where

$$\alpha_n = \frac{2\tau_{n-1}^2 - 2(1+2\theta)\tau_{n-1} + \theta(\theta+4)}{2\tau_{n-1}(\tau_{n-1}-1)},$$

and

$$\beta_n = \left(2e + \frac{2(c^2 - e^2)}{\tau_{n-1} - 1} + \frac{e^2 - a^2}{\tau_{n-1}(\tau_{n-1} - 1)}\right) \mathbf{E}[W_{n-1}] + e^2.$$

The term $\mathbf{E}[W_{n-1}]$ has been determined exactly in the proof of Proposition 10.1. The form of Eq. (10.9) is iteratable; it unwinds into

$$\mathbf{E}\left[W_n^2\right] = \sum_{j=1}^{n} \alpha_n \alpha_{n-1} \ldots \alpha_{j+1} \beta_j + \alpha_n \alpha_{n-1} \ldots \alpha_1 W_0^2.$$

After a considerable amount of simplification, via sharp asymptotics for α_n and β_n, we obtain an expression for $\mathbf{E}[W_n^2]$, in which the leading term matches that in the square of $\mathbf{E}[W_n]$ (a common phenomenon in combinatorial structures), and in which terms involving the initial condition W_0^2 die out asymptotically at the rate $O(n^{-2\theta/K})$. The variance now follows by subtracting the square of the mean from the simplified expression for $\mathbf{E}[W_n^2]$, where quadratic orders cancel out leaving at most a linear asymptotic equivalent. An exact cancelation occurs in the $n^{1-\theta/K}$ term, which is quite remarkable.

PROPOSITION 10.2

In a tenable linear urn scheme with the balanced schema $\begin{pmatrix} a & b \\ c & d \\ e & f \end{pmatrix}$ *that has constant row sum* $K \geq 1$*, the variance of the number of white balls after* n *draws is*

$$\mathbf{Var}[W_n] = \sigma_1^2 n + \sigma_2^2 + o(1),$$

as $n \to \infty$*, with*

$$\sigma_1^2 = \frac{eK[e\theta(4K + 3\theta) + 4(c^2 - e^2)(K + \theta)]}{2(K + \theta)^2(K + 2\theta)} > 0,$$

$$\sigma_2^2 = \frac{e\theta(K - a)(\theta^2 + 4(\tau_0 - 1)\theta - 2K)}{8(K + \theta)^2(K + 2\theta)}.$$

In all balanced linear cases, an application of Chebyshev's inequality gives us the following.

PROPOSITION 10.3

In a tenable linear urn scheme with the balanced schema $\begin{pmatrix} a & b \\ c & d \\ e & f \end{pmatrix}$ *that has constant row sum* $K \geq 1$*, the proportion of white balls converges in probability:*

$$\frac{W_n}{Kn} \xrightarrow{P} w := \frac{e}{K + \theta},$$

as $n \to \infty$*.*

We introduce a martingale transform. Let

$$M_n = \mu_n W_n + \nu_n,$$

for nonrandom correction factors μ_n and ν_n that will render M_n a true centered martingale. The correction factors are then obtained from recurrence equations they must satisfy.

LEMMA 10.1
The random variable

$$M_n = \frac{W_n}{g_2 \cdots g_n} - e \sum_{j=2}^{n} \frac{1}{g_2 \cdots g_j} - \mathbf{E}[W_1]$$

is a centered martingale.

PROOF Let $M_n = \mu_n W_n + \nu_n$. For M_n to be a martingale, it must satisfy

$$\begin{aligned} \mathbf{E}[M_n \mid W_{n-1}] &= \mu_n \mathbf{E}[W_n \mid W_{n-1}] + \nu_n \\ &= \mu_n [g_n W_{n-1} + e] + \nu_n \\ &= M_{n-1}. \end{aligned}$$

So, we want to enforce that

$$\mu_n g_n W_{n-1} + e\mu_n + \nu_n = \mu_{n-1} W_{n-1} + \nu_{n-1},$$

for all $n \ge 1$. This is possible if we let

$$\mu_n = \frac{\mu_{n-1}}{g_n} = \frac{\mu_{n-2}}{g_n g_{n-1}} = \cdots = \frac{\mu_1}{g_n g_{n-1} \cdots g_2}.$$

and let $e\mu_n + \nu_n = \nu_{n-1}$. This linear recurrence unfolds as

$$\begin{aligned} \nu_n &= \nu_{n-1} - e\mu_n \\ &= \nu_{n-2} - e\mu_{n-1} - e\mu_n \\ &\ \ \vdots \\ &= \nu_1 - e(\mu_2 + \cdots + \mu_n). \end{aligned}$$

We also wish to have M_n centered, that is, $\mathbf{E}[M_n] = \mathbf{E}[M_1] = 0$, or $\mu_1 \mathbf{E}[W_1] + \nu_1 = 0$. Hence

$$M_n = \frac{\mu_1 W_n}{g_2 \cdots g_n} - \mu_1 e \sum_{j=2}^{n} \frac{1}{g_2 \cdots g_j} - \mu_1 \mathbf{E}[W_1]$$

is a centered martingale, for any arbitrary constant μ_1; let us take $\mu_1 = 1$. ■

We recall a notation we used before. The symbol $\triangledown$ denotes the backward difference operator, that is, for any function h, $\triangledown h_n = h_n - h_{n-1}$. For any fixed $t > 0$, the random variable $n^{-t} \triangledown M_k$ is a martingale difference sequence (with respect to the increasing sigma-field sequence generated by W_k). The orders of magnitude for mean and variance will help us verify the martingale central limit theorem for an appropriate choice of t. It suffices to check the conditional Lindeberg condition and the conditional variance condition on the martingale differences (see Theorem 6.1, or Chapter 3 of Hall and Heyde(1980) for a broader scope).

LEMMA 10.2
Let $t > \theta/K$, and $I_{\mathcal{E}}$ be the indicator of the event $\mathcal{E}$. For any fixed $\varepsilon > 0$,

$$\sum_{j=1}^{n} \mathbf{E}\left[\left(\frac{\triangledown M_j}{n^t}\right)^2 I_{\{|n^{-t}\triangledown M_j|>\varepsilon\}} \mid W_{j-1}\right] \xrightarrow{P} 0,$$

as $n \to \infty$.

PROOF We have

$$\triangledown M_j = \frac{\triangledown W_j}{g_2 \cdots g_{j-1}} + \frac{\theta W_j}{g_2 \cdots g_{j-1}(Kj + \tau_0 - K - \theta)} - \frac{e}{g_2 \cdots g_j}.$$

Hence,

$$|\triangledown M_j| \le \frac{|\triangledown W_j|}{g_2 \cdots g_{j-1}} + \frac{\theta W_j}{g_2 \cdots g_{j-1}(Kj + \tau_0 - K - \theta)} + \frac{e}{g_2 \cdots g_j}. \quad (10.10)$$

The differential in the number of white balls is bounded from above by the constant $\max\{|a|, |c|, |e|\}$, and

$$\prod_{i=2}^{j-1} g_i = \frac{\Gamma\left(j - 1 + \frac{\tau_0 - \theta}{K}\right)\Gamma\left(1 + \frac{\tau_0}{K}\right)}{\Gamma\left(j - 1 + \frac{\tau_0}{K}\right)\Gamma\left(1 + \frac{\tau_0 - \theta}{K}\right)}$$

is of the order $j^{-\theta/K}$ by Stirling's approximation of the Gamma function (in the case $(\tau_0 - \theta)/K$ is a negative integer, say $-\ell$, the product is identically zero when the index i passes through $\ell + 1$). Hence, the first term on the right-hand side of Eq. (10.10) is $O(j^{\theta/K})$. The number of white balls after j draws cannot exceed $Kj + \tau_0$, the total number of balls in the urn. So, the second term on the right-hand side of Eq. (10.10) is also $O(j^{\theta/K})$. Likewise, the third term on the right-hand side of Eq. (10.10) is $O(j^{\theta/K})$. Hence, $(\triangledown M_j)^2 = O(j^{2\theta/K})$, and if $t > \theta/K$ the sets $\{(\triangledown M_j)^2 > \varepsilon n^{2t}\}$ are all empty for sufficiently large n, and for all $j = 1, \ldots n$. Lindeberg's conditional condition has been verified for any choice $t > \theta/K$. ■

A Z-conditional variance condition requires that, for some value of t,

$$\sum_{j=1}^{n} \mathbf{E}\left[\left(\frac{\nabla M_j}{n^t}\right)^2 \mid W_{j-1}\right] \xrightarrow{P} Z,$$

for the random variable Z.

LEMMA 10.3
For any fixed $\varepsilon > 0$,

$$\sum_{j=1}^{n} \mathbf{E}\left[\left(\frac{\nabla M_j}{n^{\theta/K+\frac{1}{2}}}\right)^2 \mid W_{j-1}\right] \xrightarrow{P} \sigma_1^2,$$

as $n \to \infty$.

According to the martingale formulation of Lemma 10.1, we have

$$\mathbf{E}[(\nabla M_j)^2 | W_{j-1}] = \frac{1}{(g_2 \dots g_{j-1})^2} \mathbf{E}\left[\left(\nabla W_j + \frac{\theta W_j}{Kj + \tau_0 - K - \theta} - \frac{e}{g_j}\right)^2 \mid W_{j-1}\right].$$

Upon squaring the variable in the conditional expectation, we get six terms. One term is the nonrandom quantity e^2/g_j^2. Each of the other random quantities has a conditional expectation that can be specified in terms of the number of white balls and the conditional first two moments as follows:

$$\mathbf{E}[(\nabla W_j)^2 \mid W_{j-1}] = W_{j-1}^2 - 2W_{j-1}\mathbf{E}[W_j \mid W_{j-1}] + \mathbf{E}[W_j^2 \mid W_{j-1}];$$

$$\mathbf{E}\left[\frac{\theta^2 W_j^2}{(Kj + \tau_0 - K - \theta)^2} \mid W_{j-1}\right] = \frac{\theta^2}{(Kj + \tau_0 - K - \theta)^2}\mathbf{E}\big[W_j^2 \mid W_{n-1}\big];$$

$$\mathbf{E}\left[2\frac{\theta W_j \nabla W_j}{Kj + \tau_0 - K - \theta} \mid W_{j-1}\right] = \frac{2\theta}{Kj + \tau_0 - K - \theta}\big(\mathbf{E}\big[W_j^2 \mid W_{j-1}\big] - W_{j-1}\mathbf{E}\big[W_j \mid W_{j-1}\big]\big);$$

$$\mathbf{E}\left[-2\frac{e \nabla W_j}{g_j} \mid W_{j-1}\right] = \frac{2e}{g_j}W_{j-1} - \frac{2e}{g_j}\mathbf{E}[W_j \mid W_{j-1}];$$

$$\mathbf{E}\left[-2\frac{e\theta}{g_j(Kj + \tau_0 - K - \theta)}W_j \mid W_{j-1}\right] = -\frac{2e\theta}{g_j(Kj + \tau_0 - K - \theta)}\mathbf{E}[W_j \mid W_{j-1}].$$

In turn, each of these is expressible in terms of conditional expectations of the indicators of the chosen multisets. Calculating all of these in a simple asymptotic form is rather tedious. We illustrate the calculation of only one of these terms:

$$\mathbf{E}[(\triangledown W_j)^2 \mid W_{j-1}] = a^2\mathbf{E}\big[I_n^{WW} \mid W_{n-1}\big] + c^2\mathbf{E}\big[I_n^{WB} \mid W_{n-1}\big] + e^2\mathbf{E}\big[I_n^{BB} \mid W_{n-1}\big];$$

the rest of the relations follow suit.

Again, what is left is lengthy and will only be sketched. Adding up the six terms, we have a relation of the form

$$\mathbf{E}\big[(\triangledown M_j)^2 \mid W_{j-1}\big] = \xi_1(j)W_{j-1}^2 + \xi_2(j)W_{j-1} + \xi_3(j),$$

where $\xi_i(j)$, for $i = 1, 2, 3$, are functions of j and the urn parameters. In this latter form, we can approximate both W_{j-1} and W_{j-1}^2 in $\mathcal{L}_1$ as follows. Of use here is the Cauchy-Schwarz inequality, which states that for any two integrable random variables X and Y

$$\mathbf{E}|XY| \le \sqrt{\mathbf{E}[X^2]\,\mathbf{E}[Y^2]}.$$

The Cauchy-Schwarz inequality gives us

$$\mathbf{E}[|W_j - wKj|] \le \sqrt{\mathbf{E}[(W_j - wKj)^2]} \tag{10.11}$$

$$= \sqrt{\mathbf{E}[(W_j - \mathbf{E}[W_j])^2] + (\mathbf{E}[W_j] - wKj)^2}$$

$$= O(\sqrt{j}), \tag{10.12}$$

following from Propositions 10.1 and 10.2. This implies that[1]

$$W_j = wKj + O_{\mathcal{L}_1}(\sqrt{j}).$$

Further, by the Cauchy-Schwarz inequality again, we have

$$\mathbf{E}\big[\big|W_j^2 - w^2K^2j^2\big|\big] \le \sqrt{\mathbf{E}[(W_j - wKj)^2]\,\mathbf{E}[(W_j + wKj)^2]}.$$

One has the bound $W_j = O(j)$, which together with and Eqs. (10.11) and (10.12) give

$$W_j^2 = w^2K^2j^2 + O_{\mathcal{L}_1}(j^{3/2}).$$

Now one has

$$\mathbf{E}\big[(\triangledown M_j)^2 \mid W_{j-1}\big] = \xi_1(j)[w^2K^2(j-1)^2 + O_{\mathcal{L}_1}(j^{3/2})] + \xi_2(j)[wK(j-1) + O_{\mathcal{L}_1}(\sqrt{j})] + \xi_3(j).$$

[1] The notation $O_{\mathcal{L}_1}(g(n))$ stands for a random quantity that is $O(g(n))$ in the $\mathcal{L}_1$ norm.

Working out the details, one reassembles the elements of σ_1^2:

$$\mathbf{E}\big[(\nabla M_j)^2 \mid W_{j-1}\big] = \left(\frac{2\theta}{K} + 1\right) \sigma_1^2 j^{2\theta/K} + O_{\mathcal{L}_1}(j^{2\theta/K-\frac{1}{2}}).$$

Therefore,

$$\sum_{j=1}^{n} \mathbf{E}\big[(n^{-t} \nabla M_j)^2 \mid W_{j-1}\big] = \frac{1}{n^{2t}} \sum_{j=1}^{n} \left[\left(\frac{2\theta}{K} + 1\right) \sigma_1^2 j^{2\theta/K} + O_{\mathcal{L}_1}(j^{2\theta/K-\frac{1}{2}})\right],$$

which, if $t = \theta/K + \frac{1}{2}$, converges to σ_1^2 (of course, what else?), in $\mathcal{L}_1$, and hence in probability too. The σ_1^2-conditional variance condition has been verified.

Lemmas 10.2 and 10.3 lead us to a main result of this section.

THEOREM 10.1

In a tenable linear urn scheme with the balanced schema $\begin{pmatrix} a & b \\ c & d \\ e & f \end{pmatrix}$ *that has constant row sum* $K \geq 1$, *the number of white balls has the central limit tendency*

$$\frac{W_n - wn}{\sqrt{n}} \xrightarrow{\mathcal{D}} \mathcal{N}\left(0, \sigma_1^2\right),$$

as $n \to \infty$; *where* $w = \frac{e}{K+e-a}$, *and* σ_1^2 *is effectively computable constants (given in Propositions 10.2 in terms of the urn parameters).*

Average-case analysis of urn schemes with 3×2 unbalanced schemata may be within reach. We have a heuristic argument suggesting that

$$\mathbf{E}[W_n] \sim wn,$$

as $n \to \infty$, where w is the positive root of the indicial equation. For example, for the scheme $\begin{pmatrix} 3 & 3 \\ 2 & 4 \\ 4 & 2 \end{pmatrix}$, the heuristic approach suggests that one may be able to rigorize an argument showing that $\mathbf{E}[W_n] \sim [\frac{5}{3} + \frac{1}{3}\sqrt{13}]n \approx 2.868517092\, n$, as $n \to \infty$. Simulation supports this hypothesis.

10.4 Application of Rectangular Schemes in Random Circuits

Various models in electro and neural sciences adopt recursive circuits as their graph-theoretic backbone. Toward a logic circuit interpretation, we call the nodes of in-degree 0 in a directed acyclic graph *inputs*, and we call the nodes of out-degree 0 *outputs*. A *recursive circuit* of size n is a directed acyclic graph, the nodes of which are labeled with $\{1, \ldots, n\}$ in such a way that labels increase along every input–output path.

The binary recursive circuit can be viewed as the second member in a hierarchy of combinatorial structures, with the recursive forest (of recursive trees) being the first in the hierarchy. The kth member of the hierarchy grows from an initial set of $r \ge 2$ nodes; at each stage, k nodes are chosen to parent a new child. A forest of recursive trees ($k = 1$) is the simplest member of the hierarchy. We have looked at the recursive tree in Section 8.2, where we used the usual (square) Pólya-urn schemata to analyze various properties.

Let $L_{n,r}$ denote the number of *outputs* in the recursive circuit having r input nodes. The number of inputs r will be held fixed throughout. So, we can drop it from all notation and think of it as implicit. For example, we shall write L_n for $L_{n,r}$, and so on.

Let us now turn to a formal definition of circuits, and the groundwork to model its growth by a rectangular urn scheme. A *recursive circuit with fan-in* k is a directed graph that starts out with $r \ge k$ isolated nodes (inputs) of in-degree and out-degree 0. The circuit evolves in stages as follows. After $n - 1$ stages, a circuit RC_{n-1} has grown. At the nth stage, k nodes are chosen *without replacement* from RC_{n-1} as parents for the new nth entrant. The new node is joined to the circuit with edges directed from the k parents to it, and is given 0 out-degree.

In view of the boolean circuit interpretation we shall refer to nodes as *gates* and borrow language from the literature of boolean circuits. We shall look at the number of outputs. In electrical engineering this may have many implications concerning the amount of output currents drawn and in boolean circuitry they stand for how many "answers" are derived from r given inputs. We shall focus on an illustration of the binary circuit $k = 2$. Figure 10.1 below shows all the possible binary circuits after two insertion steps into an initial graph of two isolated nodes (outputs are illustrated as boxes and nonoutputs as bullets).

Many models of randomness can be imposed on recursive circuits. We consider here a most natural uniform model, one in which all pairs of nodes in RC_{n-1} are equally likely candidate parents of the nth entrant. This model is equivalent to the combinatorial view that sees all labeled circuits of a given size to be equally likely.

Choosing two parents for the nth entrant from the nodes of RC_{n-1} can occur in one of three ways:

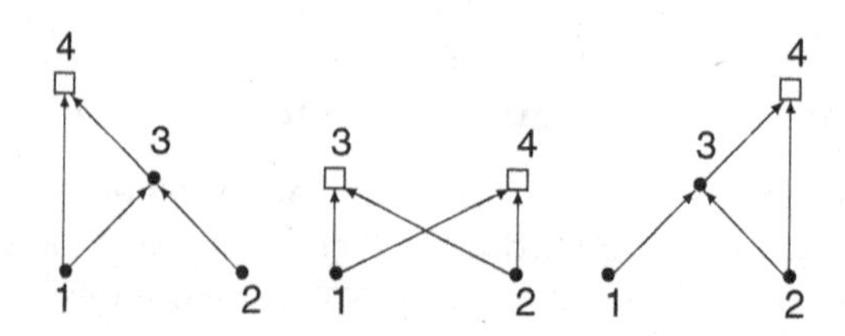

FIGURE 10.1
All binary circuits of size 4 grown from two inputs.

- The two parents are outputs in RC_{n-1}. In this case, two outputs of RC_{n-1} are turned into nonoutput gates, and a new output appears in RC_n, a net gain of -1 output.
- The two parents are nonoutputs in RC_{n-1}. In this case, one new output appears.
- The two parents are mixed, in RC_{n-1} one is output, the other is not. An output gate of RC_{n-1} is turned into a nonoutput node and one new output appears in RC_n.

There is a rectangular urn scheme that models the growth of a random circuit. One can think of the growth process in terms of addition of white and blue balls to an urn. White balls represent the outputs (squares in Figure 10.1) and blue balls represent the rest (the nonoutput gates), which are the internal nodes of the circuit, (bullets in Figure 10.1). The urn starts out with r white balls. The urn evolves as follows:

- The first of the rules of growth in the circuit is concerned with choosing two outputs as parents (picking two white balls from the urn). The response is to turn two outputs into nonoutputs and add a new output. The corresponding rule in the urn is to remove one white ball and add two blue balls.
- The second of the rules of growth in the circuit is concerned with choosing two nonoutputs as parents (picking two blue balls from the urn), in which case one new output appears. The corresponding rule in the urn is to add one white ball, while the number of blue balls stays put.
- The third of the rules of growth in the circuit is concerned with a mixed choice of one output and one nonoutput as parents (picking one white and one blue ball from the urn). In this case, one nonoutput gate appears. The corresponding rule in the urn is to add one blue ball.

The associated urn scheme is therefore

$$\begin{array}{c} \\ WW \\ WB \\ BB \end{array} \begin{array}{c} \begin{array}{cc} W & B \end{array} \\ \begin{pmatrix} a & b \\ c & d \\ e & f \end{pmatrix} \end{array} = \begin{pmatrix} -1 & 2 \\ 0 & 1 \\ 1 & 0 \end{pmatrix}.$$

In this application

$$a - 2c + e = -1 - 2 \times 0 + 1 = 0,$$

and it is a linear scheme that starts out with r white balls. The urn's white balls are the output gates, and $W_n = L_n$. Specialized to the parameters of the circuit application, the parameters of the urn scheme are

$$K = 1, \qquad \theta = e - a = 2,$$

and the recurrence (Eq. (10.8)) becomes

$$\mathbf{E}[L_n \mid L_{n-1}] = \frac{n+r-3}{n+r-1} L_{n-1} + 1;$$

the double expectation gives

$$\mathbf{E}[L_n] = \frac{n+r-3}{n+r-1} \mathbf{E}[L_{n-1}] + 1.$$

With a little work one can get the exact mean of the number of outputs by solving this recurrence; we leave this as an exercise. As is usual with variance calculation for combinatorial constructions, it takes much more labor to get the exact variance by solving Eq. (10.9). We also leave this for the exercises. The asymptotic mean and variance of the number of outputs after n random insertions follow from Propositions 10.1 and 10.2; they are

$$\mathbf{E}[L_n] = \frac{1}{3}n + \frac{1}{3}r + O\left(\frac{1}{n^2}\right);$$

$$\mathbf{Var}[L_n] = \frac{4}{45}n + \frac{1}{45}(4r-3) - \frac{2}{15n} + O\left(\frac{1}{n^2}\right).$$

As a by-product, from Proposition 10.3 we get

$$\frac{L_n}{n} \xrightarrow{P} \frac{1}{3}.$$

For circuits growing out of two gates ($r = 2$), the particular central limit tendency that ensues from Theorem 10.1 for outputs is:

$$\frac{L_n - \frac{1}{3}n}{\sqrt{n}} \xrightarrow{\mathcal{D}} \mathcal{N}\left(0, \frac{4}{45}\right).$$

Exercises

10.1 Show that the mean number of white balls after n draws (of ball pairs) from the balanced urn of white and blue balls with the scheme $\begin{pmatrix} 2 & 1 \\ 1 & 2 \\ 0 & 3 \end{pmatrix}$, when it starts with 1 white ball and 2 blue balls is

$$\mathbf{E}[W_n] = \frac{\Gamma\left(n+\frac{5}{3}\right)}{\Gamma\left(\frac{5}{3}\right)\Gamma(n+1)} \sim \frac{1}{\Gamma\left(\frac{5}{3}\right)} n^{2/3}, \qquad \text{as } n \to \infty.$$

10.2 In the general balanced scheme $\begin{pmatrix} a & b \\ c & d \\ 0 & K \end{pmatrix}$, with constant positive row sum K, and positive c and d, show that

(a) The urn is untenable if $0 < K < a$, unless it starts all blue.

(b) The mean number of white balls grows sublinearly if $K > a > 0$.

10.3 Derive the asymptotic variance of the number of white balls after n draws (of ball pairs) from the Chen-Wei urn of white and blue balls with the scheme $\begin{pmatrix} 2 & 0 \\ 1 & 1 \\ 0 & 2 \end{pmatrix}$, when it starts with 1 white and 1 blue ball.

10.4 Consider a reinforcement urn scheme of white and blue balls that grows by taking samples of size 3. The sample is taken out without replacement during the sampling, the colors are examined, we then put the sample back in the urn, and reinforce the colors in the sample by adding as many balls of any color as those that appeared in the sample—if there are w white balls and $3 - w$ blue balls in the sample, we put the sample back and add w white balls and $3 - w$ blue balls to the urn. If we denote white with W, and blue with B, we are taking into account the scheme with rectangular ball addition matrix

$$\begin{array}{c} \\ WWW \\ WWB \\ WBB \\ BBB \end{array} \begin{array}{c} \begin{array}{cc} W & B \end{array} \\ \begin{pmatrix} 3 & 0 \\ 2 & 1 \\ 1 & 2 \\ 0 & 3 \end{pmatrix} \end{array}.$$

Compute the average number of white balls after n draws.

10.5 Consider a reverse reinforcement urn scheme of white and blue balls that grows by taking triples sampled without replacement. During the sampling the colors are observed. We put the sampled triple back, and reinforce the antithetical colors: For every ball that appears in the sample we add a ball of the opposite color—if there are w white balls and $3 - w$ blue balls in the sample, we put the sample back and add w blue balls and $3 - w$ white balls to the urn. If we denote white with W, and blue with B, the scheme has the rectangular replacement matrix

$$\begin{array}{c} \\ WWW \\ WWB \\ WBB \\ BBB \end{array} \begin{array}{c} \begin{array}{cc} W & B \end{array} \\ \begin{pmatrix} 0 & 3 \\ 1 & 2 \\ 2 & 1 \\ 3 & 0 \end{pmatrix} \end{array}.$$

Compute the average number of white balls after n draws.

10.6 Consider a three-color reinforcement urn scheme of white, blue, and red balls that grows by taking sample pairs without replacement, and reinforcing the colors that appear in the sample. During the sampling the colors are observed. We put the sampled pair back, and for every ball that appears in the sample we add a ball of the same color.

Denote white with W, blue with B, and red with R. The scheme has the rectangular replacement matrix

$$\begin{array}{c} \\ WW \\ WB \\ WR \\ BB \\ BR \\ RR \end{array} \begin{array}{c} \begin{array}{ccc} W & B & R \end{array} \\ \begin{pmatrix} 2 & 0 & 0 \\ 1 & 1 & 0 \\ 1 & 0 & 1 \\ 0 & 2 & 0 \\ 0 & 1 & 1 \\ 0 & 0 & 2 \end{pmatrix} \end{array}.$$

Compute the average number of white balls after n draws.

10.7 (Tsukiji and Mahmoud, 2001). Derive the exact mean and variance for L_n, the number of outputs after n random insertions in a random circuit evolving out of two starting gates.

Answers to Exercises

Chapter 1

1.1 Suppose, toward a contradiction, that all integers are equally likely under the probability measure $\mathbf{P}$, and the probability of any integer is the fixed number $x \ge 0$. What is the probability of the whole space? We must have

$$1 = \mathbf{P}(\Omega) = \mathbf{P}\left(\bigcup_{j=1}^{\infty}\{j\}\right) = \sum_{j=1}^{\infty}\mathbf{P}(\{j\}) = \sum_{j=1}^{\infty} x.$$

If x is positive, $\sum_{j=1}^{\infty} x = \infty$, and if $x \equiv 0$, then $\sum_{j=1}^{\infty} x = 0$. Either case is a contradiction to the axioms of probability measures. Such a measure $\mathbf{P}$ cannot exist.

1.2 Verify Kolmogorov's axioms:

(i) As both $\mathbf{P}_1$ and $\mathbf{P}_2$ are probability measures, $\mathbf{P}_1(\Omega) = \mathbf{P}_2(\Omega) = 1$. Therefore,

$$\mathbf{P}(\Omega) = \alpha\mathbf{P}_1(\Omega) + (1-\alpha)\mathbf{P}_2(\Omega) = \alpha + (1-\alpha) = 1.$$

(ii) For an arbitrary measurable set A, we have $\mathbf{P}_1(A) \ge 0$, and $\mathbf{P}_2(A) \ge 0$; so, trivially, $\mathbf{P}(A) \ge 0$.

(iii) Let $A_1, A_2, \ldots$ be an infinitely countable collection of disjoint measurable sets. Then under each of the two given measures we have the infinite additivity property:

$$\mathbf{P}_i\left(\bigcup_{n=1}^{\infty} A_n\right) = \sum_{n=1}^{\infty}\mathbf{P}_i(A_n), \qquad \text{for } i = 1, 2.$$

Whence,

$$\begin{aligned}
\mathbf{P}\left(\bigcup_{n=1}^{\infty} A_n\right) &= \alpha\mathbf{P}_1\left(\bigcup_{n=1}^{\infty} A_n\right) + (1-\alpha)\mathbf{P}_2\left(\bigcup_{n=1}^{\infty} A_n\right) \\
&= \alpha\sum_{n=1}^{\infty}\mathbf{P}_1(A_n) + (1-\alpha)\sum_{n=1}^{\infty}\mathbf{P}_2(A_n) \\
&= \sum_{n=1}^{\infty}[\alpha\mathbf{P}_1(A_n) + (1-\alpha)\mathbf{P}_2(A_n)] \\
&= \sum_{n=1}^{\infty}\mathbf{P}(A_n).
\end{aligned}$$

1.3 **Remarks:** If A and B are mutually exclusive (disjoint), they *cannot* be independent. Intuitively, the condition that one of them occurs, tells us so much about the other, it simply says the other did not. From the technical definition, if both A and B have positive probabilities then their mutual exclusion gives $\mathbf{P}(A \cap B) = 0$, whereas $\mathbf{P}(A)\,\mathbf{P}(B) > 0$. This simple argument gives a clue to the answer to our problem.

The independent sets A_i and A_j cannot be disjoint. That is, the intersection of A_i and A_j must contain at least one point. More generally, $B_1 \cap B_2 \ldots \cap B_n$ must contain at least one point whenever B_j is A_j or A_j^c (because if A is independent of B, then A^c is also independent of B). For every intersection of a combination of B_j, $j = 1, \ldots, n$, we must have at least one point. There are at least 2^n such combinations.

The foregoing discussion only provides the lower bound

$$|\Omega| \ge 2^n.$$

How do we know that this bound is tight? We must further show that there is actually a space with 2^n points, over which n independent events can be defined so that none has probability 0 or 1. Consider the sample space underlying the experiment of drawing with replacement (uniformly) at random one of two balls from an urn containing one white ball and one blue ball. The history of n draws can be represented by a string of W (white) and B (Blue). For example, for $n = 3$, the string BBW stands for the history of drawing the blue ball twice, followed by drawing the white. This sample space is

$$\Omega = \{\omega_1 \ldots \omega_n \,:\, \omega_i = W \text{ or } B \text{ for } i = 1, \ldots, n\}.$$

Clearly the space has 2^n points. According to the uniform probability model, let $\mathbf{P}$ be the uniform measure that assigns probability 2^{-n}

to each sample space point. Let A_j, for $j = 1, \ldots, n$, be the event that the ball in the jth sample is white. The events $A_1, \ldots, A_n$ are independent (easy to show), and each has probability $\frac{1}{2}$.

1.4 Let W be the number of white balls in the sample, and let B be the number of the blue:

(a) $$\mathbf{P}(W = 3) = \mathbf{P}(W = 3, B = 0) = \frac{\binom{5}{3}\binom{3}{0}}{\binom{8}{3}} = \frac{10 \times 1}{56} = \frac{5}{28}.$$

(b) $$\mathbf{P}(W = 2) = \mathbf{P}(W = 2, B = 1) = \frac{\binom{5}{2}\binom{3}{1}}{\binom{8}{3}} = \frac{15}{28}.$$

(c) Let the event in question be called $A_{2,1}$, then

$$\mathbf{P}(A_{2,1}) = \frac{5}{8} \times \frac{4}{7} \times \frac{3}{6} = \frac{5}{28}.$$

1.5 Let W and B denote the colors white and blue, and let C_i be the color of the ball in draw i, $i = 1, 2$.

(a) If a blue ball is added to the second urn, its composition is changed into 4 white and 6 blue:

$$\mathbf{P}(C_2 = B \mid C_1 = B) = \frac{6}{10} = \frac{3}{5}.$$

(b) $$\begin{aligned}\mathbf{P}(C_2 = B) &= \mathbf{P}(C_2 = B \mid C_1 = W)\,\mathbf{P}(C_1 = W) \\ &\quad + \mathbf{P}(C_2 = B \mid C_1 = B)\,\mathbf{P}(C_1 = B) \\ &= \frac{5}{10} \times \frac{3}{9} + \frac{6}{10} \times \frac{6}{9} \\ &= \frac{17}{30}.\end{aligned}$$

1.6 Let N_i be the event that the balls in set $i = 1, 2$ are new. Then, by a combination view we see that

$$\mathbf{P}(N_1) = \frac{\binom{40}{6}}{\binom{91}{6}} = \frac{7030}{1220813} \approx 0.00576.$$

We can also take a permutation view of the problem and compute

$$\mathbf{P}(N_1) = \frac{40 \times 39 \times 38 \times 37 \times 36 \times 35}{91 \times 90 \times 89 \times 88 \times 87 \times 86},$$

which gives the same answer.

We figure out the balls for the second set by conditioning on the number of new balls X_1 in the first set (the event $X_1 = 6$ is N_1):

$$\begin{aligned}\mathbf{P}(N_2) &= \sum_{i=0}^{6} \mathbf{P}(N_2 \mid X_1 = i)\, \mathbf{P}(X_1 = i)\\ &= \sum_{i=0}^{6} \frac{\binom{40-i}{6}}{\binom{85}{6}} \times \frac{\binom{40}{i}\binom{51}{6-i}}{\binom{91}{6}}\\ &= \frac{7030}{1220813}\\ &= \mathbf{P}(N_1).\end{aligned}$$

1.7 We assume commercial dice to be unbiased and impose a uniform distribution on $\{1, 2, 3, 4, 5, 6\}$. Let X be the outcome of the throw, and A be the event that there are more white balls than blue in the sample. Condition A on the event $X = x$ to obtain

$$\begin{aligned}\mathbf{P}(A) &= \sum_{x=1}^{6} \mathbf{P}(A \mid X = x)\, \mathbf{P}(X = x)\\ &= \frac{1}{6} \sum_{x=1}^{6} \mathbf{P}(A \mid X = x)\\ &= \frac{1}{6} \sum_{x=1}^{6} \sum_{j > x-j} \frac{\binom{9}{j}\binom{11}{x-j}}{\binom{20}{x}}\\ &= \frac{359}{1140}.\end{aligned}$$

1.8 We first verify pairwise independence:

$$\mathbf{P}(A_1 \cap A_2) = \mathbf{P}(\text{the ball drawn is } 110) = \frac{1}{4}.$$

On the other hand,

$$\mathbf{P}(A_1) = \mathbf{P}(\text{drawing either of } 110 \text{ or } 101) = \frac{2}{4} = \frac{1}{2},$$

and

$$\mathbf{P}(A_2) = \mathbf{P}(\text{drawing either of 110 or 011}) = \frac{2}{4} = \frac{1}{2}.$$

Hence,

$$\mathbf{P}(A_1 \cap A_2) = \mathbf{P}(A_1)\,\mathbf{P}(A_2) = \frac{1}{4},$$

which verifies that the events A_1 and A_2 are independent. A similar calculation verifies that A_1 and A_3 are independent, and A_2 and A_3 are independent.

In spite of this pairwise independence, the three events are not totally independent as we can see from the calculation

$$\mathbf{P}(A_1 \cap A_2 \cap A_3) = \mathbf{P}(\phi) = 0 \neq \frac{1}{8} = \mathbf{P}(A_1)\,\mathbf{P}(A_2)\,\mathbf{P}(A_3).$$

Stepniak (2007) discusses Bernstein's example further, and shows that this is the smallest possible example (the construction must be on a sample space of size 4).

1.9 The evolution follows a three-state Markov chain. The transition matrix of this chain is

$$\mathbf{M} = \begin{array}{c} \\ A \\ B \\ C \end{array} \begin{array}{c} \begin{array}{ccc} A & B & C \end{array} \\ \begin{pmatrix} 0 & \frac{1}{2} & \frac{1}{2} \\ \frac{1}{2} & 0 & \frac{1}{2} \\ \frac{1}{2} & \frac{1}{2} & 0 \end{pmatrix} \end{array}.$$

The stationary distribution of this chain is any row vector

$$\pi = (\pi_A \quad \pi_B \quad \pi_C)$$

that solves the equation

$$\pi\mathbf{M} = \pi,$$

subject to

$$\pi_A + \pi_B + \pi_C = 1.$$

This equation has the unique solution

$$\pi_A = \pi_B = \pi_C = \frac{1}{3}.$$

The stationary distribution is uniform on the states. If we start at the states with a uniform probability, we shall be at the stationary distribution after one step (or any number of steps), and the probability of being in urn A is $\frac{1}{3}$.

Let $X_n \in \{A, B, C\}$ be the state of the chain after n transitions. The n-step transition probability from the starting state A back to A is given by the top component of the vector

$$\mathbf{M}^n \begin{pmatrix} 1 \\ 0 \\ 0 \end{pmatrix},$$

which gives

$$\mathbf{P}(X_n = A) = \frac{1}{3} + \frac{(-1)^n}{3 \times 2^{n-1}} \to \frac{1}{3}.$$

1.10 Suppose the drawing continued till the urn is completely empty, and think of the drawing process as a string (of length $2n$) of white and blue balls. We can determine a string by the structure of the n positions for the white balls (the other n positions must be for the blue balls). There are $\binom{2n}{n}$ equally likely strings. Consider the event (set of strings) where the last drawing of a white ball occurs at position k in the string. There must be a white ball at position k in the string, and all the other $n-1$ white balls occur at positions before k (there are $n-1$ white balls among the first $k-1$ positions): There are $\binom{k-1}{n-1}$ strings favoring this event. By symmetry, there is the same number of strings in which the kth drawing is the last blue balls. Let X_n denote the waiting time until an *unspecified* color is depleted first. The probability of the event $X_n = k$ is obtained by doubling the probability for the similar event when the color is specified, which is the proportion of the strings favoring that the specified color is depleted at step k (the proportion is the ratio of the count of such strings to the total count). Therefore,

$$\mathbf{P}(X_n = k) = 2\,\frac{\binom{k-1}{n-1}}{\binom{2n}{n}}, \qquad \text{for } k = n, \dots, 2n-1.$$

1.11 Let the number of white balls placed in the urn be W. Then, W is the number of successes in n independent Ber(p) experiments, or

$$W \stackrel{\mathcal{D}}{=} \text{Bin}(n, p).$$

Let $\tilde{W}$ be the number of white balls in the sample. Given the composition of the urn prior to taking the sample, the k balls are taken from a mixed populations of W white balls and $n - W$ blue balls. Given W, each drawing can be white with probability W/n, or blue with probability $1 - W/n$. Therefore, the conditional distribution of the number of white balls in the sample is like that of the number of successes in k independent Ber(W/n) experiments. And so,

$$\tilde{W} \overset{\mathcal{D}}{=} \text{Bin}\left(k, \frac{W}{n}\right) = \text{Bin}\left(k, \frac{1}{n}\text{Bin}(n, p)\right).$$

The conditional expectation is

$$\mathbf{E}[\tilde{W} \mid W] \overset{\mathcal{D}}{=} \frac{kW}{n},$$

and the average (unconditional expectation) follows by taking expectations

$$\mathbf{E}[\tilde{W}] = \mathbf{E}\left[\frac{kW}{n}\right] = \frac{k}{n}\,\mathbf{E}[\text{Bin}(n, p)] = \frac{k}{n} \times np = kp;$$

note there is no dependence on n.

1.12 If the urn is in a state where it has j white balls and k blue balls, adding a blue ball (and losing a white one) according to the sampling scheme follows a geometric distribution with $k/(j+k)$ probability of success. The rationale behind this comes from the behavior of the scheme: Nothing happens when a white ball is sampled (failure), and one awaits the success event of withdrawing a blue ball—the success rate in any individual drawing is $k/(j+k)$. We shall describe the given state by the pair (j, k).

Let X be the number of draws until total infestation, when all the balls in the urn become blue. It takes Geo$(b/(w+b))$ draws for the urn to get from the initial state (w, b) into the state $(w-1, b+1)$, then it takes and additional Geo$((b+1)/(w+b))$ draws to enter the state $(w-2, b+2)$, and so on, until it gets into the state $(1, w+b-1)$, where the urn needs an additional number of Geo$((b+w-1)/(b+w))$ draws to turn all blue. One can represent X as a sum of independent geometric random variables:

$$X \overset{\mathcal{D}}{=} \text{Geo}\left(\frac{b}{w+b}\right) + \text{Geo}\left(\frac{b+1}{w+b}\right) + \cdots + \text{Geo}\left(\frac{b+w-1}{w+b}\right).$$

Hence, for $w+b \ge 2$,

$$\mathbf{E}[X] = \frac{w+b}{b} + \frac{w+b}{b+1} + \cdots + \frac{w+b}{w+b-1} = (w+b)(H_{w+b-1} - H_{b-1}),$$

where H_j is the jth harmonic number. For $w+b=1$, we trivially have the boundary cases $w=1$, for which $X \equiv 1$, with mean 1, or $b=1$, for which $X \equiv 0$, with mean 0.

1.13 Intuitively, on average every other ball is white and the number of white balls should form a Poisson process going at half the rate of the given one. This checks out as follows. Let $W(t)$ be the number of white balls at time t, and let the total number of renewals be $X(t)$. Among the $X(t)$ balls at time t there is a random number of white balls, determined by obtaining heads in $X(t)$ independent coin flips. There are $X(t)$ identically distributed experiment, each of them "succeeds" to add a white ball with probability $\frac{1}{2}$. Hence,

$$W(t) = \text{Bin}\left(X(t), \frac{1}{2}\right).$$

We can condition on $X(t)$, which is a Poisson process (with a distribution like that of the $\text{Poi}(\lambda t)$ random variable):

$$\begin{aligned}
\mathbf{P}(W(t) = k) &= \sum_{j=0}^{\infty} \mathbf{P}(W(t) = k \mid X(t) = j)\, \mathbf{P}(X(t) = j) \\
&= \sum_{j=0}^{\infty} \mathbf{P}\left(\text{Bin}\left(j, \frac{1}{2}\right) = k\right) \times \frac{(\lambda t)^j e^{-\lambda t}}{j!} \\
&= \sum_{j=k}^{\infty} \frac{1}{2^j} \binom{j}{k} \times \frac{(\lambda t)^j e^{-\lambda t}}{j!} \\
&= \frac{e^{-\lambda t}}{k!} \sum_{j=k}^{\infty} \frac{(\lambda t/2)^j}{(j-k)!} \\
&= \frac{(\lambda t/2)^k e^{-\lambda t}}{k!} \sum_{j=k}^{\infty} \frac{(\lambda t/2)^{j-k}}{(j-k)!} \\
&= \frac{(\lambda t/2)^k e^{-\lambda t}}{k!} \times e^{\lambda t/2} \\
&= \frac{(\lambda t/2)^k e^{-\lambda t/2}}{k!}.
\end{aligned}$$

Hence,

$$W(t) = \text{Poi}\left(\frac{1}{2}\lambda t\right).$$

1.14 If a white ball is drawn in the first sample, a blue ball replaces it. The urn then contains two blue balls and it is impossible to withdraw a white ball in the second sample. The probability calculation

$$\mathbf{P}(\hat{W}_1 = 1, \hat{W}_2 = 0) = \mathbf{P}(\hat{W}_1 = 1) = \frac{1}{2},$$

summarizes this argument. Similarly, drawing a blue ball in the first sample leaves two white balls and it will be certain that a white ball appears in the second sample, that is

$$\mathbf{P}(\hat{W}_1 = 0, \hat{W}_2 = 1) = \mathbf{P}(\hat{W}_1 = 0) = \frac{1}{2}.$$

Hence

$$\mathbf{P}(\hat{W}_1 = 0, \hat{W}_2 = 1) = \mathbf{P}(\hat{W}_1 = 1, \hat{W}_2 = 0).$$

There is no need to check the joint event $\{\hat{W}_1 = 0, \hat{W}_2 = 0\}$ by virtue of the symmetry of the arguments, and similarly we need not check the event $\{\hat{W}_1 = 1, \hat{W}_2 = 1\}$. We have verified the exchangeability of $\hat{W}_2$ and $\hat{W}_2$ by demonstrating that

$$\mathbf{P}(\hat{W}_1 = \hat{w}_1, \hat{W}_2 = \hat{w}_2) = \mathbf{P}(\hat{W}_1 = \hat{w}_2, \hat{W}_2 = \hat{w}_1),$$

for all the feasible cases with $\hat{w}_1, \hat{w}_2 \in \{0, 1\}$; all probabilities for infeasible outcomes are 0.

1.15 Toward a contradiction, suppose there is a distribution function $F(x)$ that satisfies such an equality. It holds for $k = 0, 1, 2$, requiring

$$\mathbf{P}(\hat{W}_1 = 0, \hat{W}_2 = 0) = \int_0^1 (1-x)^2 \, dF(x),$$

$$\mathbf{P}(\hat{W}_1 = 1, \hat{W}_2 = 0) = \int_0^1 x(1-x) \, dF(x),$$

$$\mathbf{P}(\hat{W}_1 = 1, \hat{W}_2 = 1) = \int_0^1 x^2 \, dF(x).$$

In the preceding exercise we found the probability of the event $\{\hat{W}_1 = 1, \hat{W}_2 = 0\}$ to be $\frac{1}{2}$. By exchangeability the event $\{\hat{W}_1 = 0, \hat{W}_2 = 1\} = \frac{1}{2}$. These two events have the entire probability mass, and the joint events $\{\hat{W}_1 = 0, \hat{W}_2 = 0\}$, and $\{\hat{W}_1 = 1, \hat{W}_2 = 1\}$ have 0 probability (as is obvious from the mechanics of the ball replacement).

The third of the three equations must yield

$$\mathbf{P}(\hat{W}_1 = 1, \hat{W}_2 = 1) = \int_0^1 x^2 \, dF(x) = 0,$$

and with $F(x)$ being a distribution function, and x^2 an increasing function, $F(x)$ must have a point mass of 1 at $x = 0$. But then the first of these equations also dictates that

$$\mathbf{P}(\hat{W}_1 = 0, \hat{W}_2 = 0) = \int_0^1 (1-x)^2 \, dF(x) = 0,$$

requiring the distribution function to have a point mass of 1 at $x = 1$. This is a contradiction.

There is no contradiction with de Finetti's theorem, which requires *infinite* exchangeability. Here we have only two exchangeable random variables, for which a mixing distribution function does not exit.

1.16 Think of the outcome as a string of 1s and 0s. Any string $X_1 X_2 \dots X_{w+b}$ must contain w positions holding 1, and the rest of the positions hold 0. A particular fixed string $x_1 x_2 \dots x_{w+b}$, with 1s at positions say $i_1, i_2, \dots, i_w$ occurs with probability

$$\mathbf{P}(X_1 X_2 \dots X_{w+b} = x_1 x_2 \dots x_{w+b}) = \mathbf{P}(X_1 = x_1, \dots, X_{w+b} = x_{w+b}) = \frac{1}{\binom{w+b}{w}},$$

the denominator is the number of ways to choose w positions for 1s from among the $w + b$ positions. Note that the expression does not depend on $i_1, i_2, \dots, i_w$, and thus $X_1, X_2, \dots, X_{w+b}$ are exchangeable.

However, these variables are not independent, as we can illustrate by a simple example. The probability that a white ball is drawn first is

$$\mathbf{P}(X_1 = 1) = \frac{w}{w+b}.$$

Exchangeable random variables are identically distributed, and

$$\mathbf{P}(X_2 = 1) = \frac{w}{w+b}.$$

However,

$$\mathbf{P}(X_1 = 1)\,\mathbf{P}(X_2 = 1) = \frac{w^2}{(w+b)^2},$$

is not the same as the probability of the joint event $X_1 = 1$, and $X_2 = 1$, which is

$$\mathbf{P}(X_1 = 1, X_2 = 1) = \frac{w(w-1)}{(w+b)(w+b-1)},$$

unless there are no blue balls in the urn initially.

Chapter 2

2.1 We shall solve here a slightly more general problem, where the proportion of white balls in A's urn is $r > 0$. If B's urn has w white balls and b blue, the sampling (with replacement) at a stage gives a white ball with probability $p := w/(w+b)$, and gives a blue ball with probability $q := b/(w+b)$. The odds ratio is $w/b = p/q$, that is the ratio of the white to the blue balls, and p is the probability of drawing a white ball, and q is that of drawing a blue ball from B's urn.

The gist of the players being rich is to assume that each has an inexhaustible wealth of Ducats. Let X be A's winning (possibly zero or negative) when the game is stopped. Let us call a pair of games played first by A then followed by B a *full round*. If B wins the game after k full rounds, at that point the stakes are $2k - 1$ contributed by the two players (of which k are contributed by A), and B takes it all; A's winning is $-k$, and the probability of such an event is that of playing $k - 1$ full rounds where both players lose (each draws a blue ball from his respective urn), followed by one last full round where A draws a blue ball from his urn, then B draws a white one from his urn and wins. The probability of this event is the product of the probability of $k - 1$ failing full rounds (each has probability rq), times the probability of the last full round, which is rp.

But, if A wins after k full rounds and an extra half a round (played by A), he collects the stakes $2k$, of which k are his own contribution: He recovers his money and wins k additional Ducats. The probability of this event is that of k failing rounds times r to account for the last drawing that decided the game. The distribution of X is

$$X = \begin{cases} -k, & \text{with probability } rp(rq)^{k-1}; \\ +k, & \text{with probability } r(rq)^k. \end{cases}$$

One thinks of the game as being fair, if on average any player's winning is 0. Thus, the game is fair if

$$\mathbf{E}[X] = \sum_{k=0}^{\infty} rk(rq)^k - \sum_{k=1}^{\infty} kr^k q^{k-1} p = \left(r - \frac{p}{q}\right) \sum_{k=1}^{\infty} k(rq)^k = 0.$$

The sum is positive, and this equality can only hold if the factor before the sum is 0. The odds ratio must be r. For instance, if A's urn has one ball of each color, $r = \frac{1}{2}$, and $p : q = 1 : 2$, or in other words, the number of white balls has to be $\frac{1}{3}$ of the total number of balls in B's urn, i.e. the number of blue balls is twice that of the white balls. This agrees with an intuition that there has to be less chance for B to win an individual round to adjust for the advantage that B has the opportunity to collect first (even though A starts, if he wins at the first drawing he collects nothing). In Hudde's original problem $r = \frac{2}{3}$, and B's odds ratio must be $2 : 3$ for the game to be fair.

2.2 Let X be the number of matched pairs left in the urn after removing k balls. We look at the structure of combinations favorable to the event $X = r$. For this to happen, we have to leave behind r matched pairs: These can be determined by selecting r indexes from the set $\{1, \ldots, n\}$, and for each selected index k we leave behind the white and blue balls labeled k. The rest of the balls left should be unmatched. If there are w white balls left, we can choose them in $\binom{n-r}{w}$ ways. The rest of the unmatched balls should be $2n-k-2r-w$ blue balls with numbers not matching any of the w balls left. We can select these by taking out of the urn the w blue balls that have the same numbers as those white balls left unmatched in the urn: We choose the urn's unmatched blue balls in $\binom{n-r-w}{2n-k-2r-w}$. The number w can be made to vary over its feasible range: We let it run from 0 to ∞ as the binomial coefficients will take care of throwing out infeasible values of w. (The usual interpretation of $\binom{i}{j}$ as 0, when it is not feasible to select j objects out of i given ones will truncate irrelevant terms.) Of course, the count should be divided by the total number of unrestricted choices of k balls, which is $\binom{2n}{k}$. We have

$$\mathbf{P}(X=r) = \sum_{w=0}^{\infty} \frac{\binom{n}{r}\binom{n-r}{w}\binom{n-r-w}{2n-k-2r-w}}{\binom{2n}{k}}.$$

Using the hint, this simplifies to

$$\mathbf{P}(X=r) = \frac{\binom{n}{r}\binom{n-r}{2n-k-2r}2^{2n-k-2r}}{\binom{2n}{k}},$$

for $\max(0, n-k) \le r \le \lfloor\frac{1}{2}(2n-k)\rfloor$; the usual interpretation of the binomial coefficient as zero when one of its parameters is out of the appropriate range takes care of assigning 0 probability to infeasible cases.

2.3 Sequences of wins and losses are not equally likely. For example, the sequence of 50 consecutive wins has probability $(\frac{18}{38})^{50}$, whereas the sequence of 50 consecutive losses has probability $(\frac{20}{38})^{50}$. However, *conditionally* (given the number of wins) all sequence of 50 bets with the same number of wins are equally likely. This makes the conditional problem (given 30 wins and 20 losses) an instance of the ballot problem, with 30 votes for the winning candidate, and 20 for the loser. The probability of the winner staying ahead throughout is $\frac{30-20}{30+20} = \frac{10}{50} = \frac{1}{5}$.

2.4 Let us first discuss why the probability evaluation is correctly expressed in the case $x = n$. In this case, the event $Y_1 + Y_2 + \cdots + Y_n = x_1 + x_2 + \cdots + x_n = x < n$ cannot occur, and certainly an intersection of events involving this one is an empty set with probability 0. The expression given yields $1 - x/n = 1 - n/n = 0$.

Assume $x < n$. At $n = 1$, because of the integer constraint, $x_1 = Y_1$ must be zero, and we have

$$\mathbf{P}\left(\bigcap_{i=1}^{n}\{Y_1 + Y_2 + \cdots + Y_i < i\}\right) = \mathbf{P}(\{Y_1 < 1\}) = 1.$$

But then, $1-x/n = 1-0/1 = 1$, establishing a basis for the induction.

Let us next assume, as an induction hypothesis, that the result holds for $n-1$. Let the intersection of the n events be called $A_n(x)$. We compute its probability, while conditioning on the last ball (all balls are equally likely to be the last):

$$\begin{aligned}\mathbf{P}(A_n(x)) &= \sum_{i=1}^{n} \mathbf{P}(A_n(x) \mid Y_n = x_i)\, \mathbf{P}(Y_n = x_i) \\ &= \sum_{i=1}^{n} \mathbf{P}(A_n(x) \mid Y_n = x_i) \times \frac{1}{n}.\end{aligned}$$

However, when the ball labeled x_i appears last, the event $Y_1 + \cdots + Y_n < n$ is automatically satisfied (by the condition that $x < n$), and for $A_n(x)$ to happen, we require that the events $\{Y_1 < 1\}, \ldots, \{Y_1 + \cdots + Y_{n-1} < n-1\}$ all take place. But this is the same problem reduced to the induction hypothesis, because we are drawing $n-1$ balls, carrying a total of weights equal to $x - x_i$, that is, their simultaneous occurrence is equivalent to $A_{n-1}(x - x_i)$. Hence,

$$\begin{aligned}\mathbf{P}(A_n(x)) &= \frac{1}{n} \sum_{i=1}^{n} \mathbf{P}(A_{n-1}(x - x_i)) \\ &= \frac{1}{n} \sum_{i=1}^{n} \left(1 - \frac{x - x_i}{n-1}\right), \qquad \text{(by the induction hypothesis)} \\ &= \frac{1}{n} \sum_{i=1}^{n} \frac{n - 1 - x + x_i}{n-1} \\ &= \frac{n(n-1-x)}{n(n-1)} + \frac{1}{n(n-1)} \sum_{i=1}^{n} x_i \\ &= \frac{n(n-1-x)}{n(n-1)} + \frac{x}{n(n-1)} \\ &= 1 - \frac{x}{n}.\end{aligned}$$

2.5 Assume there are m votes for A and n for B, with $m > n$. Let $\mathcal{A}_1$ be the event that the first vote is for A. For A to remain ahead all the time, $\mathcal{A}_1$ must occur, and by the $(j+1)$st vote for A all votes for B should not exceed j, that is, we must have the event

$$\mathcal{A}_1 \bigcap \bigcap_{i=1}^{m} \{Y_1 + Y_2 + \cdots + Y_i < i\}$$

occur, where in this context Y_i is the number of votes for B between the $(i-1)$st and ith votes for A. The event $\bigcap_{i=1}^{m}\{Y_1+Y_2+\cdots+Y_i < i\}$, is an instance of the previous problem with a total of m balls carrying a sum of n. Thus, the probability that A remains ahead is

$$\begin{aligned}\mathbf{P}(A \rhd B) &= \mathbf{P}(\mathcal{A}_1) \times \mathbf{P}\left(\bigcap_{i=1}^{m}\{Y_1 + Y_2 + \cdots + Y_i < i\}\right)\\ &= \frac{m}{m+n} \times \left(1 - \frac{n}{m}\right)\\ &= \frac{m-n}{m+n}.\end{aligned}$$

2.6 The random variable X_i has a geometric distribution: The collector has obtained $i-1$ types of coupons, and he is attempting to get the ith from among n types. The waiting time X_i is the geometric random variable $\text{Geo}((n-i+1)/n)$, and

$$Y = X_1 + X_2 + \cdots + X_n,$$

with average

$$\begin{aligned}\mathbf{E}[Y] &= \mathbf{E}[X_1] + \mathbf{E}[X_2] + \cdots + \mathbf{E}[X_n]\\ &= \frac{n}{n} + \frac{n}{n-1} + \cdots + \frac{n}{1}\\ &= nH_n,\end{aligned}$$

where H_n is the nth harmonic number.

2.7 Consider the situation where winning the game requires scoring $2j+1$ points. The game lasts at least $2j+1$ points (in the case when the score at the end is $2j+1$ to zero), and at most $4j+1$ (in the case the score is $2j+1$ to $2j$). Let us call the two players A and B, and let $\mathcal{A}_r$ and $\mathcal{B}_r$, respectively, be the events that player A or B wins the game in $4j+1-r$ total points. The event $\mathcal{A}_r$ is that of having A win the game with a score $2j+1$ to $2j-r$, and $\mathcal{B}_r$ is the mirror image to have B win the game with A scoring $2j-r$ points. The event $\mathcal{A}_r$ can occur if and only if in the first $4j-r$ points, A scored $2j$ points, then wins the $(4j-r+1)$st point. These events have the probabilities

$$\mathbf{P}(\mathcal{A}_r) = \binom{4j-r}{2j} p^{2j+1} q^{2j-r},$$

and

$$\mathbf{P}(\mathcal{B}_r) = \binom{4j-r}{2j} p^{2j-r} q^{2j+1}.$$

Let X be the score of the loser. This random variable has the distribution

$$\begin{aligned}\mathbf{P}(X = 2j - r) &= \mathbf{P}(\mathcal{A}_r) + \mathbf{P}(\mathcal{B}_r)\\ &= \binom{4j-r}{2j} p^{2j+1} q^{2j-r} + \binom{4j-r}{2j} p^{2j-r} q^{2j+1},\end{aligned}$$

or

$$\mathbf{P}(X = k) = \binom{k+2j}{2j} \left(p^{2j+1} q^k + p^k q^{2j+1}\right).$$

2.8 Let $\mathcal{A}$ and $\mathcal{B}$, respectively, stand for the events that player A is ruined and the event that player B is ruined, and let $\mathcal{E}$ be the event that the betting game does not terminate. The three events $\mathcal{A}$, $\mathcal{B}$, and $\mathcal{E}$ partition the sample space, and we must have

$$\mathbf{P}(\mathcal{A}) + \mathbf{P}(\mathcal{B}) + \mathbf{P}(\mathcal{E}) = 1.$$

It was shown in the text that

$$\mathbf{P}(\mathcal{B}) = \begin{cases} \dfrac{1-(q/p)^m}{1-(q/p)^{m+n}}, & \text{if } \dfrac{q}{p} \neq 1; \\[2mm] \dfrac{m}{m+n}, & \text{if } p = q = \dfrac{1}{2}. \end{cases}$$

A symmetrical argument gives

$$\mathbf{P}(\mathcal{A}) = \begin{cases} \dfrac{1-(p/q)^n}{1-(p/q)^{m+n}}, & \text{if } \dfrac{p}{q} \neq 1; \\[2mm] \dfrac{n}{m+n}, & \text{if } p = q = \dfrac{1}{2}, \end{cases}$$

where the roles of p and q are interchanged, and so are the roles of m and n. Hence, in the case $p = q = \frac{1}{2}$, we have

$$\frac{n}{m+n} + \frac{m}{m+n} + \mathbf{P}(\mathcal{E}) = 1,$$

and $\mathbf{P}(\mathcal{E}) = 0$, and, in the case $p \neq q$, we have

$$\frac{1-(p/q)^n}{1-(p/q)^{m+n}} + \frac{1-(q/p)^m}{1-(q/p)^{m+n}} + \mathbf{P}(\mathcal{E}) = 1,$$

and $\mathbf{P}(\mathcal{E}) = 0$, too.

In all cases $\mathbf{P}(\mathcal{E}) = 0$, and the multitude of nonterminating betting sequences amounts only to a null set.

Chapter 3

3.1 We write the given form as

$$\begin{pmatrix} -a & \oplus \\ c & \oplus \end{pmatrix},$$

where a is a positive integer and c a nonnegative integer. The starting number of white balls must be a multiple of a, and so must c be, so that we always have a multiple of a white balls in the urn. Assume there is at least one blue ball. The entries for adding blue balls are 0 or positive, and there will always be at least one blue ball. Repeatedly taking out a white balls at a time drives the urn to a point where the white balls are depleted, and a blue ball will appear in the next sample, then white balls in multiples of a will appear, if $c > 0$, or the number of white balls stays put, while blue balls keep increasing, if $c = 0$.

If initially there are no blue balls, the top right entry must be positive, or else the white balls diminish in numbers to 0 and the urn remains devoid of blue balls, a state where the urn is empty.

3.2 Let us describe the state of the urn with the pair (w, b), when it contains w white and b blue balls. We call the state $(0, b)$, with $b > 0$ *critical*. If the urn is at a critical state a blue ball must be drawn, and it gets into the state $(3, b+4)$ and it returns to a critical state $(0, b+1)$ in three more draws (of white balls). We have an increase by one blue balls in every four draws.

Thus, if the urn starts at a critical $(0, B_0)$, after n draws we shall have

$$B_n = \begin{cases} B_0 + \dfrac{n}{4}, & \text{if } n \textbf{ mod } 4 = 0; \\ B_0 + \dfrac{n-1}{4} + 4, & \text{if } n \textbf{ mod } 4 = 1; \\ B_0 + \dfrac{n-2}{4} + 3, & \text{if } n \textbf{ mod } 4 = 2; \\ B_0 + \dfrac{n-3}{4} + 2, & \text{if } n \textbf{ mod } 4 = 3. \end{cases}$$

We can summarize this formula as

$$B_n = B_0 + \frac{1}{4}(n - n \textbf{ mod } 4) + (5 - n \textbf{ mod } 4)(1 - \mathbf{1}_{\{(n \textbf{ mod } 4)=0\}}).$$

If there are W_0 white balls initially, we need first to push the urn into the critical state $(0, B_0 - W_0)$, and it becomes obvious why we need $B_0 > W_0$, then from there we need $n - W_0$ additional steps from a critical state. The last formula is adjusted by replacing n with $n - W_0$.

The argument for white balls is simpler. Starting from a critical state $(0, B_0)$, we have

$$W_n = \begin{cases} 0, & \text{if } n \textbf{ mod } 4 = 0; \\ 3, & \text{if } n \textbf{ mod } 4 = 1; \\ 2, & \text{if } n \textbf{ mod } 4 = 2; \\ 1, & \text{if } n \textbf{ mod } 4 = 3. \end{cases}$$

This is summarized in

$$W_n = (4 - (n \textbf{ mod } 4))(1 - \mathbf{1}_{\{(n \textbf{ mod } 4)=0\}}),$$

and if there are W_0 white balls initially we need to draw these W_0 balls first, and adjust the formula by replacing n with $n - W_0$.

3.3 Let $\hat{W}_i$ be the indicator of the event that the ball in the ith sample is white.

(a) Given that there is a blue ball in the first sample, we are told the first draw increased the number of blue balls by two, and did not change the number of white balls. After this draw there are 4 white balls and 5 blue, and the (conditional) probability of picking white in the second draw is $\frac{4}{9}$.

(b) By exchangeability in the Pólya-Eggenberger urn, $\hat{W}_i$ has the same distribution as $\hat{W}_1$. Thus

$$\mathbf{P}(\hat{W}_i = 0) = \mathbf{P}(\hat{W}_1 = 0) = \frac{3}{7},$$

which is the probability of picking a blue ball.

(c) If this event occurs at position i, we either have two successive white balls ($\hat{W}_i = 1, \hat{W}_{i+1} = 1$), or two successive blue balls ($\hat{W}_i = 0, \hat{W}_{i+1} = 0$). The pair $(\hat{W}_i, \hat{W}_{i+1})$ has the same distribution as $(\hat{W}_1, \hat{W}_2)$. We have,

$$\mathbf{P}(\hat{W}_i = 1, \hat{W}_{i+1} = 1) = \mathbf{P}(\hat{W}_1 = 1, \hat{W}_2 = 1) = \frac{4}{7} \times \frac{6}{9} = \frac{8}{21},$$

and

$$\mathbf{P}(\hat{W}_i = 0, \hat{W}_{i+1} = 0) = \mathbf{P}(\hat{W}_1 = 0, \hat{W}_2 = 0) = \frac{3}{7} \times \frac{5}{9} = \frac{5}{21}.$$

The probability of the event $\hat{W}_i = \hat{W}_{i+1}$, is the sum of these two probabilities, which is $\frac{13}{21}$.

3.4 Let $\hat{W}_i$ be the indicator of the event of drawing a white ball in the ith sample. Then,

$$\mathbf{P}(\tilde{W}_n = k) = \sum_{\substack{\hat{w}_1+\cdots+\hat{w}_n=k \\ \hat{w}_r \in \{0,1\}, 1 \le r \le n}} \mathbf{P}(\hat{W}_1 = \hat{w}_1, \ldots, \hat{W}_n = \hat{w}_n).$$

By exchangeability, all the summed probabilities are the same as

$$\begin{aligned}\mathbf{P}(\hat{W}_1 = 1, \ldots, \hat{W}_k = 1, \hat{W}_{k+1} = 0, \ldots, \hat{W}_n = 0) \\ = \frac{W_0}{\tau_0} \times \frac{W_0 + s}{\tau_0 + s} \times \cdots \times \frac{W_0 + (k-1)s}{\tau_0 + (k-1)s} \times \frac{B_0}{\tau_0 + ks} \\ \times \frac{B_0 + s}{\tau_0 + (k+1)s} \times \cdots \times \frac{B_0 + (n-k-1)s}{\tau_0 + (n-1)s},\end{aligned}$$

and there are $\binom{n}{k}$ such summands.

3.5 It suffices to show that $\hat{W}_1$ and $\hat{W}_2$ are not exchangeable in a specific instance. Consider the Bernard Friedman's urn scheme starting with one white and two blue balls. Describe the state of the urn after n draws with the pair (W_n, B_n). If a white ball appears in the first sample (with probability $\frac{1}{3}$) the urn progresses into the state $(1, 3)$ from which the probability of sampling a blue ball is $\frac{3}{4}$, whence

$$\mathbf{P}(\hat{W}_1 = 1, \hat{W}_2 = 0) = \frac{1}{3} \times \frac{3}{4} = \frac{1}{4}.$$

On the other hand, a blue ball appearing in the first sample (with probability $\frac{2}{3}$) drives the urn into the state $(2, 2)$, from which a white ball appears in the second sample with probability $\frac{2}{4}$, whence

$$\mathbf{P}(\hat{W}_1 = 0, \hat{W}_2 = 1) = \frac{2}{3} \times \frac{2}{4} = \frac{1}{3}.$$

We see that

$$\mathbf{P}(\hat{W}_1 = 1, \hat{W}_2 = 0) \neq \mathbf{P}(\hat{W}_1 = 0, \hat{W}_2 = 1);$$

$\hat{W}_1$ and $\hat{W}_2$ are not exchangeable, and of course the entire sequence $\{\hat{W}_n\}_{n=0}^{\infty}$ is not exchangeable.

3.6 Let the small pills (10 mg) be represented by white balls, and the large pills (20 mg) be represented by blue balls in an urn. The starting state is that of the urn when it has 60 white balls and 20 blue balls. Each time a white ball is taken out (a small pill is consumed) the number of white balls in the urn is reduced by 1. Each time a blue ball is taken out, the patient breaks a large pill into the equivalent of two small pills, one is consumed and one small pill is returned to the urn (a blue ball disappears and a new white ball appears).

(a) The schema of the urn is

$$\begin{pmatrix} -1 & 0 \\ 1 & -1 \end{pmatrix};$$

this is an example of a diminishing urn.

(b) Denote by $\mathrm{NHG}_k(m, n)$ the negative hypergeometric random variable collecting from two populations of sizes m and n

until k balls of the first type (of size m) are in the sample. Let $\tilde{W}_i$ be the number of drawings (waiting time) of a white balls between the $(i-1)$st and i blue samples. We define $\tilde{W}_0 = 0$. Then $\tilde{W}_i \overset{\mathcal{D}}{=} \mathrm{NHG}_1(20, 60) - 1$. Right after the ith drawing of a blue ball, there are $20 - i$ blue balls left in the urn, and each blue sample adds one white ball to the urn, however, $\tilde{W}_1 + \tilde{W}_2 + \cdots + \tilde{W}_i$ white balls are taken before the ith blue sample, and

$$\tilde{W}_{i+1} = \mathrm{NHG}_1\left(20 - i, 60 + i - \sum_{j=1}^{i} \tilde{W}_j\right) - 1.$$

This $(i + 1)$ st waiting time has the conditional average

$$\begin{aligned}\mathbf{E}[\tilde{W}_{i+1} \mid \tilde{W}_1, \ldots, \tilde{W}_i] &= \frac{20 - i + 60 + i - \sum_{j=1}^{i} \tilde{W}_j + 1}{20 - i + 1} - 1 \\ &= \frac{81 - \sum_{j=1}^{i} \tilde{W}_j}{21 - i} - 1,\end{aligned}$$

with unconditional average

$$a_{i+1} := \mathbf{E}[\tilde{W}_{i+1}] = \frac{81 - \sum_{j=1}^{i} \mathbf{E}[\tilde{W}_j]}{21 - i} - 1.$$

We can find this average by solving the recurrence

$$(21 - i)a_{i+1} = 81 - \left(\sum_{j=1}^{i} a_j\right) - (21 - i) = 60 + i - \sum_{j=1}^{i} a_j.$$

We can difference a version of this recurrence with i from the version with $i + 1$ to get

$$a_{i+1} = a_i + \frac{1}{21 - i},$$

valid for $i \ge 1$. The solution is obtained by unwinding the recurrence

$$\begin{aligned}a_i &= \frac{1}{22 - i} + a_{i-1} \\ &= \frac{1}{22 - i} + \frac{1}{23 - i} + a_{i-2} \\ &\ \ \vdots \\ &= \frac{1}{22 - i} + \frac{1}{23 - i} + \cdots + \frac{1}{20} + a_1,\end{aligned}$$

and a_1 is the mean of $\text{NGH}_1(20, 60) - 1$, which is $\frac{60}{21}$. Hence

$$\mathbf{E}[\tilde{W}_i] = H_{20} - H_{21-i} + \frac{60}{21}.$$

where H_k is the kth harmonic number.
Then, X, the number of white balls left behind after all the blue balls are taken out is

$$60 + 20 - (\tilde{W}_1 + \tilde{W}_2 + \cdots + \tilde{W}_{20}),$$

with average

$$\begin{aligned}\mathbf{E}[X] &= 80 - \sum_{i=1}^{20} \mathbf{E}[\tilde{W}_i] \\ &= 80 - \sum_{i=1}^{20} \left(H_{20} - H_{21-i} + \frac{60}{21} \right) \\ &= \frac{100176575}{15519504} \\ &\approx 6.455.\end{aligned}$$

Remark: We obtained the rational number in the answer computationally, but more generally there is a systematic way to get such a number. Assume we wanted to solve this pills problem when w small pills (white balls) and b large pills (blue balls) are in the bottle (urn). Then $\tilde{W}_1 = \text{NHG}_1(b, w) - 1$, $\tilde{W}_2 = \text{NHG}_1(b-1, w+1-\tilde{W}_1) - 1$, and so on, till

$$\tilde{W}_{i+1} = \text{NHG}_1 \left(b - i, w + i - \sum_{j=1}^{i} \tilde{W}_j \right) - 1.$$

This gives the recurrence

$$\mathbf{E}[\tilde{W}_{i+1}] = \frac{w + b + 1 - \sum_{j=1}^{i} \mathbf{E}[\tilde{W}_j]}{b + 1 - i} - 1,$$

valid for $i \geq 1$, and with initial condition $\mathbf{E}[\tilde{W}_1] = w/(b+1)$. This recurrence has the solution

$$\mathbf{E}[W_i] = H_b - H_{b+1-i} + \frac{w}{b+1}.$$

Then, X, the number of white balls left behind after all the blue balls are taken out is

$$w + b - (\tilde{W}_1 + \tilde{W}_2 + \cdots + \tilde{W}_b),$$

with average

$$\mathbf{E}[X] = w + b - \sum_{i=1}^{b} \left(H_b - H_{b+1-i} + \frac{w}{b+1} \right)$$
$$= w + b - b\left(H_b + \frac{w}{b+1} \right) + \sum_{k=1}^{b} H_k.$$

Now,

$$\sum_{k=1}^{b} H_k = \sum_{k=1}^{b}\sum_{j=1}^{k} \frac{1}{j} = \sum_{j=1}^{b} \frac{1}{j} \sum_{k=j}^{b} 1 = \sum_{j=1}^{b} \frac{b+1-j}{j} = (b+1)H_b - b.$$

When we plug this back in, we get the simplified expression

$$\mathbf{E}[X] = H_b + \frac{w}{b+1}.$$

3.7 Let A_i be the state in which i blue balls are in the urn, and let X_i be the time it takes the urn to go from state A_i to state A_{i+1}. As in the text, we can generally represent the waiting time $Y_n(M)$ to go from the state 0 to the state n as a sum:

$$Y_n(M) = X_0 + X_1 + \cdots + X_n.$$

In the text we handled the case $Y_M(2M)$, see Eq. (3.12), which we simply called $Y(2M)$. Here we wish to deal with $Y_M(M)$.

We can write equations similar to those that gave Eq. (3.14) for the more general case and get

$$\mathbf{E}[Y_n(M)] = \frac{M}{2} \int_0^1 \frac{1}{y}(1-y)^{M-n}((1+y)^n - (1-y)^n)\, dy.$$

For the instance at hand we have

$$a_M := \mathbf{E}[Y_M(M)] = \frac{M}{2} \int_0^1 \frac{1}{y}((1+y)^M - (1-y)^M)\, dy.$$

We use a differencing technique to simplify $\mathbf{E}[Y_M(M)]/M$: Write a version with $M-1$ replacing M and take the difference

$$\frac{2a_M}{M} - \frac{2a_{M-1}}{M-1} = \int_0^1 \frac{1}{y}((1+y)^M - (1-y)^M)\, dy$$
$$- \int_0^1 \frac{1}{y}((1+y)^{M-1} - (1-y)^{M-1})\, dy$$

$$= \int_0^1 \frac{1}{y}((1+y)^M - (1+y)^{M-1})\,dy$$
$$- \int_0^1 \frac{1}{y}((1-y)^M - (1-y)^{M-1})\,dy$$
$$= \int_0^1 (1+y)^{M-1}\,dy + \int_0^1 (1-y)^{M-1}\,dy$$
$$= \frac{(1+y)^M}{M}\bigg|_{y=0}^{1} - \frac{(1-y)^M}{M}\bigg|_{y=0}^{1}$$
$$= \frac{2^M}{M}.$$

Hence,

$$\frac{2a_M}{M} = \frac{2a_{M-1}}{M-1} + \frac{2^M}{M} = \frac{2a_{M-2}}{M-2} + \frac{2^{M-1}}{M-1} + \frac{2^M}{M} = \sum_{i=1}^{M} \frac{2^i}{i}.$$

And finally,

$$\mathbf{E}[Y_M(M)] = \frac{M}{2}\sum_{i=1}^{M}\frac{2^i}{i}.$$

Chapter 4

4.1(a) $W(0)$ must be a multiple of α, and $B(0)$ must be a multiple of β.

(b) Suppose the next renewal time after t is t'. There is no change in the scheme between time t and t', that is,

$$\frac{W(t)}{\alpha} + \frac{B(t)}{\beta} = \frac{W(t'')}{\alpha} + \frac{B(t'')}{\beta}, \qquad \text{for all } t \le t'' < t'.$$

At time t' there is one renewal (almost surely). The renewal either comes from a white process, in which case the number of white balls goes down by α, and the number of blue balls goes up by β, or comes from a blue process, in which case the number of white balls goes up by α, and that of blue balls goes down by β. In the case of a renewal from a white process

$$\frac{W(t')}{\alpha} + \frac{B(t')}{\beta} = \frac{W(t) - \alpha}{\alpha} + \frac{B(t) + \beta}{\beta} = \frac{W(t)}{\alpha} + \frac{B(t)}{\beta}.$$

The argument is quite similar for a renewal from a blue process.

In either case we have

$$\frac{W(t')}{\alpha} + \frac{B(t')}{\beta} = \frac{W(t)}{\alpha} + \frac{B(t)}{\beta}.$$

At all times $\frac{W(t)}{\alpha} + \frac{B(t)}{\beta}$ remains invariant, and equal to its value at time 0.

The argument for the event of multiple simultaneous renewals (a null set) is essentially the same: For instance, if there are i renewals from white processes, and j from blue processes, the net result is to add to the Ehrenfest process $(j - i)\alpha$ white balls and add $(i - j)\beta$ blue balls, and again $\frac{W(t)}{\alpha} + \frac{B(t)}{\beta}$ remains invariant.

At time 0, the invariance dictates that

$$C := \frac{W(0)}{\alpha} + \frac{B(0)}{\beta}.$$

However, for tenability $W(0)$ is a multiple of α, and $B(0)$ is a multiple of β; the invariant C is an integer.

(c) Let $\mathbf{B} = \mathbf{A}^T$. The matrix $\mathbf{B}$ has the two distinct eigenvalues 0 and $-(\alpha + \beta)$, and the exponential expansion

$$e^{\mathbf{B}t} = \frac{1}{\alpha + \beta}\left(\begin{pmatrix} \beta & \alpha \\ \beta & \alpha \end{pmatrix} + e^{-(\alpha+\beta)t}\begin{pmatrix} \alpha & -\alpha \\ -\beta & \beta \end{pmatrix}\right).$$

According to Theorem 4.1

$$\begin{pmatrix} \mathbf{E}[W(t)] \\ \mathbf{E}[B(t)] \end{pmatrix} = e^{\mathbf{B}t}\begin{pmatrix} W(0) \\ B(0) \end{pmatrix}.$$

Extracting the first component we find

$$\mathbf{E}[W(t)] = \frac{1}{\alpha + \beta}(\beta W(0) + \alpha B(0) + e^{-(\alpha+\beta)t}(\alpha W(0) - \alpha B(0)).$$

4.2 Here $\tau = 3$, and by solving Eq. (4.11) we get

$$\mathbf{E}[e^{W(t)u}] = \frac{1}{8}(1 - e^{-2t} - e^{-4t} + e^{-6t} + (3 - e^{-2t} + e^{-4t} - 3e^{-6t})e^{u}$$
$$+(3 + e^{-2t} + e^{-4t} + 3e^{-6t})e^{2u} + (1 + e^{-2t} - e^{-4t} - e^{-6t})e^{3u}).$$

Set $z = e^u$ to get the probability generating function

$$\sum_{k=0}^{3}\mathbf{P}(W(t) = k)z^k = \frac{1}{8}(1 - e^{-2t} - e^{-4t} + e^{-6t} + (3 - e^{-2t} + e^{-4t} - 3e^{-6t})z$$
$$+(3 + e^{-2t} + e^{-4t} + 3e^{-6t})z^2 + (1 + e^{-2t} - e^{-4t} - e^{-6t})z^3).$$

The probability $\mathbf{P}(W(t) = k)$, follows upon extracting the coefficient of z^k, for $k = 0, 1, 2, 3$.

For the mean and variance we can either calculate sums like

$$\sum_{k=0}^{3} k\, \mathbf{P}(W(t) = k),$$

(for the mean), or compute directly from the moment generating function:

$$\mathbf{E}[W(t)] = \frac{\partial}{\partial u}\phi(t,u)\Big|_{u=0} = \frac{3}{2} + \frac{1}{2}e^{-2t} \to \frac{3}{2},$$

$$\begin{aligned}\mathbf{Var}[W(t)] &= \left(\frac{\partial^2}{\partial u^2}\phi(t,u) - \left(\frac{\partial}{\partial u}\phi(t,u)\right)^2\right)_{u=0}\\ &= \frac{3}{2} + \frac{1}{2}e^{-2t}\\ &\to \frac{3}{2};\end{aligned}$$

the asymptotics are taken as $t \to \infty$.

Note that, as $t \to \infty$, the moment generating function converges to

$$\frac{1}{8} + \frac{3}{8}e^{u} + \frac{3}{8}e^{2u} + \frac{1}{8}e^{3u},$$

which is that of $\text{Bin}(3, \frac{1}{2})$.

4.3 Let $\mathbf{B} = \mathbf{A}^T$. If $W(t)$ and $B(t)$ are, respectively, the number of white and blue balls at time t, their averages are given by Theorem 4.1:

$$\begin{pmatrix}\mathbf{E}[W(t)]\\ \mathbf{E}[B(t)]\end{pmatrix} = e^{\mathbf{B}t}\begin{pmatrix}W(0)\\ B(0)\end{pmatrix}.$$

To facilitate the computation of $e^{\mathbf{B}t}$, we seek a representation in terms of the two eigenvalues of $\mathbf{B}$ (which are 5 and 1), and the 2×2 idempotents of $\mathbf{B}$. One finds

$$e^{\mathbf{B}t} = \frac{1}{4}e^{5t}\begin{pmatrix}1 & 1\\ 3 & 3\end{pmatrix} + \frac{1}{4}e^{t}\begin{pmatrix}3 & -1\\ -3 & 1\end{pmatrix}.$$

Completing the matrix multiplication, we get

$$\mathbf{E}[W(t)] = \left(\frac{1}{4}e^{5t} + \frac{3}{4}e^{t}\right)W(0) + \left(\frac{1}{4}e^{5t} - \frac{1}{4}e^{t}\right)B(0).$$

4.4 Let $\mathbf{B} = \mathbf{A}^T$. Let $W(t)$, $B(t)$, and $R(t)$ be the respective number of white, blue, and red balls at time t, and denote by $\mathbf{X}(t)$ the vector with these three components. As in the multicolor extension of (see Section 4.2) Theorem 4.1, the average of $\mathbf{X}(t)$ is given by

$$\mathbf{E}[\mathbf{X}(t)] = e^{\mathbf{B}t}\mathbf{X}(0).$$

The matrix $\mathbf{B}$ has a repeated eigenvalue; the eigenvalues are 2, 2, and 1. When an eigenvalue λ is repeated once in a 3×3 matrix, the exponential expansion involves $e^{\lambda t}$, and $te^{\lambda t}$, with coefficients that are 3×3 fixed matrices:

$$e^{\mathbf{B}t} = te^{2t}\mathcal{E}_1 + e^{2t}\mathcal{E}_2 + e^{t}\mathcal{E}_3.$$

The three matrices $\mathcal{E}_1, \mathcal{E}_2, \mathcal{E}_3$ can be obtained by a variety of methods, like for example matching the coefficients of both sides at $t = 0$, and the derivatives once and twice at $t = 0$. We find the expansion as given in the hint. We then find

$$\begin{aligned}\mathbf{E}\big[\mathbf{X}(t)\big] &= te^{2t}\begin{pmatrix}0 & 0 & 1\\ 0 & 0 & 1\\ 0 & 0 & 0\end{pmatrix}\mathbf{X}(0) + e^{2t}\begin{pmatrix}0 & 1 & 0\\ 0 & 1 & 0\\ 0 & 0 & 1\end{pmatrix}\mathbf{X}(0)\\ &\quad + e^{t}\begin{pmatrix}1 & -1 & 0\\ 0 & 0 & 0\\ 0 & 0 & 0\end{pmatrix}\mathbf{X}(0)\\ &= \begin{pmatrix}R(0)te^{2t} + B(0)e^{2t} + \big(W(0) - B(0)\big)e^{t}\\ R(0)te^{2t} + B(0)e^{2t}\\ R(0)e^{2t}\end{pmatrix}.\end{aligned}$$

4.5 Suppose the given generator is

$$\begin{pmatrix}U & X\\ Y & Z\end{pmatrix},$$

and let $\psi_{U,X}(u, v)$ be the joint moment generating function of U and X, and $\psi_{Y,Z}(u, v)$ be the joint moment generating function of Y and Z. Differentiate the partial differential equation in Lemma 4.1 to get

$$\begin{aligned}\frac{\partial}{\partial u}\left(\frac{\partial \phi}{\partial t}\right) &+ (1 - \psi_{U,X}(u, v))\frac{\partial^2 \phi}{\partial u^2} - \frac{\partial \psi_{U,X}}{\partial u} \times \frac{\partial \phi}{\partial u}\\ &+ (1 - \psi_{Y,Z}(u, v))\frac{\partial^2 \phi}{\partial u \partial v} - \frac{\partial \psi_{Y,Z}}{\partial u} \times \frac{\partial \phi}{\partial v} = 0.\end{aligned}$$

Evaluate at $u = v = 0$ to get

$$\frac{\partial}{\partial t}\left(\frac{\partial \phi}{\partial u}\right)\Big|_{u=v=0} = \left[\mathbf{E}[U]\frac{\partial \phi}{\partial u} + \mathbf{E}[Y]\frac{\partial \phi}{\partial v}\right]_{u=v=0},$$

which gives

$$\frac{d}{dt}\,\mathbf{E}[W(t)] = \mathbf{E}[U]\,\mathbf{E}[W(t)] + \mathbf{E}[Y]\,\mathbf{E}[B(t)].$$

Symmetrically, by carrying out the operation on v, by first taking the partial derivative with respect to v, then evaluating at $u = v = 0$, we obtain the mirror image equation:

$$\frac{d}{dt}\,\mathbf{E}[B(t)] = \mathbf{E}[X]\,\mathbf{E}[W(t)] + \mathbf{E}[Z]\,\mathbf{E}[B(t)].$$

Writing these individual ordinary differential equations as a system in matrix form, we have the representation

$$\frac{d}{dt}\begin{pmatrix}\mathbf{E}[W(t)]\\ \mathbf{E}[B(t)]\end{pmatrix} = \begin{pmatrix}\mathbf{E}[U] & \mathbf{E}[Y]\\ \mathbf{E}[X] & \mathbf{E}[Z]\end{pmatrix}\begin{pmatrix}\mathbf{E}[W(t)]\\ \mathbf{E}[B(t)]\end{pmatrix} = \mathbf{E}[\mathbf{A}^T]\begin{pmatrix}\mathbf{E}[W(t)]\\ \mathbf{E}[B(t)]\end{pmatrix},$$

where $\mathbf{A}^T$ is the transpose of $\mathbf{A}$. This system of first-order ordinary differential equations has the solution

$$\begin{pmatrix}\mathbf{E}[W(t)]\\ \mathbf{E}[B(t)]\end{pmatrix} = e^{\mathbf{E}[\mathbf{A}^T]t}\begin{pmatrix}W(0)\\ B(0)\end{pmatrix}.$$

4.6 Let $q := 1 - p$. From Exercise 4.5, we need only the average generator:

$$\begin{pmatrix}\mathbf{E}[X] & \mathbf{E}[1-X]\\ \mathbf{E}[1-X] & \mathbf{E}[X]\end{pmatrix} = \begin{pmatrix}p & q\\ q & p\end{pmatrix}.$$

It then follows that

$$\begin{pmatrix}\mathbf{E}[W(t)]\\ \mathbf{E}[B(t)]\end{pmatrix} = e^{\begin{pmatrix}p & q\\ q & p\end{pmatrix}t}\begin{pmatrix}W(0)\\ B(0)\end{pmatrix}$$

$$= \frac{1}{2}\begin{pmatrix}(e^t + e^{(2p-1)t})W(0) + (e^t - e^{(2p-1)t}))B(0)]\\ (e^t - e^{(2p-1)t})W(0) + (e^t + e^{(2p-1)t})B(0))\end{pmatrix}.$$

Chapter 5

5.1 In the Pólya-Eggenberger scheme the ball replacement matrix is

$$\mathbf{A} = \begin{pmatrix}s & 0\\ 0 & s\end{pmatrix},$$

for some positive integer s. So,

$$\mathbf{A} = \mathbf{A}^T =: \mathbf{B} = s\mathbf{I}_2,$$

where $\mathbf{I}_2$ is the identity matrix.Thus,

$$\begin{aligned} e^{\mathbf{B}t} &= e^{s\mathbf{I}_2 t} \\ &= \mathbf{I}_2 + \frac{st}{1!}\mathbf{I}_2 + \frac{s^2t^2}{2!}\mathbf{I}_2^2 + \cdots \\ &= \left(1 + \frac{st}{1!} + \frac{s^2t^2}{2!} + \cdots\right)\mathbf{I}_2 \\ &= e^{st}\mathbf{I}_2 \\ &=: e^{st}\boldsymbol{\mathcal{E}}_1. \end{aligned}$$

The Pólya-Eggenberger schema has the eigenvalues $\lambda_1 = \lambda_2 = s$ (repeated).

Assume the two colors are white and blue, and the number of white and blue balls in the urn after n draws are, respectively, W_n and B_n. Let $W(t)$ and $B(t)$ be the poissonized counterparts of W_n and B_n. Hence, with $\mathbf{J} = (1\ \ 1)$,

$$\boldsymbol{\mathcal{E}}_1 \begin{pmatrix} W(0) \\ B(0) \end{pmatrix} = \mathbf{I}_2 \begin{pmatrix} W(0) \\ B(0) \end{pmatrix} = \begin{pmatrix} W(0) \\ B(0) \end{pmatrix};$$

$$\mathbf{J}\mathbf{B}^{-1}\boldsymbol{\mathcal{E}}_1 \begin{pmatrix} W(0) \\ B(0) \end{pmatrix} = \frac{1}{s}\mathbf{J}\mathbf{I}_2\mathbf{I}_2 \begin{pmatrix} W(0) \\ B(0) \end{pmatrix} = \frac{1}{s}(W(0) + B(0)).$$

Plugging these calculations in Eq. (5.5), we obtain

$$\begin{pmatrix} \mathbf{E}[W_n] \\ \mathbf{E}[B_n] \end{pmatrix} \approx \frac{s}{W_0 + B_0} \begin{pmatrix} W_0 \\ B_0 \end{pmatrix} n,$$

in accord with Corollary 3.1. Note that in the final presentation we replaced $W(0)$ and $B(0)$ with W_0 and B_0 to recognize the fact that the poissonized process coincides with the discrete process at the points of renewal, only if the two processes start with the same initial conditions.

5.2 The depoissonization of hyper Pólya-Eggenberger schemes follows essentially the same steps as for the bivariate Pólya-Eggenberger scheme (pursued in Exercise 5.1), only the linear algebra is done in k dimensions. We have

$$\mathbf{A} = \mathbf{A}^T =: \mathbf{B} = s\mathbf{I}_k,$$

and

$$e^{\mathbf{B}t} = e^{st}\mathbf{I}_k =: e^{st}\boldsymbol{\mathcal{E}}_1,$$

where $\boldsymbol{\mathcal{E}}_1 = \mathbf{I}_k$ is the $k \times k$ identity matrix.

Assume the colors are numbered $1, \ldots, k$, and the number of balls of color i in the urn after n draws is $X_n^{(i)}$. Let $\mathbf{R}_n = (X_n^{(1)}, \ldots, X_n^{(k)})^T$, and $\mathbf{R}(t)$ be its poissonized counterpart. Hence, with $\mathbf{J} = (1\ 1\ \dots 1)$ being the row vector of k ones,

$$\boldsymbol{\mathcal{E}}_1\mathbf{R}(0) = \mathbf{I}_k\mathbf{R}(0) = \mathbf{R}(0);$$

$$\mathbf{J}\mathbf{B}^{-1}\boldsymbol{\mathcal{E}}_1\mathbf{R}(0) = \mathbf{J}\mathbf{I}_k\mathbf{I}_k\mathbf{R}(0) = X_0^{(1)} + \dots + X_0^{(k)}.$$

Plugging these calculations in Eq. (5.5), we obtain

$$\begin{pmatrix} X_n^{(1)} \\ \vdots \\ X_n^{(k)} \end{pmatrix} \approx \frac{1}{X_0^{(1)} + \dots + X_0^{(k)}} \begin{pmatrix} X_0^{(1)} \\ \vdots \\ X_0^{(k)} \end{pmatrix} n.$$

5.3 The Pólya-Eggenberger-like schema

$$\begin{pmatrix} a & 0 \\ 0 & d \end{pmatrix},$$

has the two distinct eigenvalues $\lambda_1 = a$, $\lambda_2 = d$, and $\mathbf{A} = \mathbf{A}^T =: \mathbf{B}$.

Let the number of white and blue balls in the urn after n draws be, respectively, W_n and B_n. The urn must start nonempty. If $W_0 = 0$, we only draw blue balls and add blue balls, and W_n remains zero at all times, and $B_n = dn + B_0$. Likewise, if $B_0 = 0$, then $W_n = an + W_0$, and $B_n = 0$.

In the sequel, we assume $W_0 \ge 1$ and $B_0 \ge 1$. We have

$$\begin{aligned} e^{\mathbf{B}t} &= \boldsymbol{\mathcal{E}}_1 e^{at} + \boldsymbol{\mathcal{E}}_2 e^{dt} \\ &=: \begin{pmatrix} 1 & 0 \\ 0 & 0 \end{pmatrix} e^{at} + \begin{pmatrix} 0 & 0 \\ 0 & 1 \end{pmatrix} e^{dt}. \end{aligned}$$

The averaging measure $\bar{t}_n$ is defined by

$$n \approx \mathbf{J}\mathbf{B}^{-1}\boldsymbol{\mathcal{E}}_1 \begin{pmatrix} W_0 \\ B_0 \end{pmatrix} e^{a\bar{t}_n} = \frac{W_0}{a} e^{a\bar{t}_n},$$

and $\bar{t}_n \approx \frac{1}{a} \ln n$. By the depoissonization heuristic

$$\begin{aligned} \begin{pmatrix} \mathbf{E}[W_n] \\ \mathbf{E}[B_n] \end{pmatrix} &\approx e^{\mathbf{B}\bar{t}_n} \begin{pmatrix} W_0 \\ B_0 \end{pmatrix} \\ &= \begin{pmatrix} e^{a\bar{t}_n} & 0 \\ 0 & e^{d\bar{t}_n} \end{pmatrix} \begin{pmatrix} W_0 \\ B_0 \end{pmatrix} \\ &= \begin{pmatrix} W_0 e^{a\bar{t}_n} \\ B_0 e^{d\bar{t}_n} \end{pmatrix} \end{aligned}$$

$$= \left(\frac{an}{B_0(e^{a\bar{t}_n})^{d/a}} \right)$$

$$= \left(\frac{an}{B_0\left(\frac{an}{W_0}\right)^{d/a}} \right).$$

The total number of balls after n draws, τ_n, cannot exceed an asymptotically. Hence, the white balls dominate asymptotically and the difference $\tau_n - W_n$ is equal to B_n, hence the lower-order term in W_n must be a negative function of n that is approximately $O(n^{d/a})$.

5.4 Let $\mathbf{B} = \mathbf{A}^T$. Let $W(t)$, $B(t)$, and $R(t)$ be the respective number of white, blue, and red balls at time t, and denote by $\mathbf{X}(t)$ the vector with these three components. If we find the approximating measure $\bar{t}_n$, then according to the calculation in Exercise 4.4 and the depoissonization heuristic

$$\mathbf{E}[\mathbf{X}(\bar{t}_n)] = \begin{pmatrix} R(0)\bar{t}_n e^{2\bar{t}_n} + B(0)e^{2\bar{t}_n} + \big(W(0) - B(0)\big)e^{\bar{t}_n} \\ R(0)\bar{t}_n e^{2\bar{t}_n} + B(0)e^{2\bar{t}_n} \\ R(0)e^{2\bar{t}_n} \end{pmatrix}.$$

If $R(0) = 0$, red balls never appear in the urn and we can reduce the problem to one on two colors, with the ball addition matrix $\begin{pmatrix} 1 & 0 \\ 1 & 2 \end{pmatrix}$. Assume $R(0) > 0$. We obtain $\bar{t}_n$ from the relation

$$n \approx \bar{t}_n e^{2\bar{t}_n} \mathbf{J}\mathbf{B}^{-1}\boldsymbol{\mathcal{E}}_1 \begin{pmatrix} W(0) \\ B(0) \\ R(0) \end{pmatrix}$$

$$= \bar{t}_n e^{2\bar{t}_n} (1\ \ 1\ \ 1) \begin{pmatrix} 1 & -\frac{1}{2} & -\frac{1}{4} \\ 0 & \frac{1}{2} & -\frac{1}{4} \\ 0 & 0 & \frac{1}{2} \end{pmatrix} \begin{pmatrix} 0 & 0 & 1 \\ 0 & 0 & 1 \\ 0 & 0 & 0 \end{pmatrix} \begin{pmatrix} W(0) \\ B(0) \\ R(0) \end{pmatrix}$$

$$= \bar{t}_n e^{2\bar{t}_n} R(0).$$

Hence,

$$\mathbf{E}[\mathbf{X}(\bar{t}_n)] \approx \begin{pmatrix} n + O\Big(\dfrac{n}{\ln n}\Big) \\ n + O\Big(\dfrac{n}{\ln n}\Big) \\ \dfrac{n}{\ln n} \end{pmatrix}.$$

Chapter 6

6.1 Suppose the schema is $\begin{pmatrix} a & b \\ c & d \end{pmatrix}$. Let the principal eigenvalue of the schema be $\lambda_1 > 0$, and the corresponding left row eigenvector be $(v_1 \ \ v_2)$. Let W_n and B_n be, respectively, the number of white and blue balls in the urn after n draws. Every time we sample a white ball, we add a white balls, and every time we sample a blue ball we add c white balls. Then,

$$W_n = a\tilde{W}_n + c\tilde{B}_n + W_0 = a\tilde{W}_n + c(n - \tilde{W}_n) + W_0,$$

which we rearrange as

$$\tilde{W}_n = \frac{W_n - W_0 - cn}{a - c}.$$

Scaling, we write

$$\frac{\tilde{W}_n}{n} = \frac{W_n - W_0}{(a - c)n} - \frac{c}{(a - c)}.$$

By Theorem 6.2,

$$\frac{W_n}{n} \xrightarrow{P} \lambda_1 v_1.$$

We also have

$$\frac{W_0}{n} \xrightarrow{P} 0.$$

Hence,

$$\frac{\tilde{W}_n}{n} \xrightarrow{P} \frac{\lambda_1 v_1 - c}{a - c} = v_1.$$

Similarly,

$$\frac{\tilde{B}_n}{n} \xrightarrow{P} v_2.$$

6.2 Let W_n be the number of white balls after n draws. If we sample a white ball (with replacement), we add (an independent copy of) X white balls, and if we sample a blue ball (with replacement), we add (an independent copy of) X white balls. That is, we always return the sampled ball to the urn then add (an independent copy of) X white balls after each draw, and the number of white balls has the representation

$$W_n \stackrel{\mathcal{D}}{=} X_1 + X_2 + \cdots + X_n + W_0,$$

where the X_i's are independent and identically distributed Bernoulli Ber(p) random variables.

(a) The sum of n independent and identically distributed Ber(p) random variables is the binomial Bin(n, p), and

$$W_n \overset{\mathcal{D}}{=} \text{Bin}(n, p) + W_0.$$

Consequently,

$$\begin{aligned}\mathbf{P}(W_n = k) &= \mathbf{P}(\text{Bin}(n, p) = k - W_0)\\ &= \binom{n}{k} p^{k-W_0} q^{n-k+W_0}, \qquad \text{for } W_0 \le k \le n + W_0.\end{aligned}$$

(b) From the standard approximation of binomial distribution by the normal we find

$$\frac{W_n - pn}{\sqrt{n}} \xrightarrow{\mathcal{D}} \mathcal{N}(0, p(1-p)).$$

6.3 The average generator is

$$\begin{pmatrix} \frac{1}{3} & \frac{2}{3} \\ \frac{2}{3} & \frac{1}{3} \end{pmatrix},$$

with eigenvalues $\lambda_1 = 1$, and $\lambda_2 = -\frac{1}{3} < \frac{1}{2}\lambda_1$, so the urn scheme is from extended family, and we always add exactly one ball. The principal eigenvalue is λ_1, with corresponding left eigenvector $\frac{1}{2}(1\ 1)$.

Let W_n be the number of white balls after n draws, and let $\mathbf{1}_n^W$ be the indicator of sampling a white ball in the nth draw. Theorem 6.5 gives us the form of the limiting distribution:

$$\frac{W_n - \frac{1}{2}n}{\sqrt{n}} \xrightarrow{\mathcal{D}} \mathcal{N}(0, \sigma^2).$$

To completely characterize the limiting normal distribution, we still need to work out the variance.

The exact stochastic recurrence is

$$W_n = W_{n-1} + X\mathbf{1}_n^W + (1 - X)(1 - \mathbf{1}_n^W),$$

with conditional expectation (over the sigma field $\mathcal{F}_{n-1}$, generated by the first $n - 1$ draws):

$$\mathbf{E}[W_n \mid \mathcal{F}_{n-1}] = W_{n-1} + \mathbf{E}[X]\,\mathbf{E}\big[\mathbf{1}_n^W \mid \mathcal{F}_{n-1}\big] + \mathbf{E}[1 - X]\,\mathbf{E}\big[1 - \mathbf{1}_n^W \mid \mathcal{F}_{n-1}\big];$$

for this representation we used the independence of X from all the previous sampling steps. The indicator $\mathbf{1}_n^W$ has the conditional expectation

$$\mathbf{E}\big[\mathbf{1}_n^W \mid \mathcal{F}_{n-1}\big] = \frac{W_{n-1}}{\tau_0 + n - 1}.$$

Plugging in, taking expectations, and simplifying, we get the recurrence

$$\mathbf{E}[W_n] = \left(1 + \frac{2p-1}{\tau_0 + n - 1}\right) \mathbf{E}[W_{n-1}] + 1 - p.$$

The solution is computed by iteration:

$$\mathbf{E}[W_n] = \frac{1}{2}n + \frac{\tau_0}{2} - \frac{(\tau_0 - 2W_0)\,\Gamma\left(n + \tau_0 - \frac{1}{3}\right)\Gamma(\tau_0)}{2\Gamma\left(\tau_0 - \frac{1}{3}\right)\Gamma(n + \tau_0)} \sim \frac{1}{2}n.$$

The square of the stochastic recurrence gives

$$W_n^2 = W_{n-1}^2 + X\mathbf{1}_n^W + (1-X)\left(1 - \mathbf{1}_n^W\right) + 2XW_{n-1}\mathbf{1}_n^W + 2(1-X)W_{n-1}\left(1 - \mathbf{1}_n^W\right),$$

(the square of an indicator is itself). The conditional expectation is

$$\begin{aligned}\mathbf{E}\left[W_n^2 \mid \mathcal{F}_{n-1}\right] &= W_{n-1}^2 + \mathbf{E}[X]\,\mathbf{E}\left[\mathbf{1}_n^W \mid \mathcal{F}_{n-1}\right] + \mathbf{E}[1-X]\,\mathbf{E}\left[1 - \mathbf{1}_n^W \mid \mathcal{F}_{n-1}\right]\\ &\quad + 2\mathbf{E}[X]\,\mathbf{E}\left[W_{n-1}\mathbf{1}_n^W \mid \mathcal{F}_{n-1}\right]\\ &\quad + 2\mathbf{E}[1-X]\,\mathbf{E}\left[W_{n-1}\left(1 - \mathbf{1}_n^W\right) \mid \mathcal{F}_{n-1}\right];\end{aligned}$$

a term like $W_{n-1}\mathbf{1}_n^W$ has the conditional expectation

$$\mathbf{E}\left[W_{n-1}\mathbf{1}_n^W \mid \mathcal{F}_{n-1}\right] = \frac{W_{n-1}^2}{\tau_0 + n - 1}.$$

Plugging in, taking expectations, and simplifying, we get the recurrence

$$\begin{aligned}\mathbf{E}\left[W_n^2\right] &= \left(1 + \frac{4p-2}{\tau_0 + n - 1}\right)\mathbf{E}\left[W_{n-1}^2\right]\\ &\quad + \left(2(1-p) + \frac{2p-1}{\tau_0 + n - 1}\right)\mathbf{E}[W_{n-1}] + 1 - p.\end{aligned}$$

The solution is computed by iteration, then the variance is obtained from $\mathbf{E}[W_n^2] - \mathbf{E}^2[W_n]$, and we find

$$\begin{aligned}\mathbf{Var}[W_n] &= \frac{3}{20}n + \frac{3}{20} + \frac{21\sqrt{3}\,\Gamma\left(\frac{2}{3}\right)\Gamma\left(n + \frac{1}{3}\right)}{\pi\Gamma(n+1)} - \left(\frac{3\Gamma\left(n + \frac{2}{3}\right)}{2\Gamma\left(\frac{2}{3}\right)\Gamma(n+1)}\right)^2\\ &\sim \frac{3}{20}n.\end{aligned}$$

We have determine the limit:

$$\frac{W_n - \frac{1}{2}n}{\sqrt{n}} \xrightarrow{\mathcal{D}} \mathcal{N}\left(0, \frac{3}{20}\right).$$

Chapter 7

7.1 When $K = 0$, the number of balls remains the same throughout any number of draws. As in the text, if τ_n is the total number of balls, and h_n is the number of histories after n draws, then

$$h_n = \tau_0 \tau_1 \dots \tau_{n-1} = \tau_0^n.$$

Therefore, The histories generating function is

$$\mathcal{H}(z) = \sum_{n=0}^{\infty} h_n \frac{z^n}{n!} = \sum_{n=0}^{\infty} \frac{(\tau_0 z)^n}{n!} = e^{\tau_0 z}.$$

7.2 When $K < 0$ one adds to the urn a negative number of balls after each drawing. That is, the urn is untenable and diminishing, and in time it gets to a point where it is no longer possible to take out balls. In $\lfloor \frac{\tau_0}{|K|} \rfloor$ steps the sampling must stop, and all histories are finite. One has

$$\tau_n = \tau_0 - n|K|,$$

and h_n, the number of histories after n draws, is

$$h_n = \tau_0 \tau_1 \dots \tau_{n-1} = \tau_0(\tau_0 - |K|) \dots (\tau_0 - (n-1)|K|), \qquad \text{for } n \le \left\lfloor \frac{\tau_0}{|K|} \right\rfloor,$$

and $h_n = 0$, if $n > \lfloor \frac{\tau_0}{|K|} \rfloor$. Its generating function is

$$\begin{aligned}
\mathcal{H}(z) &= \sum_{n=0}^{\infty} h_n \frac{z^n}{n!} \\
&= \sum_{n=0}^{\lfloor \tau_0/|K| \rfloor} \left(\frac{\tau_0}{|K|} \right)_n \frac{|K|^n z^n}{n!} \\
&= \sum_{n=0}^{\lfloor \tau_0/|K| \rfloor} \binom{\frac{\tau_0}{|K|}}{n = 0} |K|^n z^n \\
&= (1 + |K| z)^{\lfloor \tau_0/|K| \rfloor},
\end{aligned}$$

where $(x)_k$ is the k-times falling factorial of x. Note that the generating function is only a polynomial, to account for the stoppage of the sampling.

7.3 The associated differential system is

$$x'(t) = 1,$$
$$y'(t) = 1,$$

with solution

$$x(t) = t + x(0),$$
$$y(t) = t + y(0).$$

According to Theorem 7.1, the generating function is

$$H(x(0), y(0), z) = (z + x(0))^{W_0}(z + y(0))^{B_0}.$$

Extracting the coefficient $u^w\ v^b\ z^n$ in $H(u, v, z)$ and coefficient of z^n in $H(1, 1, z)$:

$$\mathbf{P}(W_n = w,\ B_n = b) = \frac{\binom{W_0}{w}\binom{B_0}{b}}{\binom{W_0 + B_0}{w + b}},$$

for $w + b = n \le W_0 + B_0$.

7.4 Let the three colors be white, blue, and red, and let W_n, B_n, and R_n, respectively, stand for the number of white, blue, and red balls after n draws. We introduce the three parametric functions $x(t)$, $y(t)$, and $z(t)$ (one for each color) and mimic the proof of Theorem 7.1 to get the generating function of the urn histories:

$$\mathcal{H}\big(x(0), y(0), z(0), u\big) = x^{W_0}(u)\, y^{B_0}(u) z^{R_0}(u),$$

where the parametric system $(x(t), y(t), z(t))$ is the solution to the differential system of equations

$$x'(t) = x^2(t),$$
$$y'(t) = y^2(t),$$
$$z'(t) = z^2(t).$$

Each of these equations is a Riccati equation, with solutions similar to the two-color case, and we have

$$x(t) = \frac{x(0)}{1 - x(0)t},$$
$$y(t) = \frac{y(0)}{1 - y(0)t},$$
$$z(t) = \frac{z(0)}{1 - z(0)t}.$$

The generating function is

$$\mathcal{H}(x(0), y(0), z(0), u) = \left(\frac{x(0)}{1 - x(0)u}\right)^{W_0}\left(\frac{y(0)}{1 - y(0)u}\right)^{B_0}\left(\frac{z(0)}{1 - z(0)u}\right)^{R_0}.$$

Extracting coefficients, we get the exact distribution:

$$\mathbf{P}(W_n = w, B_n = b, R_n = r) = \frac{\binom{w-1}{W_0-1}\binom{b-1}{B_0-1}\binom{r-1}{R_0-1}}{\binom{n+\tau_0-1}{\tau_0-1}},$$

at every feasible (w, b, r) satisfying

$$w + b + r = \tau_0 + n,$$

and $\tau_0 := W_0 + B_0 + R_0$ is the total number of balls at the start.

Chapter 8

8.1 Think of the polymer as an extended ternary tree, where each node is supplied with enough external nodes to make the out-degree of each node of the polymer exactly equal to 4. For example, a leaf (necessarily of degree 1) receives three external nodes in the extension. Color the external nodes joined to the leaves with white, the external nodes joined to nodes of degree 2 with blue, and the external nodes joined to nodes of degree 3 with red, and leave the saturated nodes of degree 4 colorless. Put the colored balls in an urn and grow the urn under the schema

$$\mathbf{A} = \begin{pmatrix} 0 & 2 & 0 \\ 3 & -2 & 1 \\ 3 & 0 & -1 \end{pmatrix};$$

the indexing in both rows and columns comes in the order white-blue-red.

This is an extended urn scheme with eigenvalues $\lambda_1 = 2$, $\lambda_2 = -2$, and $\lambda_3 = -3$, and a principal eigenvector

$$\mathbf{v} := \begin{pmatrix} v_1 \\ v_2 \\ v_3 \end{pmatrix} = \begin{pmatrix} 0.6 \\ 0.3 \\ 0.1 \end{pmatrix}.$$

Let W_n be the number of white balls after n molecules have been added. Initially, $W_0 = 6$. From Theorem 6.2,

$$\frac{W_n}{n} \xrightarrow{P} \lambda_1 v_1 = \frac{6}{5}.$$

Three white balls correspond to one leaf; the asymptotic proportion of leaves is $\frac{1}{3} \times \frac{6}{5} = \frac{2}{5}$ (in probability).

8.2 Let L_n and I_n be, respectively, the number of leaves and internal nodes after n key insertions. Each leaf contains a key of four bytes, and each internal node contains a key and two pointers, that is 12 bytes in total. The size of the tree is

$$S_n = 4L_n + 12I_n.$$

Refer to the urn representation of binary search trees in Theorem 8.1. By Theorem 6.2,

$$\frac{L_n}{n} \xrightarrow{P} \frac{1}{3}.$$

The proportion of internal nodes must then follow the weak law

$$\frac{I_n}{n} \xrightarrow{P} \frac{2}{3}.$$

Then,

$$\frac{S_n}{n} = 4\frac{L_n}{n} + 12\frac{I_n}{n} \xrightarrow{P} 4 \times \frac{1}{3} + 12 \times \frac{2}{3} = \frac{28}{3}.$$

8.3 Consider the evolution of the extended binary search tree. The tree starts out empty (one white external node). After interarrival times that are distributed like Exp(1), two external nodes replace an existing one. Thinking of the external nodes as balls in an urn, the study of the size is reduced to the monochromatic urn embedded in real time, with the schema $\mathbf{A} = [1]$. The moment generating function of the total number of balls by time t is given by Eq. (4.6) with $W(0) = 1$, and $a = 1$. Let $W(t)$ be the number of white balls by time t. As discussed after Eq. (4.6),

$$W(t) \stackrel{\mathcal{D}}{=} \text{Geo}(e^{-t}).$$

(a) The number of actual nodes in a binary tree with $n + 1$ external nodes is n, and $W(t)$ and $S(t)$ are related by $W(t) = S(t) + 1$. Hence,

$$S(t) \stackrel{\mathcal{D}}{=} \text{Geo}(e^{-t}) - 1.$$

(b) Generally, when a random variable X has moment generating function $\phi_X(u)$, the scaled random variable bX has moment generating function $\phi_X(bu)$. Hence, for $u < 1$,

$$\phi_{e^{-t}W(t)}(u) = \frac{e^{-t}e^{ue^{-t}}}{1 - (1 - e^{-t})e^{ue^{-t}}} \to \frac{1}{1-u}.$$

The limit is the moment generating function of the Exp(1) random variable. So,

$$\frac{S(t)}{e^t} = \frac{W(t)}{e^t} - \frac{1}{e^t} \xrightarrow{\mathcal{D}} \text{Exp}(1),$$

as $e^{-t} \to 0$.

8.4 The recursive tree embedded in real time is a special case of the sprout, with Bernard Friedman's scheme on white and blue balls with the schema $\begin{pmatrix} 0 & 1 \\ 1 & 0 \end{pmatrix}$. The eigenvalues of this matrix are 1 and -1. Color the leaves with white, and the internal nodes with blue to complete the correspondence with Bernard Friedman's urn scheme. Let $W(t)$ and $B(t)$ be the number of white and blue balls, respectively, in the urn by time t. Then $W(t)$ is the number of leaves, and the total size is a Yule process. By Theorem 4.1, on average we have

$$\begin{pmatrix} \mathbf{E}[W(t)] \\ \mathbf{E}[B(t)] \end{pmatrix} = e^{\begin{pmatrix} 0 & 1 \\ 1 & 0 \end{pmatrix} t} \begin{pmatrix} 1 \\ 0 \end{pmatrix}.$$

Hence,

$$\mathbf{E}[W(t)] \sim \frac{1}{2} e^t.$$

Remark: Upon completing the depoissonization heuristic we get a result that coincides with the asymptotic average in Theorems 3.4 and 8.6.

8.5 Let the leaves be colored with white, the rest of the unsaturated nodes with blue, and leave the saturated nodes colorless. As in the discrete case, and by the same reasoning, the schema underlying the growth is $\begin{pmatrix} 0 & 1 \\ 1 & -1 \end{pmatrix}$.

The eigenvalues of this matrix are the golden ratio ϕ and its conjugate ϕ^*, which are namely

$$\phi := \frac{\sqrt{5} - 1}{2}, \qquad \phi^* := -\frac{\sqrt{5} + 1}{2}.$$

Let $W(t)$ and $B(t)$ be the number of white and blue balls, respectively, in the urn by time t. By Theorem 4.1, on average we have

$$\begin{pmatrix} \mathbf{E}[W(t)] \\ \mathbf{E}[B(t)] \end{pmatrix} = e^{\begin{pmatrix} 0 & 1 \\ 1 & -1 \end{pmatrix} t} \begin{pmatrix} 1 \\ 0 \end{pmatrix}.$$

Hence,

$$\mathbf{E}[W(t)] \sim \frac{\phi}{\sqrt{5}} e^{\phi t}.$$

Remark: Upon completing the depoissonization heuristic we get a result that coincides with the asymptotic mean in Theorem 8.10.

Chapter 9

9.1 The content of the hint is easily argued, as the size of the oldest species after n evolutionary steps is what it was after $n-1$ steps, plus a possible increase by 1 (a Bernoulli random variable) with success probability equal to the proportion of individuals of the oldest species in the population. It is then routine to compute (for $n \ge 2$)

$$\begin{aligned}
\mathbf{E}[X_n] &= \mathbf{E}[X_{n-1}] + \mathbf{E}\left[\mathrm{Ber}\left(\frac{X_{n-1}}{n+\theta-1}\right)\right] \\
&= \mathbf{E}[X_{n-1}] + \sum_{k=0}^{\infty} \mathbf{E}\left[\mathrm{Ber}\left(\frac{X_{n-1}}{n+\theta-1}\right) \mid X_{n-1} = k\right] \mathbf{P}(X_{n-1} = k) \\
&= \mathbf{E}[X_{n-1}] + \sum_{k=0}^{\infty} \mathbf{E}\left[\mathrm{Ber}\left(\frac{k}{n+\theta-1}\right)\right] \mathbf{P}(X_{n-1} = k) \\
&= \mathbf{E}[X_{n-1}] + \sum_{k=0}^{\infty} \frac{k}{n+\theta-1} \mathbf{P}(X_{n-1} = k) \\
&= \mathbf{E}[X_{n-1}] + \frac{1}{n+\theta-1} \sum_{k=0}^{\infty} k\, \mathbf{P}(X_{n-1} = k) \\
&= \mathbf{E}[X_{n-1}] + \frac{1}{n+\theta-1} \mathbf{E}[X_{n-1}] \\
&= \frac{n+\theta}{n+\theta-1} \mathbf{E}[X_{n-1}] \\
&= \frac{n+\theta}{n+\theta-1} \times \frac{n+\theta-1}{n+\theta-2} \mathbf{E}[X_{n-2}] \\
&= \frac{n+\theta}{n+\theta-1} \times \frac{n+\theta-1}{n+\theta-2} \times \cdots \frac{2+\theta}{1+\theta} \mathbf{E}[X_1] \\
&= \frac{n+\theta}{\theta+1}.
\end{aligned}$$

9.2 In the text we found the representation

$$X_{3,n} \stackrel{\mathcal{D}}{=} 1 + \text{Geo}\left(\frac{n-1}{n}\right) + \text{Geo}\left(\frac{n-2}{n}\right),$$

and the two geometric random variables are independent. Subsequently,

$$\begin{aligned}
\mathbf{P}(X_{3,n} = k) &= \mathbf{P}\left(\text{Geo}\left(\frac{n-1}{n}\right) + \text{Geo}\left(\frac{n-2}{n}\right) = k-1\right) \\
&= \sum_{\substack{i+j=k-1 \\ i,j\ge 1}} \mathbf{P}\left(\text{Geo}\left(\frac{n-1}{n}\right) = i, \text{Geo}\left(\frac{n-2}{n}\right) = j\right) \\
&= \sum_{\substack{i+j=k-1 \\ i,j\ge 1}} \mathbf{P}\left(\text{Geo}\left(\frac{n-1}{n}\right) = i\right) \mathbf{P}\left(\text{Geo}\left(\frac{n-2}{n}\right) = j\right) \\
&= \sum_{\substack{i+j=k-1 \\ i,j\ge 1}} \left[\frac{n-1}{n} \times \left(\frac{1}{n}\right)^{i-1}\right] \left[\frac{n-2}{n} \times \left(\frac{2}{n}\right)^{j-1}\right] \\
&= \frac{(n-1)(n-2)}{n^2} \sum_{i=1}^{k-2} \frac{2^{k-i-2}}{n^{k-3}} \\
&= \frac{(n-1)(n-2)(2^k-4)}{4n^{k-1}}.
\end{aligned}$$

9.3 Let L_n be the number of leaves after n edge splits. The tree starts with one edge and two leaves, and every edge splitting adds two edges and two nodes. The number of nodes after n splits is $2n+2$.

We refer to the color scheme in the proof of Theorem 9.4. It was shown in the proof of that theorem that the white edges, with count W_n, satisfy

$$\frac{W_n}{n} \xrightarrow{P} \lambda_1 v_1 = \frac{1}{2}.$$

Each white edge has a leaf of a cherry at only one of its ends. The rest of the leaves are joined to blue edges with count B_n, satisfying

$$\frac{B_n}{n} \xrightarrow{P} \lambda_1 v_2 = \frac{1}{2}.$$

Then,

$$\frac{W_n}{2n+2} \xrightarrow{P} \frac{1}{4}, \qquad \frac{B_n}{2n+2} \xrightarrow{P} \frac{1}{4};$$

in probability, one fourth of the nodes are leaves not in cherries.

Further,

$$L_n = W_n + B_n,$$

and

$$\frac{L_n}{2n+2} = \frac{W_n}{2n+2} + \frac{B_n}{2n+2} \xrightarrow{P} \frac{1}{2};$$

in probability, one half of the nodes are leaves, and the other half is internal nodes.

Chapter 10

10.1 This instance of balanced urns is degenerate ($e = 0$). The parameters of this scheme yield $W_0 = 1$, $\tau_0 = 3$, $K = 3$, and $\theta = e - a = -2$. The exact formula developed in Proposition 10.1 can be applied; though the proof was written for a linear nondegenerate schemes, it only requires $K + \theta \neq 0$, which holds here. The asymptotic formula follows from Stirling's approximation to the ratio of Gamma functions.

10.2 This instance of balanced urns is degenerate ($e = 0$).

(a) Suppose $K < a$, and initially there are white balls in the urn. The top row must add up to K in a balanced scheme. Therefore, b must be negative. Starting with any number of white balls greater than 1, draws of pairs of white balls increase the white balls (and it is possible to continue drawing pairs of white balls), but reduce the number of blue balls. There comes a time when $|b|$ blue balls should be taken out, but there are not enough balls of this color in the urn. If the initial number of white balls is just 1, there must be at least a blue ball initially. Because the entry c is positive, a drawing of a white-blue pair increases the number of white balls, and after the first draw there are at least two white balls, and we can follow a path of drawing pairs of white balls, which depletes the blue color. On the other hand, starting with an all blue urn, we keep adding blue balls (K balls after each pair drawing), and this can be sustained indefinitely.

(b) Here $\theta = -a < 0$, and $K + \theta = K - a > 0$. As discussed in Exercise 10.1, the exact formula developed in Proposition 10.1 can be applied to get

$$\mathbf{E}[W_n] = W_0 \frac{\left\langle \frac{\tau_0 - \theta}{K} \right\rangle_n}{\left\langle \frac{\tau_0}{K} \right\rangle_n}.$$

If we do not have any white balls at the beginning, the urn is all blue and any drawing adds K blue balls; $\mathbf{E}[W_n] = 0$ (which is sublinear), and if we do

$$\mathbf{E}[W_n] = W_0 \frac{\Gamma\left(n + \frac{\tau_0 - \theta}{K}\right)}{\Gamma\left(n + \frac{\tau_0}{K}\right)} \times \frac{\Gamma\left(\frac{\tau_0}{K}\right)}{\Gamma\left(\frac{\tau_0 - \theta}{K}\right)}.$$

By Stirling's approximation to the Gamma function,, as $n \to \infty$, we write

$$\mathbf{E}[W_n] \sim W_0 \frac{\Gamma\left(\frac{\tau_0}{K}\right)}{\Gamma\left(\frac{\tau_0 - \theta}{K}\right)} n^{-\theta/K} = n^{a/K},$$

and with $0 < a < K$, this is sublinear growth.

10.3 Let W_n be the number of white balls in the urn after n draws. The variance can be computed from the first two moments. The first moment in a Chen-Wei urn is $2C W_0 n/\tau_0 + W_0$. We get the second moment by solving the linear recurrence in Eq. (10.9). We find the asymptotic variance

$$\mathbf{Var}[W_n] \sim \left(7 + \frac{16\sqrt{\pi}}{\Gamma\left(\frac{3}{4} + \frac{\sqrt{17}}{4}\right)\Gamma\left(\frac{3}{4} - \frac{\sqrt{17}}{4}\right)}\right) n^2.$$

10.4 Let W_n and B_n be the number of white and blue balls in the urn after n draws, and let τ_n be the total. We set up a recurrence:

$$\mathbf{E}[W_n \mid W_{n-1}] = W_{n-1} + 3\frac{\binom{W_{n-1}}{3}}{\binom{\tau_{n-1}}{3}} + 2\frac{\binom{W_{n-1}}{2} B_{n-1}}{\binom{\tau_{n-1}}{3}} + \frac{W_{n-1}\binom{B_{n-1}}{2}}{\binom{\tau_{n-1}}{3}}.$$

Substitute for B_{n-1} with $\tau_{n-1} - W_{n-1}$. The recurrence is simplified, only linear terms stay: Quadratic terms are eliminated and so are all the terms involving W_{n-1}^3. After simplification and taking expectations, the recurrence takes the form

$$\mathbf{E}[W_n] = \frac{3n + \tau_0}{3n + \tau_0 - 3}\,\mathbf{E}[W_{n-1}],$$

which can be iterated to give

$$\mathbf{E}[W_n] = 3\frac{W_0}{\tau_0} n + W_0.$$

10.5 Let W_n and B_n be the number of white and blue balls in the urn after n draws, and let τ_n be the total. We set up a recurrence:

$$\mathbf{E}[W_n \mid W_{n-1}] = W_{n-1} + \frac{\binom{W_{n-1}}{2} B_{n-1}}{\binom{\tau_{n-1}}{3}} + 2\frac{W_{n-1}\binom{B_{n-1}}{2}}{\binom{\tau_{n-1}}{3}} + 3\frac{\binom{B_{n-1}}{3}}{\binom{\tau_{n-1}}{3}}.$$

Substitute for B_{n-1} with $\tau_{n-1} - W_{n-1}$. The recurrence is simplified to a linear one: Quadratic terms are eliminated and so are all the terms involving W_{n-1}^3. After simplification and taking expectations, the recurrence takes the form

$$\mathbf{E}[W_n] = \frac{n-1}{n}\,\mathbf{E}[W_{n-1}] + 3,$$

which can be iterated to give, for $n \ge 1$

$$\mathbf{E}[W_n] = \frac{3}{2}n + \frac{3}{2}.$$

10.6 Let W_n, B_n, and R_n be the number of white, blue, and red balls in the urn after n draws, and let τ_n be the total. We set up a recurrence:

$$\mathbf{E}[W_n \mid W_{n-1}] = W_{n-1} + 2\frac{\binom{W_{n-1}}{2}}{\binom{\tau_{n-1}}{2}} + \frac{W_{n-1}B_{n-1}}{\binom{\tau_{n-1}}{2}} + \frac{W_{n-1}R_{n-1}}{\binom{\tau_{n-1}}{2}}.$$

Substitute for R_{n-1} with $\tau_{n-1} - W_{n-1} - B_{n-1}$. The simplified recurrence contains only W_{n-1} and is further reduced to a linear one:

$$\mathbf{E}[W_n] = \frac{2n+\tau_0}{2n+\tau_0-2}\,\mathbf{E}[W_{n-1}],$$

which can be iterated to give, for $n \ge 1$,

$$\mathbf{E}[W_n] = 2\frac{W_0}{\tau_0}n + W_0.$$

10.7 As shown in the text, the recurrence equation for the mean is

$$\mathbf{E}[L_n] = \frac{n-1}{n+1}\,\mathbf{E}[L_{n-1}] + 1.$$

This can be iterated as follows

$$\begin{aligned}
\mathbf{E}[L_n] &= 1 + \frac{n-1}{n+1}\,\mathbf{E}[L_{n-1}] \\
&= 1 + \frac{n-1}{n+1}\left(1 + \frac{n-2}{n}\,\mathbf{E}[L_{n-2}]\right) \\
&= 1 + \frac{n-1}{n+1} + \frac{(n-1)(n-2)}{(n+1)n}\left(1 + \frac{n-3}{n-1}\,\mathbf{E}[L_{n-3}]\right) \\
&= \frac{(n+1)n}{(n+1)n} + \frac{n(n-1)}{(n+1)n} + \frac{(n-1)(n-2)}{(n+1)n} \\
&\quad + \frac{(n-2)(n-3)}{(n+1)n}\left(1 + \frac{n-4}{n-2}\,\mathbf{E}[L_{n-4}]\right) \\
&\ \ \vdots
\end{aligned}$$

$$= \frac{1}{(n+1)n} \sum_{j=1}^{n} (j+1)j$$

$$= \frac{n^3 + 3n^2 + 2n}{3(n+1)n}$$

$$= \frac{n+2}{3}.$$

The variance takes quite a bit of tedious but straightforward work, as is typical in random structures. We shall only outline this lengthy calculation. The rectangular urn considered has $a = -1$, $c = 0$, and $e = 1$. Starting with squaring both sides of the recurrence in Eq. (10.1), one derives

$$\begin{aligned} \mathbf{E}\left[L_n^2 \,|\, \mathcal{F}_{n-1}\right] &= L_{n-1}^2 + \mathbf{E}\left[(I_n^{WW})^2 \,|\, \mathcal{F}_{n-1}\right] + \mathbf{E}\left[(I_n^{BB})^2 \,|\, \mathcal{F}_{n-1}\right] \\ &\quad + 2L_{n-1}\mathbf{E}\left[I_n^{BB} - I_n^{WW} \,|\, \mathcal{F}_{n-1}\right] - 2\mathbf{E}\left[I_n^{WW} I_n^{BB} \,|\, \mathcal{F}_{n-1}\right]. \end{aligned}$$

Apply the relations $(I_n^{WW})^2 = I_n^{WW}$, $(I_n^{BB})^2 = I_n^{BB}$ and $I_n^{WW} I_n^{BB} = 0$, take expectations and simplify to get

$$\mathbf{E}\left[L_n^2\right] = \frac{(n-1)(n-2)}{(n+1)n}\mathbf{E}\left[L_{n-1}^2\right] + \frac{2(n-1)}{n}\mathbf{E}[L_{n-1}] + 1.$$

Use the now-known exact average to get the exact second moment then the variance

$$\mathbf{Var}[L_n] = \frac{(n+2)(4n-3)}{45n}.$$

Notation

π	circle's circumference to diameter ratio 3.141592...
γ	Euler's constant 0.577215...
e	base of the natural logarithm 2.718281...
∞	greater than any real number
$-\infty$	less than any real number
i	$\sqrt{-1}$
ϕ	empty set
$\mathbb{R}$	real numbers
$\{x : P(x)\}$	set of all x satisfying the predicate P
$A \bigcap B$	intersection of the sets A and B
$\bigcap_{i=1}^{n} A_i$	intersection of the sets $A_1, A_2, \ldots, A_n$
$\bigcap_{i=1}^{\infty} A_i$	intersection of the sets $A_1, A_2, \ldots$
$A \bigcup B$	union of the sets A and B
$\bigcup_{i=1}^{n} A_i$	union of the sets $A_1, A_2, \ldots, A_n$
$\bigcup_{i=1}^{\infty} A_i$	union of the sets $A_1, A_2, \ldots$
$A \subseteq B$	A is a subset of B
$x \in A$	x belongs to the set A
A^c	complement of the set A with respect to the universal set
$\Longleftrightarrow$	if and only if
$\lvert x \rvert$	absolute value of the real number x
$x = y$	x is equal to y
$x \equiv y$	x is identically equal to y
$x \approx y$	x is approximately equal to y
$x := y$	x is equal to y by definition
$x =: y$	y is equal to x by definition
$x \neq y$	x is not equal to y
$x < y$	x is less than y
$x \leq y$	x is less than or equal to y
$x > y$	x is greater than y

$x \geq y$	x is greater than or equal to y
$-x$	negative of x
$x + y$	x plus y
$x - y$	x minus y
xy, $x \times y$	x times y
$\frac{x}{y}$, x/y	x divided by y
x^y	x raised to the power y
$\sqrt{x}$	square root of x
$\sqrt[n]{x}$	nth root of x
$\lfloor x \rfloor$	floor of the real number x, that is, the largest integer less than or equal to x
$\lceil x \rceil$	ceiling of the real number x, that is, the smallest integer greater than or equal to x
$n!$	n factorial
k **mod** m	the remainder obtained after dividing integer k by integer m
$\triangledown$	backward difference operator: $\triangledown h_n = h_n - h_{n-1}$
$\diamond$	end of example
$\blacksquare$	end of proof
H_n	nth harmonic number ($H_n = 1 + 1/2 + \cdots + 1/n$)
$H_n^{(p)}$	nth harmonic number of order p, integer $p \geq 1$, ($H_n^{(p)} = 1 + 1/2^p + \cdots + 1/n^p$)
$\binom{n}{k}$	binomial coefficient (n choose k)
$\binom{n}{k_1, \ldots, k_r}$	multinomial coefficient ($n!/(k_1! \ldots k_r!)$), $k_1 + \cdots + k_r = n$)
$[z^n]f(z)$	coefficient of z^n in f
$(z)_n$	Pochhammer's symbol or the falling factorial $z(z-1)\ldots(z-n+1)$
$\langle z \rangle_n$	rising factorial $z(z+1)\ldots(z+n-1)$
$\left[{n \atop k}\right]$	the kth signless Stirling number of the first kind ($[z^n]\langle z \rangle_n$)
$\left\{{n \atop k}\right\}$	the kth Stirling number of order n of the second kind
$\left\langle{n \atop k}\right\rangle$	the kth Eulerian number of order n
e^x, $\exp(x)$	exponential of x
$\sin x$	sine of x
$\cos x$	cosine of x
$\Gamma(x)$	gamma function ($\int_0^\infty u^{x-1}e^{-u}\,du$, $x > 0$)
$\ln x$	natural logarithm (base e) of z

$\mathbf{det}(\mathbf{A})$	determinant of the matrix $\mathbf{A}$
(a, b)	the open interval $\{x : a < x < b\}$
$[a, b)$	the semi-open interval $\{x : a \leq x < b\}$
$(a, b]$	the semi-open interval $\{x : a < x \leq b\}$
$[a, b]$	the closed interval $\{x : a \leq x \leq b\}$
$\max(x_1, \ldots, x_k)$	maximum value of x_i, $1 \leq i \leq k$
$\min(x_1, \ldots, x_k)$	minimum value of x_i, $1 \leq i \leq k$
$\mathbf{P}(A)$	probability of the event A
$\mathbf{E}[X]$	expected value or average of X
$\mathbf{Var}[X]$	variance of X
$\mathbf{P}(A \mid B)$	conditional probability of A given that B occurred
$\mathbf{E}[X \mid B]$	conditional expectation of X given that B occurred
$\mathbf{E}[X \mid \mathcal{F}]$	conditional expectation of X given the sigma field $\mathcal{F}$
$\mathbf{E}[X \mid Y]$	conditional expectation of X given Y, or more specifically the sigma field generated by Y
$\phi_X(t)$	moment generating function or characteristic function of the random variable X, that is $\mathbf{E}[e^{tX}]$ or $\mathbf{E}[e^{itX}]$
$f_X(x)$	density of the continuous random variable X
$F_X(x)$	distribution function of the random variable X at x, or $\mathbf{P}(X \leq x)$
$\lim_{n\to\infty} x_n$	the limit of the sequence of numbers $x_1, x_2, \ldots$
$\lim_{x\to x_0} y(x)$	the limit of the function $y(x)$ as x approaches x_0
$x_n \to a$	the sequence of numbers $x_1, x_2, \ldots$ converges to a
$X \stackrel{\mathcal{D}}{=} Y$	the random variable X is equal in distribution to Y
$X \stackrel{\text{a.s.}}{=} Y$	the random variable X is almost surely equal to Y
$X_n \xrightarrow{\mathcal{D}} X$	the sequence of random variables $X_1, X_2, \ldots$ converges to the random variable X in distribution or in law
$X_n \xrightarrow{P} X$	the sequence of random variables $X_1, X_2, \ldots$ converges to the random variable X in probability
$X_n \xrightarrow{a.s.} X$	the sequence of random variables $X_1, X_2, \ldots$ converges to the random variable X almost surely

Ber(p)	Bernoulli random variable with probability of success p
Bin(n, p)	Binomial random variable on n independent trials and probability of success p per trial
Exp(λ)	exponential random variable with mean λ
Gamma(a, b)	gamma random variable with parameters a and b
Geo(p)	geometric random variable with probability of success p per trial
$\mathcal{N}(\mu, \sigma^2)$	normal random variable with mean μ and variance σ^2
$\mathcal{N}_k(\boldsymbol{\mu}_k, \boldsymbol{\Sigma}_k^2)$	k-dimensional multivariate normal random vector with the k-dimensional mean vector $\boldsymbol{\mu}_k$ and the $k \times k$ variance-covariance matrix $\boldsymbol{\Sigma}_k^2$
NB(r, p)	negative binomial random variable with success probability p per trial to reach the rth success
Uniform(a, b)	continuous uniform distribution on the interval (a, b)
Uniform[$a\,..\,b$]	discrete uniform distribution on the set $\{a, a+1, \ldots, b\}$ of integers
Poi(λ)	Poisson random variable with rate λ
$\beta(a, b)$	beta random variable with parameters a and b
$\chi^2(\nu)$	chi-squared random varible with ν degrees of freedom
$\frac{df(z)}{dz}$	first derivative of f with respect to z
$\frac{df(z)}{dz}\Big\vert_{z=z_0}$	first derivative of f evaluated at z_0
$f', f'', f''', \ldots$	first, second, third, ..., derivatives of $f(z)$ with respect to z
$\frac{d^n f(z)}{dz^n}$	nth derivative of f with respect to z
$\frac{\partial f(z_1, \ldots, z_k)}{\partial z_i}$	partial derivative of f with respect to z_i
$\frac{\partial^n f(z_1, \ldots, z_k)}{\partial z_i^n}$	nth partial derivative of f with respect to z_i
$\sum_{i \in R(i)} x_i$	sum of the terms x_i over every index i satisfying the relation R (interpreted as 0, if R is empty)
$\prod_{i \in R(i)} x_i$	product of the terms x_i over every index i satisfying the relation R (interpreted as 1, if R is empty)
$\int_a^b f(x)\, dx$	integral of $f(x)$ from a to b

$f(n) = O(g(n))$	$f(n)$ is big Oh $g(n)$ ($\lvert f(n)/g(n)\rvert$ is bounded from above, as $n \to \infty$)
$f(n) = \Omega(g(n))$	$f(n)$ is Omega $g(n)$ ($\lvert f(n)/g(n)\rvert$ is bounded from below, as $n \to \infty$)
$f(n) = o(g(n))$	$f(n)$ is little oh of $g(n)$ ($f(n)/g(n)$ tends to zero as $n \to \infty$)
$f(n) \sim g(n)$	$f(n)$ is asymptotically equivalent to $g(n)$ ($f(n)/g(n)$ tends to one, as $n \to \infty$)
$f(x) = O(g(x))$ over S	$f(x)$ is big Oh of $g(x)$ over S (for some positive constant C, $\lvert f(x)\rvert \le C\lvert g(x)\rvert$ for all $x \in S$)
$f(x) = o(g(x))$ as $x \to x_0$	$f(x)$ is little oh of $g(x)$ when x approaches x_0 ($f(x)/g(x)$ tends to zero, as $x \to x_0$)
$f(x) \sim g(x)$ as $x \to x_0$	$f(x)$ is asymptotic to $g(x)$ when x approaches x_0 ($f(x)/g(x)$ tends to one, as $x \to x_0$)

Bibliographic Notes

The writing of this book was influenced by several sources, among which is *Urn Models and Their Applications*, the classic magnum opus on urn models by Johnson and Kotz (1977b). The present book is meant to get into advances in the subject that were made after Johnson and Kotz (1977b), and into numerous applications not considered in that book. Among other sources of impact on this book is Blom, Holst, and Sandell's book *Problems and Snapshots from the World of Probability*, which takes a problem-driven approach of some classical probability topics presented in the format of short articles, including a few on urns. The little presentation we made on exchangeability (cf. Example 1.1) is based on material taken from this source. The discussion on waiting times in Ehrenfest urns in Section 3.5 is based on that book, too. Shedlon Ross' books, particularly *A First Course in Probability and Stochastic Processes* provided high inspiration. A reader wishing to consult on basic theory can look up either of these two books. A book that remains influential in the field of discrete probability is Feller's (1968) *An Introduction to Probability Theory and Its Applications.* It has material on Pólya urns interspersed throughout.

We highly recommend all the above books as first supplemental readings and sources for basic theory, other topics of interest in urn models, as well as more general areas of probability. The huge body of published material on urn models and applications in the form of original research papers was an indispensable source of various material.

Bibliography

1. Aebly, J. (1923). Démonstration du problème du scrutin par des considérations géométriques. *L' Enseignement Mathématique*, **23**, 185–186.
2. Aldous, D. (1995). Probability distributions on cladograms. In *Random Discrete Structures*, Eds. Aldous, D. and Pemantle, R., *IMA Volumes in Mathematics and its Applications*, **76**, 1–18.
3. Aldous, D., Flannery, B., and Palacios, J. (1988). Two applications of urn processes: The fringe analysis of search trees and the simulation of quasi-stationary distributions of Markov chains. *Probability in Engineering and Informational Sciences*, **2**, 293–307.
4. André, D. (1887). Solution direct du problème résolu par M. Bertrand. *Comptes Rendus de l' Académie des Sciences*, Paris, **105**, 436–437.
5. Arthur, B., Ermoliev, Y., and Kaniovski, Y. (1963). On generalized urn schemes of the Pólya kind. *Cybernetics*, **19**, 61–71 (English translation).
6. Athreya, K. and Karlin, S. (1968). Embedding of urn schemes into continuous time Markov branching process and related limit theorems. *The Annals of Mathematical Statistics*, **39**, 1801–1817.
7. Athreya, K. and Ney, P. (1972). *Branching Processes*. Springer-Verlag, New York.
8. Bagchi, A. and Pal, A. (1985). Asymptotic normality in the generalized Pólya-Eggenberger urn model with applications to computer data structures. *SIAM Journal on Algebraic and Discrete Methods*, **6**, 394–405.
9. Bai, Z. and Hu, F., (1999). Asymptotic theorems for urn models with nonhomogeneous generating matrices. *Stochastic Processes and their Applications*, **80**, 87–101.
10. Bai, Z. and Hu, F. (2005). Strong consistency and asymptotic normality for urn models. *The Annals of Applied Probability*, **15**, 914–940.
11. Bai, Z., Hu, F., and Rosenberger, W. (2002). Asymptotic properties of adaptive designs for clinical trials with delayed response. *The Annals of Statistics*, **30**, 122–139.
12. Bailey, R. (1972). *A Montage of Diversity*. Ph.D. Dissertation, Emory University.
13. Balaji, S. and Mahmoud, H. (2003). Problem 762. *The College Mathematics Journal*, **34**, 405–406.
14. Balaji, S. and Mahmoud, H. (2006). Exact and limiting distributions in diagonal Pólya processes. *The Annals of the Institute of Statistical Mathematics*, **58**, 171–185.
15. Balaji, S., Mahmoud, H., and Watanabe, O. (2006). Distributions in the Ehrenfest process. *Statistics and Probability Letters*, **76**, 666–674.
16. Barbier, É. (1887). Généralisation du problème résolu par M. Bertrand. *Comptes Rendus des Séances de l'Académie des Sciences*, Paris, **105**, 407.
17. Baum, L. and Billingsley, P. (1965). Asymptotic distributions for the coupon collector's problem. *The Annals of Mathematical Statistics*, **36**, 1835–1839.
18. Bellman, R. and Harris, T. (1951). Recurrence times for the Ehrenfest model. *Pacific Journal of Mathematics*, **1**, 184–188.

19. Benaïm, M., Schreiber, S., and Tarrés, P. (2004). Generalized urn models of evolutionary processes *The Annals of Applied Probability*, **14**, 1455–1478.
20. Bentley, J. (1975). Multidimensional binary search trees used for associative searching. *Communications of the ACM*, **18**, 509–517.
21. Bergeron, F., Flajolet, P., and Salvy, B. (1992). Varieties of increasing trees. *Lecture Notes in Computer Science*, Ed. J. Raoult, **581**, 24–48.
22. Bernoulli, J. (1713). *Ars Conjectandi*. Reprinted in *Die Werke von Jakob Bernoulli* (1975). Birkhäuser, Basel.
23. Bernstein, S. (1940). Sur un problème du schéme des urnes à composition variable. *C. R. (Dokl.) Acad. Sci. URSS*, **28**, 5–7.
24. Bernstein, S. (1946). *Theory of Probability*. Gostechizdat, Moscow-Leningrad.
25. Bertrand, J. (1887). Solution d'un problème. *Comptes Rendus des Séances de l'Académie des Sciences*, Paris, **105**, 369.
26. Billingsley, P. (1995). *Probability and Measure*. Wiley, New York.
27. Blackwell, D. and McQueen, J. (1973). Ferguson distributions via Pólya urn schemes. *The Annals of Statistics*, **1**, 353–355.
28. Blom, G. (1989). Mean transition times for the Ehrenfest urn model. *Advances in Applied Probability*, **21**, 479–480.
29. Blom, G., Holst, L., and Sandell, D. (1994). *Problems and Snapshots from the World of Probability*. Springer-Verlag, New York.
30. Brennan, C. and Prodinger, H. (2003). The pills problem revisited. *Quaestiones Mathematicae*, **26**, 427–439.
31. Burge, W. (1958). Sorting, trees, and measures of order. *Information and Control*, **1**, 181–197.
32. Chauvin, B. and Pouyanne, P. (2004). m–ary Search trees when $m \geq 27$: A strong asymptotics for the space requirements. *Random Structures and Algorithms*, **24**, 133–154.
33. Chen, M. and Wei, C. (2005). A new urn model. *Journal of Applied Probability*, **42**, 964–976.
34. Chern, H. and Hwang, H. (2001). Phase change in random m-ary search trees and generalized quicksort. *Random Structures and Algorithms*, **19**, 316–358.
35. Chern, H., Hwang, H., and Tsai, T. (2002). An asymptotic theory for Cauchy-Euler differential equations with applications to the analysis of algorithms. *Journal of Algorithms*, **44**, 177–225.
36. Chow, Y. and Teicher, H. (1978). *Probability Theory*. Springer-Verlag, New York.
37. Cohen, J. (1966). *A Model of Simple Competition*. Harvard University Press, Cambridge, Massachusetts.
38. David, F. and Barton, D. (1962). *Combinatorial Chance*. Griffin, London.
39. De Finetti, B. (1937). La prévision: ses lois logiques, ses sources subjectives. *Annales de l'Institut Henri Poincaré*, **7**, 1–68.
40. De Moivre, A. (1712). *The Doctrine of Chances*. Millar, London. Reprinted in 1967 by Chelsea, New York.
41. Devroye, L. (1991). Limit laws for local counters in random binary search trees. *Random Structures and Algorithms*, **2**, 303–316.
42. Devroye, L. and Vivas, A. (1998). Random fringe-balanced quadtrees (manuscript).
43. Dirienzo, G. (2000). Using urn models for the design of clinical trials. *Indian Journal of Statistics*, **62**, 43–69.
44. Donnelly, P. (1986). Partition structures, Pólya urns, the Ewens sampling formula, and the ages of alleles. *Theoretical Population Biology*, **30**, 271–288.

45. Donnelly, P. and Tavaré, S. (1986). The ages of alleles and a coalescent. *Advances in Applied Probability*, **18**, 1–19.
46. Donnelly, P., Kurtz, T., and Tavaré, S. (1991). On the functional central limit theorem for the Ewens sampling formula. *The Annals of Applied Probability*, **1**, 539–545.
47. Dvortzky, A. and Motzkin, T. (1947). A problem of arrangements. *Duke Mathematical Journal*, **14**, 305–313.
48. Eggenberger, F. and Pólya, G. (1923). Über die statistik verketteter vorgäge. *Zeitschrift für Angewandte Mathematik und Mechanik*, **3**, 279–289.
49. Ehrenfest, P. and Ehrenfest, T. (1907). Über zwei bekannte einwände gegen das Boltzmannsche H-theorem. *Physikalische Zeitschrift*, **8**, 311–314.
50. Erdős, P. and Rényi, A. (1961). On a classical problem of probability theory. *Magyar Tud. Akad. Mat. Kutató Int. Kőzl.*, **3**, 79–112.
51. Ewens, W. (1972). The sampling theory of selectively neutral alleles. *Theoretical Population Biology*, **3**, 87–112.
52. Ewens, W. (2004). *Mathematical Population Genetics*. Springer, New York.
53. Feller, W. (1968). *An Introduction to Probability Theory and Its Applications, Volume I*. Wiley, New York.
54. Feller, W. (1971). *An Introduction to Probability Theory and Its Applications, Volume II*. Wiley, New York.
55. Fill, J. and Kapur, N. (2005). Transfer theorems and asymptotic distributional results for m–ary search trees. *Random Structures and Algorithms*, **26**, 359–391.
56. Finkel, R. and Bentley, J. (1974). Quad trees: a data structure for retrieval on composite keys. *Acta Informatica*, **4**, 1–9.
57. Fisher, R. (1922). On the dominance ratio. *Proceedings of the Royal Society of Edinburgh*, **42**, 321–341.
58. Fisher, R. (1930). *The Genetical Theory of Natural Selection*. Clarendon Press, Oxford; Dover Publications, New York (1958).
59. Flajolet, P. and Sedgewick, R. (2008+). *Analytic Combinatorics*. Addison-Wesley, Reading, Massachusetts.
60. Flajolet, P., Dumas, P., and Puyhaubert, V. (2006). Some exactly solvable models of urn process theory. In *Discrete Mathematics and Computer Science Proceedings*, Ed. Philippe Chassaing, **AG**, 59–118.
61. Flajolet, P., Gabarró, J., and Pekari, H. (2005). Analytic urns. *The Annals of Probability*, **33**, 1200–1233.
62. Fréchet, M. (1943). *Les Probabilités Associées à un Système d' Evenements Compatibles et Dépendants*. Herman, Paris.
63. Freedman, D. (1965). Bernard Friedman's urn. *The Annals of Mathematical Statistics*, **36**, 956–970.
64. Friedman, B. (1949). A simple urn model. *Communications of Pure and Applied Mathematics*, **2**, 59–70.
65. Gastwirth, J. (1977). A probability model of pyramid schemes. *The American Statistician*, **31**, 79–82.
66. Gastwirth, J. and Bhattacharya, P. (1984). Two probability models of pyramids or chain letter schemes demonstrating that their promotional claims are unreliable. *Operations Research*, **32**, 527–536.
67. Gouet, R. (1989). A martingale approach to strong convergence in a generalized Pólya-Eggenberger urn model. *Statistics and Probability Letters*, **8**, 225–228.
68. Gouet, R. (1993). Martingale functional central limit theorems for a generalized Pólya urn. *The Annals of Probability*, **21**, 1624–1639.

69. Gouet, R. (1997). Strong convergence of proportions in a multicolor Pólya urn. *Journal of Applied Probability*, **34**, 426–435.
70. Graham, R., Knuth, D., and Patashnik, O. (1994). *Concrete Mathematics: A Foundation for Computer Science*. Addison-Wesley, Reading, Massachusetts.
71. Hald, A. (2003). *History of Probability and Statistics and Their Applications before 1750*. Wiley, New York.
72. Hall, P. and Heyde, C. (1980). *Martingale Limit Theory and Its Applications*. Academic Press, New York.
73. Heath, D. and Sudderth, W. (1976). De Finetti's theorem on exchangeable random variables. *The American Statistician*, **30**, 188–189.
74. Hermosilla, L. and Olivos, J. (1985). A bijective approach to single rotation trees. Presented at *The Fifth Conferencia Internacional en Ciencia de la Computacion*, Santiago, Chile.
75. Hill, B., Lane, D., and Sudderth, W. (1980). A strong law for some generalized urn processes. *The Annals of Probability*, **8**, 214–226.
76. Hoppe, F. (1984). Pólya-like urns and the Ewens' sampling formula. *Journal of Mathematical Biology*, **20**, 91–94.
77. Hoppe, F. (1987). The sampling theory of neutral alleles and an urn model in population genetics. *Journal of Mathematical Biology*, **25**, 123–159.
78. Hoshi, M. and Flajolet, P. (1992). Page usage in a quadtree index. *BIT*, **32**, 384–402.
79. Hurley, C. and Mahmoud, H. (1994). Analysis of an algorithm for random sampling. *Probability in the Engineering and Informational Sciences*, **8**, 153–168.
80. Huygens, C. (1657). *De Ratiociniis in Ludo Alea*. Reprinted in *Œuvres*, **14**, Paris (1920).
81. Inoue, K. and Aki, S. (2001). Pólya urn model under general replacement schemes. *Journal of Japan Statistical Society*, **31**, 193–205.
82. Ivanova, A. (2003). A play-the-winner-type urn model with reduced variability. *Metrika*, **58**, 1–13.
83. Ivanova, A. (2006). Urn designs with immigration: Useful connection with continuous time stochastic processes. *Journal of Statistical Planning and Inference*, **136**, 1836–1844.
84. Ivanova, A., Rosenberger, W., Durham, S., and Flournoy, N. (2000). A birth and death urn for randomized clinical trials: asymptotic methods. *Sankhya*, **B 62**, 104–118.
85. Janson, S. (2004). Functional limit theorems for multitype branching processes and generalized Pólya urns. *Stochastic Processes and Applications*, **110**, 177–245.
86. Janson, S. (2005a). Limit theorems for triangular urn schemes. *Probability Theory and Related Fields*, **134**, 417–452.
87. Janson, S. (2005b). Asymptotic degree distribution in random recursive trees. *Random Structures and Algorithms*, **26**, 69–83.
88. Johnson, N. and Kotz, S. (1969). *Distributions in Statistics: Discrete Distributions*. Wiley, New York.
89. Johnson, N. and Kotz, S. (1976). Two variants of Pólya' s urn models (towards a rapprochement between combinatorics and probability theory). *The American Statistician*, **13**, 186–188.
90. Johnson, N. and Kotz, S. (1977a). On a multivariate generalized occupancy model. *Journal of Applied Probability*, **13**, 392–399.
91. Johnson, N. and Kotz, S. (1977b). *Urn models and Their Applications: An Approach to Modern Discrete Probability Theory*. Wiley, New York.

92. Johnson, N., Kotz, S., and Mahmoud, H. (2004). Pólya-type urn models with multiple drawings. *Journal of the Iranian Statistical Society*, **3**, 165–173.
93. Kac, M. (1949). On deviations between theoretical and empirical distributions. *Proceedings of the National Academy of Sciences*, **35**, 252–257.
94. Karamata, J. (1935). Théorèmes sur la sommabilité exponentielle et d'autres sommabilités rattachant. *Mathematica (Cluj)*, **9**, 164–178.
95. Karlin, S. and McGregor, J. (1965). Ehrenfest urn models. *Journal of Applied Probability*, **2**, 352–376.
96. Karlin, S. and McGregor, J. (1972). Addendum to a paper of W. Ewens. *Theoretical Population Biology*, **3**, 113–116.
97. Karlin, S. and Taylor, H. (1974). *A First Course in Stochastic Processes*. Academic Press, New York.
98. Kemeny, J. and Snell, J. (1960). *Finite Markov Chains*. Wiley, New York.
99. Kemp, R. (1984). *Fundamentals of the Average Case Analysis of Particular Algorithms*. Wiley-Teubner Series in Computer Science, John Wiley, New York.
100. Kingman, J. and Volkov, S. (2003). Solution to the OK Corral problem via decoupling of Friedman's urn. *Journal of Theoretical Probability*, **16**, 267–276.
101. Kirschenhofer, P. and Prodinger, H. (1998). Comparisons in Hoare's Find algorithm. *Combinatorics, Probability, and Computing*, **7**, 111–120.
102. Knuth, D. (1992). Two notes on notation. *The American Mathematical Monthly*, **99**, 403–422.
103. Knuth, D. (1998). *The Art of Computer Programming, Vol. 3: Sorting and Searching*, 2nd ed. Addison-Wesley, Reading, Massachusetts.
104. Knuth, D. and McCarthy, J. (1991). Problem E3429: Big pills and little pills. *The American Mathematical Monthly*, **98**, 264.
105. Kolchin, V. (1986). *Random Mappings*. Optimization Software, Inc., Publications Division, New York.
106. Kolcin, V., Sevastyanov, B., Cistyakov, V., and Blakrishnn, A. (1978). *Random Allocations*. Winston, Great Falls, Montana (Translated from Russian).
107. Kotz, S. and Balakrishnan, N. (1997). Advances in urn models during the past two decades. In *Advances in Combinatorial Methods and Applications to Probability and Statistics*, **49**, 203–257. Birkhäuser, Boston.
108. Kotz, S., Mahmoud, H., and Robert, P. (2000). On generalized Pólya urn models. *Statistics and Probability Letters*, **49**, 163–173.
109. Kriz, J. (1972). Die PMP-vertielung. *Statistique Hefte*, **13**, 211–224.
110. Laplace, P. (1812). *Théorie Analytique des Probabilités*. Reprinted in *Œuvres*, **7**, Paris (1886).
111. Levine, H. (1997). *Partial Differential Equations*. American Mathematical Society and International Press, Providence, Rhode Island.
112. Mahmoud, H. (1992). *Evolution of Random Search Trees*. Wiley, New York.
113. Mahmoud, H. (1998). On rotations in fringe-balanced binary trees. *Information Processing Letters*, **65**, 41–46.
114. Mahmoud, H. (2000). *Sorting: A Distribution Theory*. Wiley, New York.
115. Mahmoud, H. (2002). The size of random bucket trees via urn models. *Acta Informatica*, **38**, 813–838 (Erratum in *Acta Informatica*, **41**, 63 (2004)).
116. Mahmoud, H. (2003). Pólya urn models and connections to random trees: A review. *Journal of the Iranian Statistical Society*, **2**, 53–114.
117. Mahmoud, H. (2004). Random sprouts as Internet models and Pólya processes. *Acta Informatica*, **41**, 1–18.

118. Mahmoud, H. and Pittel, B. (1989). Analysis of the space of search trees under the random insertion algorithm. *Journal of Algorithms*, **10**, 52–75.
119. Mahmoud, H. and Smythe, R. (1991). On the distribution of leaves in rooted subtrees of recursive trees. *The Annals of Applied Probability*, **1**, 406–418.
120. Mahmoud, H. and Smythe, R. (1992). Asymptotic joint normality of outdegrees of nodes in random recursive trees. *Random Structures and Algorithms*, **3**, 255–266.
121. Mahmoud, H. and Smythe, R. (1995). Probabilistic analysis of bucket recursive trees. *Theoretical Computer Science*, **144**, 221–249.
122. Mahmoud, H., Smythe, R., and Szymański, J. (1993). On the structure of plane-oriented recursive trees and their branches. *Random Structures and Algorithms*, **4**, 151–176.
123. Maistrov, L. (1974). *Probability Theory: A Historical Sketch*, Academic Press, New York.
124. Markov, A. (1917). Generalization of a problem on a sequential exchange of balls (in Russian). *Collected Works*, Meeting of PhysicoMathematical Society of the Academy of Sciences.
125. Matthews, P. and Rosenberger, W. (1997). Variance in randomized play-the-winner clinical trials. *Statistics and Probability Letters*, **35**, 233–240.
126. McKenzie, A. and Steele, M. (2000). Distributions of cherries for two models of trees. *Mathematical Biosciences*, **164**, 81–92.
127. Meir, A. and Moon, J. (1988). Recursive trees with no nodes of out-degree one. *Congressus Numerantium*, **66**, 49–62.
128. Milenkovic, O. and Compton, K. (2004). Probabilistic transforms for combinatorial urn models. *Combinatorics, Probability and Computing*, **13**, 645–675.
129. Mirimanoff, D. (1923). A propos de l' interprétation géométrique du problème du scrutin. *L' Enseignement Mathématique*, **23**, 187–189.
130. Na, H. and Rapoport, A. (1970). Distribution of nodes of a tree by degree. *Mathematical Biosciences*, **6**, 313–329.
131. Najock, D. and Heyde, C. (1982). On the number of terminal vertices in certain random trees with an application to stemma construction in philology. *Journal of Applied Probability*, **19**, 675–680.
132. Panholzer, A. and Prodinger, H. (1998). An analytic approach for the analysis of rotations in fringe-balanced binary search trees. *The Annals of Combinatorics*, **2**, 173–184.
133. Pemantle, R. (1990). A time-dependent version of Pólya's urn. *Journal of Theoretical Probability*, **3**, 627–637.
134. Pemantle, R. (2007). A survey of random processes with reinforcement. *Probability Surveys*, **4**, 1–79 (electronic).
135. Pittel, B. (1987). An urn model for cannibal behavior. *Journal of Applied Probability*, **24**, 522–526.
136. Poblete, P. and Munro, J. (1985). The analysis of a fringe heuristic for binary search trees. *Journal of Algorithms*, **6**, 336–350.
137. Pólya, G. (1931). Sur quelques points de la théorie des probabilités. *Annales de l'Institut Henri Poincaré*, **1**, 117–161.
138. Quintas, L. and Szymański J. (1992). Nonuniform random recursive trees with bounded degree. In *Sets, Graphs, and Numbers: Colloquia Mathematica Societas Janos Bolyai*, **60**, 611–620.
139. Rényi, A. (1962). Three new proofs and generalizations of a theorem of I. Weiss. *Magyar Tud. Akad. Mat. Kutató Int. Kőzl.*, **7**, 203–214.

140. Riordan, J. (1968). *Combinatorial Identities*. Wiley, New York.
141. Robbins, H. (1952). Some aspects of the sequential design of experiments. *Bulletin of the American Mathematical Society*, **58**, 527–535.
142. Rosenberger, W. (2002). Randomized urn models and sequential design. *Sequential Analysis*, **21**, 1–41.
143. Rosenberger, W. and Lachin, J. (2002). *Randomization in Clinical Trials: Theory and Practice*. Wiley, New York.
144. Rosenblatt, A. (1940). Sur le concept de contagion de M. G. Pólya dans le calcul des probabilités. *Proc. Acad. Nac. Cien. Exactas, Fis. Nat.*, Peru (Lima), **3**, 186–204.
145. Ross, S. (1983). *Stochastic Processes*. Wiley, New York.
146. Ross, S. (2006). *A First Course in Probability And Stochastic Processes*. Pearson Prentice Hall, Upper Saddle River, New Jersey.
147. Samet, H. (1990). *Applications of Spatial Data Structures*. Addison-Wesley, Reading, Massachusetts.
148. Savkevich, V. (1940). Sur le schéma des urnes à composition variable. *C. R. (Dokl.) Acad. Sci. URSS*, **28**, 8–12.
149. Sedgewick, R. and Flajolet, P. (1996). *An Introduction to the Analysis of Algorithms*. Addison-Wesley, Reading, Massachusetts.
150. Smiley, M. (1965). *Algebra of Matrices*. Allyn and Bacon, Inc., Boston, Massachusetts.
151. Smythe, R. (1996). Central limit theorems for urn models. *Stochastic Processes and Their Applications*, **65**, 115–137.
152. Smythe, R. and Mahmoud, H. (1996). A survey of recursive trees. *Theory of Probability and Mathematical Statistics*, **51**, 1–29 (appeared in Ukrainian in (1994)).
153. Stepniak, C. (2007). Bernstein's examples on independent events. *The College Mathematics Journal*, **38**, 140–142.
154. Styve, B. (1965). A variant of Pólya urn model. *Nordisk Mat. Tidsk.*, **13**, 147–152 (in Norwegian).
155. Takács, L. (1961). A generalization of the ballot problem and its application in the theory of queues. *Journal of the American Statistical Association*, **57**, 327–337.
156. Takács, L. (1997). On the ballot theorems. *Advances in Combinatorial Methods and Applications to Probability and Statistics*. Birkhäuser, Boston.
157. Tchuprov, A. (1922). Ist die normale stabilität empirisch nachweisbar? *Nord Stat. Tidskr.*, **B/I**, 373–378.
158. Trieb, G. (1992). A Pólya urn model and the coalescent. *Journal of Applied Probability*, **29**, 1–10.
159. Tsukiji, T. and Mahmoud, H. (2001). A limit law for outputs in random circuits. *Algorithmica*, **31**, 403–412.
160. Uppuluri, V. and Carpenter, J. (1971). A generalization of the classical occupancy problem. *Journal of Mathematical Analysis and Its Applications*, **34**, 316–324.
161. Wei, L. (1977). A class of designs for sequential clinical trials. *Journal of the American Statistical Association*, **72**, 382–386.
162. Wei, L. (1978). An application of an urn model to the design of sequential controlled clinical trials. *Journal of the American Statistical Association*, **73**, 559–563.
163. Wei, L. (1978). The adaptive biased coin design for sequential experiments. *The Annals of Statistics*, **6**, 92–100.
164. Wei, L. (1979). The generalized Pólya's urn design for sequential medical trials. *The Annals of Statistics*, **7**, 291–296.
165. Wei, L. and Durham, S. (1978). The randomized play-the-winner rule in medical trials. *Journal of the American Statistical Association*, **73**, 840–843.

166. Wei, L. and Lachin, J. (1988). Properties of the urn randomization in clinical trials. *Controlled Clinical Trials*, **9**, 345–364.
167. Whitworth, W. (1878). Arrangements of m things of one sort and n things of another sort under certain conditions of priority. *Messenger of Mathematics*, **8**, 105–114.
168. Whitworth, W. (1886). *Choice and Chance*. Deighton Bell, Cambridge.
169. Woodbury. M. (1949). On a probability distribution. *The Annals of Mathematical Statistics*, **20**, 311–313.
170. Wright. S. (1929). Fisher's theory of dominance. *Animal Naturalist*, **63**, 274–279.
171. Wright. S. (1929). Evolution in a Mendelian population. *Anatomical Record*, **44**, 287.
172. Wright. S. (1930). Review of the genetical theory of natural selection by R. A. Fisher. *Journal of Heredity*, **21**, 349–356.
173. Wright. S. (1931). Evolution in Mendelian populations. *Genetics*, **16**, 97–159.
174. Wright. S. (1932). The roles of mutation, inbreeding, crossbreeding and selection in evolution. *Proceedings of the Sixth International Congress of Genetics*, **1**, 356–366.
175. Yao, A. (1978). On random 2–3 trees. *Acta Informatica*, **9**, 159–170.
176. Yao, F. (1990). Computational Geometry. In *Handbook of Theoretical Computer Science, Volume A: Algorithms and Complexity*, Eds. van Leeuwen, J., 343–389. MIT Press, Amsterdam.
177. Zelen, M. (1969). Play-the-winner rule and the controlled clinical trial. *Journal of the American Statistical Association*, **64**, 131–146.

Name Index

J

K

L

M

N

O

P

Q

R

S

T

V

W

Y

Z

Subject Index

N

O

P

Q

R